The McGraw-Hill
Essential ISDN
Sourcebook

The McGraw-Hill Series on Computer Communications (Selected Titles)

*To order or receive additional information on these or any other
McGraw-Hill titles, in the United States please call 1-800-822-8158.
In other countries, contact your local McGraw-Hill representative.*

KEY = WM16XXA

The McGraw-Hill Essential ISDN Sourcebook

Martin Nemzow

McGraw-Hill

New York   San Francisco   Washington, D.C.   Auckland   Bogotá
Caracas   Lisbon   London   Madrid   Mexico City   Milan
Montreal   New Delhi   San Juan   Singapore
Sydney   Tokyo   Toronto

Library of Congress Cataloging-in-Publication Data

Nemzow, Martin A. W.
    The McGraw-Hill essential ISDN sourcebook / Martin Nemzow.
       p.   cm.
    McGraw-Hill series on computer communications
    Includes index.
    ISBN 0-07-046384-0 (pb)
    1. Integrated services digital networks.   I. Title.
TK5103.75.N45   1997
621.382—dc20   96-34059
                                                       CIP

McGraw-Hill

A Division of The McGraw·Hill Companies

ISBN 0-07-046384-0

The sponsoring editor for this book was Steven Elliot and the production supervisor was Donald F. Schmidt. It was set in New Times Roman by the author.

Printed and bound by R. R. Donnelley & Sons Company.

This book is printed on recycled, acid-free paper containing
a minimum of 50% recycled de-inked fiber.

McGraw-Hill books are available at special quantity discounts to use as premiums and sales promotions, or for use in corporate training programs. For more information, please write to the Director of Special Sales, McGraw-Hill, 11 West 19th Street, New York, NY 10011. Or contact your local bookstore.

Contents

Preface

Introduction

It is easy to relegate the importance of Integrated Services Digital Network (ISDN), a product some 20 years in rollout, to something that is an interim product both too little and too late. ISDN comes in several "flavors" that include Narrowband ISDN (N-ISDN) and Broadband ISDN (B-ISDN). Although this book is about N-ISDN, the common format for ISDN for business, telecommuting, Internet access, and home-office connectivity, this same technology is closely related to B-ISDN, now standardized as Asynchronous Transfer Mode (ATM). All references in this book to ISDN refer to N-ISDN unless specifically designated as B-ISDN or ATM.

What most people do not realize is that ISDN was already a significant product in most European and Asian industrialized nations before vendors and trade magazines recognized its significance in the United States; that scramble was only within the last 2 years. Even as vendors and the press begin to describe ATM as the next wave of networking power and, with asynchronous bandwidth services and modems from the CATV industry, as the replacement for ISDN, realize that all these technologies are based upon fiber optics, synchronous optical networks (SONET), and digital switching technology. Fiber optics and digital switching makes high-speed optical networks possible. It is easy to miss the subtle and important significance of ISDN. *The McGraw-Hill Essential ISDN Sourcebook* is really about this digital switching technology and its benefits on normal, everyday communications. Although ATM promises extraordinary gigabit (Gbps) bandwidths, ISDN really represents a new wave in communications, more profound to everyday workflows, processes, businesses, and people than just a bigger bandwidth pipe.

The Real Value of ISDN

ISDN is a very important product, but for reasons different from the World Wide Web, Internet, and remote Local Area Network/Wide Area Network (LAN/WAN) interconnections. In fact, ISDN represents the very significant transition from the analog phone system invented and implemented by Alexander Graham Bell over 120 years ago to the next step in digital transmission. The regions with the most advanced digital telephone infrastuctures available were the first ones to upgrade from antiquated and inefficient technology. We are really playing "catch up" in the United States, Russia, the Middle East, and Asia. The real significance of ISDN is not that it provides fast Internet connections and cheaper point-to-point links; that is just a simple byproduct of the new technology. Rather, ISDN represents the transition from analog telephone infrastructures to a single interchangeable and interoperable digital mesh.

It is important to realize that ISDN does *not* stand for independent switched data networking or some of the other many variations of the official name which appear so frequently in other books, in the copious ISDN information on the Internet, or even in vendor literature. The *I* is for *integrated* in that the new service connects into existing voice and data services. The *S* is for *services* to accentuate the range and flexibility of this technology. The *D* stresses the technology migration to an all-*digital* infrastructure. Finally, the *N* is for *network*, as in friends, family, peers, colleagues, but not necessarily computer networking, data networking, or even anything to do with information and computers; it could be just a voice communication network.

ISDN is the foundation for enhanced telephony, Centrex phone systems, public branch exchanges (PBXs), analog key systems, computer telephony, fax-on-demand, fax back services, call centers, and circuit-switched connections. It is also a solution for telemarketing operations that have been stymied by noisy analog lines and the limited availability of new phone number exchanges. The North American Number Plan (NANP) transition to area codes in formats other than (X0X) and narrow bandwidths create some problems with older telephone systems, but in general they create many new phone numbers. Digital connections with switched infrastructures create new opportunities. Although ISDN is basically a two-wire delivery, it provides from 1 to 30 different and simultaneous channels with data rates from 56 to 1984 kbps. B-ISDN is mentioned in the same sentences with ATM, but provides realistic migration paths for N-ISDN way beyond 1984 kbps. Although each ISDN channel can carry digital computer data, it can also carry voice, sound, video, and packet data at the same time. Each channel can be connected to a different destination or combined with others for greater capacity.

ISDN Integration

Although new hardware seems to be the driving force behind ISDN, the real innovative features and benefits of ISDN come from software and process workflow integration. The operative feature of ISDN is that each channel is a switched digital connection; as such, it can do many things. The Basic Rate Interface (BRI), the small office/home office (SOHO) offering, is three discrete channels and it costs about the same as a single plain old telephone service (POTS) phone line. Simple hardware provides a single 56- or 64-kbps connection to replace the slower V.34 analog modem line, which is trying hard to yield 28.8 kbps. Complex software provides multiple links, packet-data delivery, and a real phone line with distinctive rings for fax, data, voice, and extensions. This software is not all in the ISDN terminal adapter; a lot of it is controlled by the service provider in their central office switches. If you think this makes ordering ISDN from a provider like BellSouth difficult, you are right. The process of ordering and configuring ISDN is called *provisioning* and it is the single most confusing part of ISDN. A number of vendors, such as Intel with their ProShare videoconferencing *Intel Blue* provisioning definition, simplify installation of worldwide videoconferencing capability. However, this minimizes the power of ISDN by making fewer options, features, and functions available for computer telephone integration (CTI). Although the North American ISDN Users' Forum (NIUF) and vendors have created a national ISDN specification called NT-1 and provisioning simplification, there are already glimmers of specifications for NT-2, NT-3, et cetera. ISDN will not get simpler, but rather more powerful with more options and more features.

While ISDN capability packages are intended to simplify ISDN ordering and installation for the average user, ISDN is more than just a faster modem connection. Do not overlook the inherent power in this radically new technology. ISDN represents the beginning of our transition from an analog communication infrastructure to a switched digital wave.

Martin Nemzow

The McGraw-Hill Essential ISDN Sourcebook

1

Overview

Introduction

Integrated Services Digital Network (ISDN) represents the beginning of a transition from analog telephone services to a digital infrastructure. The new digital services provide raw data bandwidths greater than 33.6 kbps, currently the most available with analog modems and standard plain old telephone service (POTS). ISDN does provide remote connectivity which is faster than that, starting anywhere from 56, 64, 112, or 128 kbps to capacities as great as 4 Gbps with B-ISDN (and ATM). ISDN is interoperable with both switched digital services and analog phone services, but it is also a complex product with very advanced conferencing and telephone processing features. The current headlong rush to ISDN for faster Internet access is representative of just one of the many features available with ISDN. While this book shows you in mechanical detail how to hook into the Internet with ISDN and sidestep many of the implementation problems of ISDN, it also defines many other ISDN features and the technology behind it because the most exciting benefits from ISDN are hidden as workflow applications, advanced communications services, and increased flexibility brought about by digital communications.

Purpose of Book

The McGraw-Hill Essential ISDN Sourcebook presents practical information about installation, service, usage, utility, and troubleshooting. ISDN is like an onion; it has made some people cry and might have the same effect on you too. It has been described for years in the press as a solution in search of a problem and a product without hardware, software, or carriers. While this is partly true, now it is a real product. You can touch it, feel it, smell it. Like that onion, there are numerous layers to it.

Figure 1.1 You can expose ISDN services like layers of an onion.

The purpose of this book is to show you how to pick ISDN when it is ripe and ready and cook it into your stew of enterprise networking and telecommunications so that it benefits your big picture. ISDN is effective for Internet connections, telecommuters, site-to-site interconnections, and supplemental bandwidth on demand. ISDN can provide bandwidth as low as 56 kbps but goes as high as multiple gigabits per second. Just as an onion isn't always an ingredient—for example, in Toll House cookies®—ISDN is not always the solution you need for networking and telephony.

This book is like that onion also. You can use this Sourcebook as a cookbook to get ISDN up and running quickly for fast access to Internet web sites, for faster connectivity between remote LANs, for faster access to the corporate network from home, and for videoconferencing. The material in this book is simple and straightforward. Behind the simple facade of ISDN underlies a complex technology suitable for application development, complex workflow optimization, and complex telephony applications. This book provides the technical information for more complex and involved applications.

The McGraw-Hill Essential ISDN Sourcebook is designed primarily to answer questions facing a telecommuter or enterprise network team plotting network redesign and upgrades to the wiring infrastructure. Much of the material in this book relates to physical wiring, cabling, and connectivity issues, if only because the success of ISDN transmission hinges on the physical cabling plant. Although physical issues seem to be "beneath" the threshold of most network managers, initial lack of attention to cabling and plant issues will just raise it later.

Intended Audience

The intended audience includes people who work from home, telecommuters, MIS directors, industry consultants, network managers, workgroup managers, application developers building client/server and distributed networking applications, support technicians, computer professionals, and anyone who uses the Internet, provides site-to-site or multisite connectivity, requires greater data bandwidth, or wants to learn about the worldwide digital communication infrastructure. Anyone who is intending to migrate from analog modems to faster digital services, is actively planning to replace a PBX or key telephone system for more flexible and capable services, seeks to integrate traditional data communications with phone services, or is designing advanced telephony application development will find this material valuable for its insight and performance caveats. Anyone redesigning the work environment, specifically with regard to networking workflow, can benefit from the strategic information and practical know-how. This book is also a useful teaching tool; it covers many telephone and network environments from the vantage point of specific technical details, but also with the perspective of practical ISDN implementation.

Book Content

The content of *The McGraw-Hill Essential ISDN Sourcebook* is straightforward. This book provides some case studies for faster networks requirements and transitions, gives a brief history on ISDN, demonstrates the utility of ISDN, shows you how to design, order, install, resolve service provider provisioning issues, test, manage, and debug ISDN services.

Structure of Book

This chapter presents the organization of *The McGraw-Hill Essential ISDN Sourcebook*. As an acknowledgment of the reader's valuable time and the many layers of ISDN, Figure 1.2 illustrates the design and flow of the knowledge contained within this book.

Chapter 1, this chapter, provides a general overview of the ISDN book explaining the purpose, audience, book content, and structure of book.

Chapter 2 presents the gritty fundamentals of ISDN, including useful operational definitions of ISDN, how to specify the options and services required to create a viable connection, and how to bypass the pitfalls of the technology. Since ISDN is not a homogeneous or universal service, this chapter shows (by map and chart) where services are available, what these services are, and how to interconnect.

Chapter 3 shows what ISDN in terms of its many provisioning options (approximately 130 by voice, 20 for data, and various compound services for Automatic Number Identification [ANI], Public Branch Exchange [PBX], messaging, and mail). This chapter elaborates on the options and the utility for each one.

Chapter 4 details the physical installation process, mostly showing what the end-user (both for a residential or business user) would need to do or supervise for the successful completion of an ISDN facility. This chapter shows how to install an ISDN terminal adapter, the hub, the router, and multidrop network terminations and how to deal with the myriad of physical and software issues relating to installation.

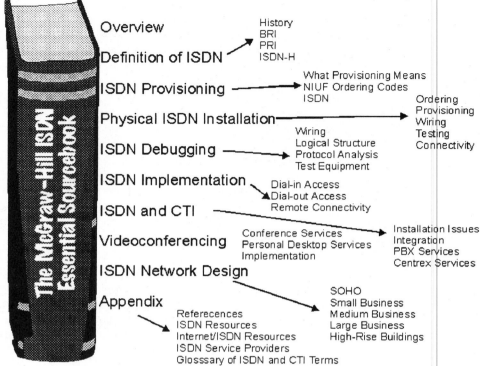

Figure 1.2 The organization of *The McGraw-Hill Essential ISDN Sourcebook.*

Chapter 5 shows what to do when ISDN doesn't work. Although, ISDN may seem as easy as a phone service because it is a copper line connection, site problems, punch-down blocks and jumpers, or equipment connectivity problems often necessitate a fair degree of analytical and diagnostic skills to keep the data flowing and people talking.

Chapter 6 shows what is needed and what to do to actually use the ISDN channel. As previously stated, ISDN is just a communication medium without any inherent purpose until it is applied to remote LAN access, dial-in or dial-out services, telecommuting, or Internet access.

Chapter 7 addresses another facility with ISDN (at least in North America), that it is just a fast data service. ISDN also includes voice services and ancillary communication options than can be integrated into an organization for significant cost savings and strategic advantages.

Chapter 8 shows how to use videoconferencing with ISDN. Although PictureTel, ProShare, and other like systems have been available for about 10 years, ISDN simpli-fies the establishment and durability of a two-way connection (over lease or dedicated lines). This seems like an important complement to ISDN, and along with white board services is an important application to profile in terms of costs, ease of use, installation, problems, and benefits.

Chapter 9 shows the floor plans, site plans, and connectivity designs for actually planning, ordering, and installing ISDN. Issues of site-to-site lengths, internal wiring distances, and end uses are discussed. Also, since ISDN is likely to be employed for both voice and data, routing of lines through telecommunication closets, customer service multiplexers, and splitters to switchboards, PBXs, or computer rooms is relevant as well to the overall design and planning process.

Appendix A includes useful Internet and web site sources, news groups, and some other technical documents about ISDN. It includes reference to cited materials and sources for statistics, models, software, notes, standards, photographs, and addresses of useful ISDN-related Internet sites.

There is also a glossary of terms specific to ISDN, data networking, and telephony defined and cross-referenced by the all-too-common acronyms. There are many terms in the glossary not directly defined in this book but used by ISDN vendors, service providers, and in the general provisioning or configuration process of ISDN for dial-up data transmissions, packet data connectivity, or computer telephony integration with ISDN telephones.

2

ISDN Implementation Fundamentals

Introduction

Integrated Services Digital Network (ISDN) is an exciting technology, not because of the attention given to it in the press, but rather because it represents a cost-effective, scalable, reliable, and flexible wide area networking (WAN) technology. ISDN is a telecommunications service *package* turned all digital end to end, using existing or modified switches and existing wiring, for the most part upgraded so that the basic "call" is at least a 56-kbps end-to-end channel. When analog circuits are converted to ISDN, the switches need to be upgraded, modified, or tested, but calls from ISDN to standard analog circuits switch through existing analog switches. ISDN and plain old telephone service (POTS) are interoperable.

Packet and frame modes are thrown in for good measure, too, in some places. ISDN represents the transition from analog telephony to digital communications. ISDN is strictly a wide area communications technology very much like the current telephone system, although it does have both local and long-distance components. The transition from analog to digital service has occurred faster where older and insufficient telephone systems stymied development. Technically, the shift is from analog frequency-division multiplexing and mechanical circuit switching to digital time-division multiplexing with packet multiplexing and fast-packet electronic switching.

Digital signaling is important because digital signals ignore the static and noise that often affect analog transmissions, especially over long-distances and older telephone lines. Digital signaling produces connections of the highest possible quality and carries digital data at significantly higher speeds than analog conversion. Digital services promise even greater speeds as digital compression and other techniques become more sophisticated and available. ISDN merely offers transmission capacity of a specified type, with a standard set of call-control and capacity-management functions. You really do not need to know any of that to benefit from ISDN technology. The benefits include bandwidth-on-demand and any-to-any connectivity. Tantamount above anything else, even above the technical benefits of ISDN, is the financial advantage of ISDN connectivity in that you get more lines and more bandwidth for the same cost. In general the cost of a dial-up ISDN connection is identical to that of an analog connection and ISDN provides at least twice the true throughput.

This transition is now unifying the worldwide communication structure. ISDN is as easy to use as any standard analog phone or modem. In fact, many of the services you pay extra for on analog phone circuits are available with ISDN during initial provisioning without extra cost. ISDN connects calls faster than does analog, within a second versus many seconds. ISDN provides greater basic bandwidth and scales to megabits for growth. ISDN is available in most locations in the world, although some developing nations lacking in basic phone services also lack access to ISDN. In fact, many hotel halls, conference centers, and even hotel rooms have installed ISDN services. Figure 2.1 illustrates the flexibility of ISDN and digital connections.

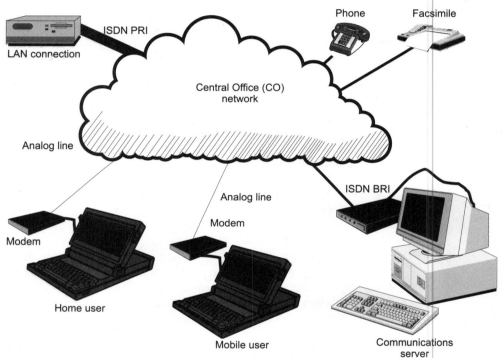

Figure 2.1 ISDN provides a flexible data communications connection.

To many users, especially individuals and those in smaller companies, ISDN is by far the most important of these new digital technologies because it is available in smaller packages at reasonable prices. For millions of users, ISDN offers inexpensive dialed service, high-speed data transmission, and the ability to send and receive voice, data, and still and moving images through the same completely digital connections. ISDN is also a communications standard accepted throughout the world, which means that voice and high-speed data connections to most of the major business centers of Europe and the Pacific Rim are literally no more than a dialed phone call away.

It is important to recognize that ISDN is not a one-size-fits-all solution. It is not a one-size service either since various ISDN service bundles exist in different configurations and capacities. However, the viability for ISDN breaks down into four main attributes:

- Price
- Performance
- Reliability
- Digital

ISDN compares favorably in some locations with a comparable number of analog phone lines, with frame relay, with switched digital lines, and with dedicated lease lines. It provides greater performance than analog connections, faster dialed connections, fewer bit errors, and greater bandwidth. It can match other digital service offerings, such as X.25, frame relay, switched multimegabit data service (SMDS), T-1, E-1, and most current asynchronous transfer mode (ATM) offerings in terms of bandwidth. ISDN is likely to be more reliable than analog lines because the central office switch hardware is likely to be 15 years newer. You see this reliability disparity in regions where the telephone equipment is indeed older. ISDN is also likely to provide more reliable data streams because it produces fewer bit errors than analog.

The pervasiveness of personal computing, inexpensive PCs, and Internet or on-line access generate and consume increasing amounts of bandwidth. Although specific applications require more bandwidth than is normally available with analog lines, as shown Figure 2.2. the requirement for increased bandwidth is really more of the same—a lot more of the same—as listed in Figure 2.3.

Common schedule for implementation

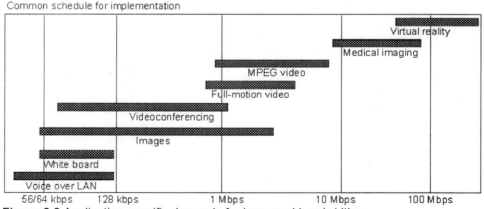

Figure 2.2 Application-specific demands for increased bandwidths.

ATM, digital simultaneous voice-and-data (DVSD), asymmetric digital subscriber lines (ADSL), high bit-rate digital subscriber loop (HDSL), a mid-term solution called rate adaptive digital subscriber lines (RADSL) more likely to be implemented, and community antenna television, which is better known as cable TV (CATV), with cable modems are not ready for wide-scale deployment as an alternative to ISDN. The phone companies also have an installed wiring base that provides greater coverage and penetration than CATV providers. ISDN is now widely deployed. Despite the "digital" in its name. DVSD runs on POTS, and it is not really a viable competitor with ISDN except at very low bandwidths and minimal computer telephone integration (CTI) functionality.

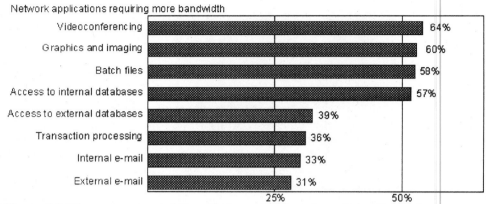

Figure 2.3 Where more bandwidth really is applied (courtesy of Idea Factory).

Voice over data or voice concurrent with data is a service support with ISDN. Voice mixed with data (conferencing, video, fax, voice, and e-mail) is also provided as an analog service with special DSVD modems. However, DSVD is wholly analog (as evidenced by its need for modified modems) and is not to be confused with ISDN. DSVD works with analog POTS lines, while ISDN requires special switched digital or provisioned ISDN lines. DSVD services, most readily available with Radish Communication's VoiceView interface, are shown in Figure 2.4.

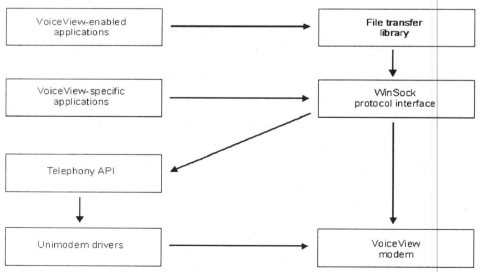

Figure 2.4 DSVD and VoiceView are analog products that are inferior to ISDN.

You are likely to see this same DVSD functionality programmed into new ISDN equipment for interoperability with analog DVSD. Net Commander is one of the first ISDN products for small office/home office (SOHO) services which is compatible with or comparable to the analog DVSD products. DVSD really is a low-end solution which is not as flexible or scalable as ISDN. In time, it might become a good tool to integrate for telecommuters when there are more products and acceptance of this technology.

When you compare ISDN against analog, leased lines, very small aperture terminal (VSAT, better known as a satellite connection), and other delivery services, it can be cheaper per hour and per megabit. In addition, adaptation of ATM, multipoint, multidistribution subscriber systems (MMDS), and CATV cable modems, or ADSL and HDSL into standard telephony services is not yet defined. Once the necessary interfaces are defined, implementation may require several years. ISDN is here now. DVSD (which is limited to 28.8 kbps) seems to be an interim solution better served by ISDN with its faster base transmission speed and better concurrent integration of voice and data services. The wireless MMDS—much like a 45 cm (18 in) satellite TV subscriber service—providing 90 channels for service could pose more significant future competition to ISDN delivery than cable services. In addition, recognize that cable architectures are ring-based, rather than the star-based design of telephone systems, and are thus prone to hacking, eavesdropping, and tapping. ISDN is part of the standard telephony infrastructure and operates as part of that structure. It will be a long time before ADSL, HDSL, cable TV services, and ATM are integrated into the infrastructure. It took at least 10 years for ISDN to interoperate within the existing network.

Furthermore, ISDN at its slowest is at least 2 times faster than a 28.8- or 33.6- kbps modem on an analog line. This will be important as Internet usage grows and Internet services include more images, video, even more data, and virtual reality (VR) presentations. ISDN provides enough bandwidth for real-time videoconferencing. The technology of bandwidth-on-demand will soon make video calls and soon-to-be video messaging possible. Broadband ISDN (B-ISDN) is available in metropolitan areas, and bandwidths will grow as ATM and SONET are implemented as telecommunication backbones.

ISDN is available in 80 percent of North America and is installed in about 1 million sites already. There are 1 million active ISDN lines in Germany, 400,000 lines in the British Isles, and 600,000 in France. About 500,000 ISDN lines are deployed in Japan, with an equal number throughout the rest of southeast Asia.

Japan has established the Tokyo Teleport Town (TTT), a model city with high-speed communication links for 63,000 residents and 106,000 additional daily workers. This trial project of Nippon Telephone and Telegraph (NTT) demonstrates the Japanese government's interest in building new digital infrastructures, including ISDN, as part of the new information highway. That is a small quantity compared to the 800 million analog phone lines in use in the world, but represents the explosive transition from analog telephony to digital communications. ISDN is real and useful, but more than that, and the reason of greatest importance, is that ISDN is a cost-effective solution for on-demand high-bandwidth connectivity.

It really is easy to install and utilize ISDN. In most cases, installation goes easily and quickly. Making it work is also easy. Getting an ISDN circuit is a matter of a phone call to the local telephone provider, asking for the ISDN line, and matching central office switch equipment with your local ISDN terminal adapter and your usage requirements. Because ISDN is not available everywhere yet, you may have complications or experience a long installation wait. Custom installations and complex usage requirements often complicate the process. Mission-critical applications raise the ante for maintaining backup, switchovers, and complex telephony integration. Notwithstanding, ISDN is a feasible solution to many telecommunication and data communication needs. You can usually work through any ISDN problem in an hour or two in the same technical support phone call. Worst case, you can resolve ISDN problems within a couple hours spread over a week or so.

The primary benefit of ISDN is cost-effectiveness. There are always other services that will provide the same results. In fact, you should explore them before committing to ISDN because these options may be better. ISDN provides phone service, fax service, and communication channels for video and data. But so does standard analog service, switched or leased lines, and T-1. You can split T-1 services with multiplexing units just as you do with ISDN voice routers or special public branch exchange- (PBX-) to-PBX connections. In fact, you can provision a T-1 for ISDN, analog voice, and frame relay. The ISDN decision distills to services versus cost. The table in Figure 2.5 illustrates the comparisons.

Service	Installation	Monthly	Per Minute	Long-distance	Bandwidth
POTS	62	29	0.035	0.14	33.4
Switched 9.6	576	176	--	--	9.6
Switched 56	576	214	--	--	56
Switched 64	576	228	--	--	64
ISDN	217	78	--	0.14 per B	to 128
Frame Relay 56	1256	314	0.17	0.14 per B	to 56
T-1	1784	1254	--	0.14 per B	to 1534

Figure 2.5 Telecommunication options cost comparison (courtesy of BellSouth and Networks, Inc.).

If you are contemplating a leased line or digital line and the local telecommunications provider supports ISDN and the costs (including flat rate local access) are comparable to these BellSouth quotations, consider ISDN. It is cheaper for all but 7×24 operation when there is a per-minute charge for ISDN. If you already have a leased or switched digital line installed, the pay back on all hardware and switchover is 4 months. The transition is a no-brain decision. There is no comparison between ISDN and frame relay in terms of cost, although the tariff is coming down for frame relay. If you can convince the provider that they do not even have to install a new wire since the existing copper is already in place and thus can prorate or waive installation charges, the pay back is just 2 months. If the ISDN service provider charges per minute for ISDN usage (PacBell and Bell Atlantic do), the comparison between POTS, ISDN, and leased line is a little more complex. For example, the quoted rate for a plain telephone line does not include Caller ID, call forwarding, automatic call-back, call announcement, and conference capabilities. When you add those services into the monthly charges, the cost for a single BellSouth phone line could be as high as $68. ISDN with (at least) two phone numbers, more bandwidth, more flexible use of bandwidth, and all those optional services becomes a good deal. If you compare ISDN with T-1, ISDN costs $0.54/kbps per month while T-1 costs $1.22/kbps per month. Figure 2.6 contrasts the fixed and marginal costs for various services with marginal costs for long-distance service.

Contrast service costs in terms of the circuit, the channels actually used, the usage per hour, or the fixed plus marginal cost per kilobits per second. Choose the comparison that fits your service usage. Simple comparison is not completely correct in that a basic

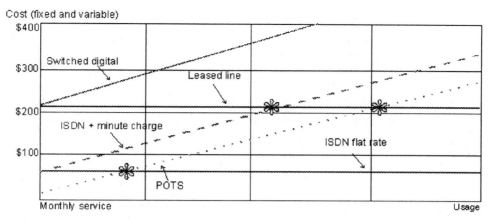

Figure 2.6 Fixed and marginal telecommunication options cost comparison.

ISDN package can carry 4 times the data that the analog line can carry and that some local services have different installation, monthly, and per-minute costs. For example, at $0.012 per minute, the current PacBell and Bell Atlantic ISDN per-minute rates, the break-even for a leased or switched BellSouth "Synchronet" 56-kbps line is 13,200 minutes (220 hours). This means ISDN is cheaper unless the connection is maintained almost constantly. Do not think of ISDN as something special; it is just a bundle of services that can be provided similarly as a mix of analog, frame relay, packet services, and digital lines.

Reading Network Diagrams

ISDN is about networks—data and voice. Since ISDN also supports sound and video, it is a communications infrastructure. This infrastructure is described as wires called the *local loop*. the service provider central office (CO), and "clouds" of services often represented by simplistic diagrams. This is a quick lesson on reading diagrams for both data and voice networks. In general, lines indicate wiring connections between points, and boxes represent hardware devices such as computers, switches, PBXs, or key phone systems. The graphics are metaphors for the physical components. Circles represent local area network rings while ladders indicate backbones and buses. A network diagram is precise in terms of displaying purpose and connectivity. This is shown in Figure 2.7.

Since ISDN includes data and voice components, you might need to read data and voice networks and separate the phone from the data from the integrated voice and data networks. Thick lines indicate a network backbone to connect all network devices. Network devices. or nodes, include PCs, workstations, printer ports, dial-in and dial-out communication concentrators, servers, and all intermediate network devices, such as wiring hubs. repeaters. bridges, routers, switches, and gateways, are indicated by boxes. Cabling between a device and a backbone or hub is indicated by thin lines. Thin lines represent coaxial network cables, serial cables, or the typical twisted-pair wiring commonly used now for both data and voice.

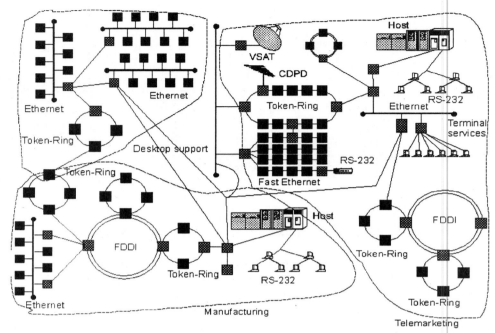

Figure 2.7 A simple network diagram.

A network diagram can be either an exact physical representation showing component placement or actual paths for the wiring. This is called a *physical* diagram and is usually a layered floor plan. Each layer represents a different function, from basic structure designs, to heating, ventilation and air conditioning (HVAC) ductwork, water and sewer lines, power lines, fixed partitioning, movable partitions, and data wiring. You will see floor plans throughout this book. Figure 2.8 illustrates the wiring floor plan for a typical modern office building. Such a diagram might be imported to Windows from AutoCAD as a bitmap.

One of the hallmarks of a physical diagram is the ambiguity in the purpose. You cannot tell from Figure 2.8 whether the wiring is for an analog phone system, ISDN, or a local area network. Sometimes, a blueprint shows too much. When layers are extracted to highlight key points, a blueprint can still be too big. The network design is therefore condensed to its salient points, which are typically network devices, the wire between them, and the logical way these devices are attached together. This stick-figure diagram is called a *logical* representation. It is the condensed diagram for the network. Issues of scale, color, placement, and measurement are unimportant. The logical relationship between devices is important. You will see logical designs throughout this book because they convert complex relationships into stick figures. They are cartoons, but are useful for distilling complex concepts into simple images, as shown by Figure 2.9.

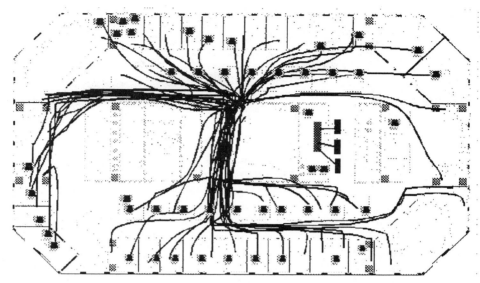

Figure 2.8 Physical network diagram.

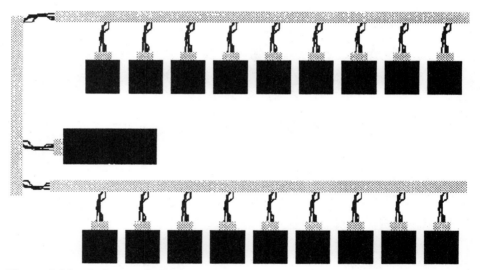

Figure 2.9 Logical network diagram.

Because telephone service is more complex than just a point-to-point connection, many of these complexities are minimized to a cloud. In effect, you can take for granted the services offered through the central office ISDN switch just as you do with standard POTS. This ISDN service cloud is shown in Figure 2.10.

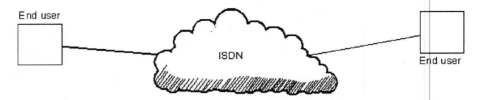

Figure 2.10 ISDN service cloud.

The meaning of the cloud is that the user expects a standard service and doesn't need to know how it works or where it really exists. Access to it is not an issue, but having an access line to it somewhere is an important concern. This is called *availability*. However, there are many more layers to the ISDN onion than just that. Because you may need to know about these layers (of services, channels, features, and functions), Figure 2.11 is an expansion of the network diagram. This figure doesn't include all the detail because that is too much at this point. The book does get there and describes it all. The difference in Figure 2.11 is the inclusion of the channel bank (the local access switch), which is the fast-packet switching technology at the local carrier's central office. The ISDN switch at the central office is called the *point-of-presence* (POP). Notice that the end-user devices which connect the central office switch are generically called *customer premise equipment* (CPE) because they are installed at the telecommunication service provider's customers' sites.

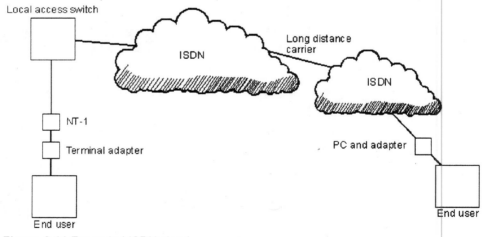

Figure 2.11 Expanded ISDN cloud.

ISDN equipment includes a carrier line breakout called the *network terminator* (NT-1 or NT-2) and the actual data translator, officially called a *terminal adapter* (TA). The network terminator is a digital version of the familiar analog channel service unit/digital service unit (CSU/DSU) used to terminate a digital or T-1 line from a carrier for proper electrical characteristics, perform line conditioning, and signal equalization.

Most network terminators, particularly the ISDN types, include loopback tests and line diagnostics. The terminal adapter is variously referred to as a digital modem (which is incorrect because it does not *mo*dulate or *dem*odulate signals, voltages, or current levels), digital adapter, data service unit (DSU), or an ISDN user interface. The terminal adapter literally adapts the data streams to the network terminator and the computer equipment interfaces. This might include RS-232, V.35, or a Peripheral Component Interface (PCI) bus. As ISDN represents a transformation from the old and outmoded analog phone systems, the ISDN terminal adapter represents a transformation from old data delivery to a new more expedient method; you no longer need to transform digital data or voice into analog signals and back again.

Figure 2.12 illustrates the physical hardware and wiring within a logical overview. It is a good illustrative view of what ISDN really is.

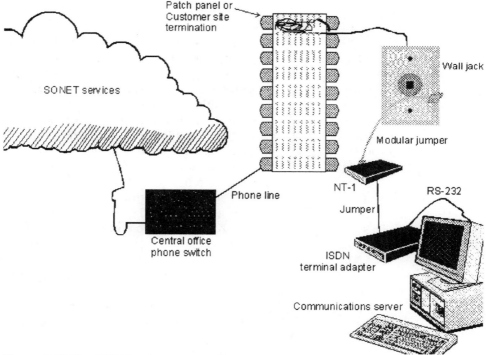

Figure 2.12 The ISDN hardware connection.

The Internet is not ISDN, nor is ISDN the Internet. If you are exploring the use of ISDN to connect into the Internet, realize that the Internet is a logical connection of various services by Synchronous Optical Network (SONET), T-3, T-1, SMDS, dedicated digital lines, analog modems, and wiring. ISDN is just another service to connect into that logical cloud of services which define the network connectivity that is the Internet. Similarly, ISDN is just a cloud of logical connectivity apart from and separate from the Internet, which is also a cloud of services, as Figure 2.13 illustrates.

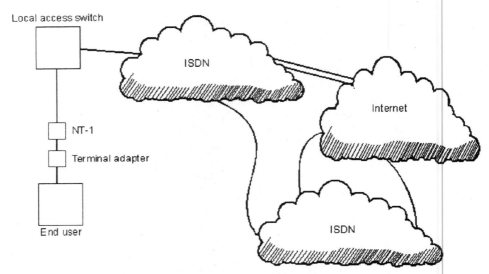

Figure 2.13 The connection of the ISDN cloud to the Internet cloud.

Another important issue between data communications and telecommunications is the difference between channel width and channel length. Conceptualize channel width as the thickness in wires and hence how much data it can carry at a time. In reality, each communication wire carries only a single data signal at a time within an electrical wave format. However, a signal clock (and the limiting ability of the wire to maintain the quality of that wave) determines the frequency of that wave. In general, a single wire pair can maintain up to about 1.984 Mbps, although the telephone service providers are suggesting that ADSL, HDSL, or RADSL (variously termed *xDSL*) can carry data at speeds up to 6 Mbps unidirectionally with 384 kbps on the return on two pairs.

There is one more complication in that the signal on that wire may contain 24 (American T-1 format) or 30 (European E-1 format) separate *channels* multiplexed together. The signal width is called *bandwidth* and is the useful data-carrying capacity. The signal bandwidth is the useful data carrying capacity of a single wire, whereas each channel has the carrying capacity of just one of those multiplexed channels. For example, T-1 signaling supports 1.536 Mbps in total bandwidth, but each of the 24 channels supports either 56, 57.6, or 64 kbps. Channel length is a function of the wire length point-to-point, and the propagation speed of the signal factored by the wire length creates a signal latency. Since the speed of the signal in copper wire is about 55 percent of the speed of light, signal latency around the world should be about 0.14 s. In reality, switch delays, signal boosters, and various queues slow delivery to about 3 s. The difference between bandwidth and latency is best illustrated in Figure 2.14 by analogy with the diameter and length of a pipe and the actual data packets or streams fitted to the available pipe shape. It is important to differentiate between bandwidth and latency because it is a difference of capacity over time. There is always a semifixed latency whatever the bandwidth and protocol. Adding bandwidth increases your ability to send more data, but the latency rarely changes. The end-to-end delivery time (that is, latency) for a phone call, for various configurations of ISDN, for a modem connection, for frame relay, or for ATM are all the same; it takes about 0.1s to send a signal across the United States, Asia, Europe, or Africa. You can send about 14 KB with a modem in the same time you can deliver about 155 MB with ATM or Broadband ISDN. Although the differences in data delivery are a function of bandwidth, the time to get either signal from the source to the destination remains exactly the same. That is the latency and there is no known way to circumvent it.

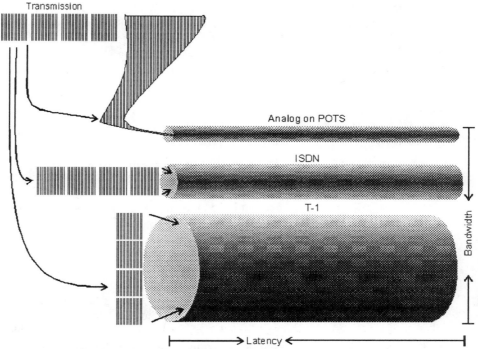

Figure 2.14 Difference between bandwidth and latency.

History of ISDN

AT&T developed ISDN based on research with IBM on digital data transmission requirements. It became a pet project of the United Nations (UN) and the International Telecommunication Union (ITU although formerly known as the CCITT) when bureaucrats realized it could form the basis of a new, unified, worldwide communications system. Research began as early as 1968. ISDN became a viable AT&T product in 1981. The forced divestiture of AT&T and breakup into the Baby Bells, the regional Bell operating companies (RBOC), and a separate long-distance carrier, in 1984 created new monopolistic opportunities for these new companies.

Following the divestiture of AT&T, 161 areas in the United States were geographically defined as Local Access and Transport Areas (LATA). Currently, within one LATA (intra-LATA) a Local Exchange Carrier (LEC) may handle an entire call if the call originates and ends within a LATA. A Local Exchange Carrier can be a Regional Bell operating company, such as Ameritech, Bell Atlantic, Bell South, NYNEX, Pacific Telesis. US West, and others, or an independent telephone company called an independent operating company (IOC), such as GTE or Southern New England Telephone, that provides customers with their local service.

Divestiture created opportunities with hardware, pricing, services, and software. As a result, the RBOCs exercised these choices. Some deployed ISDN with their own proprietary implementations and central office switches from AT&T, Northern Telecom, Siemens, Fujitsu, Bull, and Stromberg-Carlson. The postdivestiture environment did not stress interoperability. Since local service worked very well with long-distance carriers, there was no incentive to create anything new or expand the digital services. However, Europe had a tradition of poor analog phone service and it needed something new to support the growing interdependence among countries and the formation of the European Common Market. Not only was phone service a monopoly in Europe, it was also controlled by governments. Therefore, it was a simple legislative matter to create incentives for a common digital network. ISDN is widely available in Europe. At the same time, the research arm of AT&T, Bellcore (now called Lucent Technologies) began to investigate interoperability and the potential for ISDN services. This research and interaction with RBOCs and independent equipment manufacturers lead to the National ISDN initiative.

Bellcore had viable ISDN consumer products as early as 1990. Most of these products are being phased out due to marketing "obsolescence" because any service provided on digital lines, such as call forwarding, Caller ID, and conference calling, is available on analog lines as part of the in-band signaling overhead. You might also note that voice and fax conversations require only 4 kbps of the 33.6 kbps available on analog circuits. Any extra bandwidth was irrelevant until V.35 could push modulated data streams above that rate. Therefore, there has been no particular reason to convert analog PBX and key systems to digital and ISDN until recently.

In addition, digital services have been more expensive, less reliable, and less robust than most working phone systems until very recently. However, a desperate need for more phone numbers and the need to update the North American Numbering Plan (NANP) drove many local providers to create new area codes to provide new exchanges and more lines. Although digital connections provide twice the data capacity of analog in terms of circuits and bandwidth (64 versus 33.6 kbps), lack of availability, equipment, and users wanting ISDN and digital circuits, high costs, and unconvincing rationale for a shift to new technology delayed deployment in the United States. Elsewhere, the story is different.

Until 1995 there was limited need for data transmission bandwidths greater than 9.6-kbps modem speeds. Organizations that needed it, could (and still do) contract for leased lines or switched services at bandwidths from 300 bps to T-1. T-1 is the Bell technology for a 1.544-Mbps communications circuit provided by long-distance carriers for voice or data transmission through the telephone hierarchy. Since the required framing bits do not carry data, actual T-1 throughput is 1.536 Mbps. T-1 lines may be divided into 24 separate 64-kbps channels. This circuit is common in North America. Elsewhere, the T-1 is superseded by the ITU-TTS designation DS-1. Satellite communications in the form of VSAT provides from 24 to 384 kbps. VSAT also provides a cheaper mesh than telephone lines. Packet-data networks, such as X.25, provided cheap data transmission when application latency was not critical. You might compare X.25 to the postal system and ISDN or dial-up modem service to next-day courier service. Organizations in need of anything greater than that public bandwidth created their own lines or leased SONET and fiber optical service. It really wasn't until the HTTP mark-up language (HTML) used for web pages increased the simplicity and viable of Internet file systems that the Internet started to attract attention. In 1995, the World Wide Web browser and the utility of home pages took off. As more and more HTTP sites included graphic images, the need for greater bandwidth drove modem speeds from 9.6 to 14.4 to 28.8 kbps.

Analog POTS typically provides a telephone channel with 3500- to 4200-hertz bandwidth between 300 and 4200 Hz for normal voice conversations. This corresponds to the previously mentioned 4 kbps. Each hertz corresponds to a bit that can be carried by a fluctuation in the frequency of sound. However, because these circuits can actually carry a wider range of sound than can be heard by the human ear, there is a lot more range with the 56 kHz in each circuit. In general, an analog circuit provides less than 33.6 kbps of converted data stream bandwidth because inefficiencies in converting digital to analog signals. That 33.6 kbps is the functional upper limit. Data compression provides from 2 to 7 times this throughput depending on the compressibility and latency characteristics. The lack of modem reliability for anything above 33.6 kbps made it clear that some other technology was necessary.

ISDN has become the right product at the right time. All the service providers (regional providers and long-distance carriers) in the United States have agreed to support National ISDN in their central office switches for uniform interoperability. The fact that ISDN was widely available through PacBell in metropolitan areas in California accelerated its trial, success, and wider application. The real delay with ISDN has been that it requires a basic infrastructure so that users can connect remote sites, access the Internet, call up CompuServe or America Online, or share videoconferencing. That was placed into service during 1995 and 1996. Because ISDN is now cheaper than two analog lines just about anywhere in the United States, this will accelerate its acceptance and utility. Furthermore, events like the 1994 Northridge earthquake in Los Angeles and other natural disasters splitting communities and complicating commutes have advanced the state of the art for telecommuting. Organizations have found it to be effective, cost-effective, and efficient.

Ethernet also supports ISDN in another odd twist of technological bundling. IsoEthernet, for *Isochronous Ethernet*, also known by the Institute for Electrical and Electronics Engineers (IEEE) 802.9a designation, supports standard 10Base-T and 96 ISDN channels on standard unshielded twisted-pair wiring. The concept for this Ethernet enhancement was real-time support for videoconferencing within the office network environment. While National Television Systems Committee (NTSC) analog video is more easily transported on coaxial cable and is more efficient than digitized video, the integration of Ethernet and ISDN adds to the infrastructure supporting digital telephony, videoconferencing, and data networking.

Seventeen Years of Hyperbole

It's true! ISDN died in the rhetorical overstatement over the years as the product which jaundiced users proclaimed "It Still Does Nothing." ISDN was a dead product for IBM mainframe connectivity, late to the table, and without mass vendor support. ISDN may still be a product too late for mass acceptance, except I think the pricing issues, general availability, and flexibility finally make it viable. It really is a question on price, needs, and other options. The bottom line for ISDN, if you are considering it and have not committed to this package of services, is current availability, integration with standard POTS, interconnectivity with existing infrastructures, and your requirements.

Although ADSL, HDSL, ATM, MMDS, and cable modems seem to foretell the end to ISDN, contrast this with the assumption from the inventor of Ethernet that his 1975 creation would be superseded by 1988. Ethernet is still going strong and the number of installations has increased by 4500 percent since that terminal date of 1988. Cable providers are promising interactive communications via cable modems with bandwidths from 1 to 40 Mbps. Some vendors of ISDN equipment, including Motorola, Nortel (Northern Telecom), and Intel have produced working cable modems with bandwidths to 10 Mbps. This bandwidth and hence speed is misleading because these devices access a bus network where bandwidth is *shared* communally. In addition, the network backbone is only one of many factors in performance. Noise, packet overhead, host server loading, Internet service provider performance, and bandwidth between modem and computer are also significant issues. I think the utility for cable modem connectivity will find its niche as on-demand movie delivery, video gaming, and interactive TV rather than as an alternative Internet, data delivery pipe, or high-speed infrastructure.

Similarly, ATM is promising greater interoperability and bandwidths. The forecasts are irrelevant. If you need the flexibility of ISDN or its bandwidth, it is a cost-effective delivery method available now. Equipment is cheap. Software is functional and becoming more flexible. I use it. I like it. You can use it from DOS, Windows, Unix, or host mainframes. Some ISDN hardware supports the Hayes modem command set, TCP/IP, SLIP, and PPP with the result that ISDN can be a compatible upgrade for analog modem sets. ISDN can be more reliable, less prone to latency, and cheaper to use than existing or future modems, frame relay, ATM, SMDS, and dedicated or switched leased lines.

In contrast, the two-way infrastructure for cable modems is not widely available, has yet to be standardized, and connectivity to POTS and other services has not been demonstrated. ATM works, only sort of, because the adaptation layers for telephony, video, and data conversions are not fully defined yet. Security has not been defined for either the cable modems or ATM. These products are promising, but that is all at this time. ISDN is promising too. However, it is widely available, and so long as you accept the available service (not hoping or waiting for more implementations) and use it where cost-effective, it is a good solution. Realize that hardware and particularly software are developing and are not fully complete. I think the hyperbole of ISDN is now history, resolved by its widespread availability and efficient pricing.

Viability

National ISDN standards 1 and 2 are already accepted and widely implemented. The question for most telecommunication providers is whether the central office switches will support National ISDN-3 (NI-3). NI-3 will complete interface uniformity (for all ISDN equipment), sharing of directory information and numbers over multipole terminals, and support for contention. Other features may include noninitialization of an ISDN device, automated configuration and provisioning, and uniform assignment of features on a directory number and call-type basis. This is not directly a consumer concern, but rather a carrier and equipment vendor concern. However, I have included this information as a way of showing the commitment of regional and long-distance carriers to ISDN and the importance of ISDN as a "consumer" product. NI-3 merely shows the commitment to the increased ISDN uniformity worldwide.

ISDN provides more service than comparable analog lines, and as such, represents a significant enhancement for all organizations with many phone lines. Furthermore, because it is digital, many supplemental services that cost extra (such as Caller ID, call forwarding, conference calling, and many other services) are built into ISDN. Anyone and any organization hoping to compete with internetworking and remote-access services will benefit from ISDN now and will see this prerequisite for business in the future.

Definition of ISDN

ISDN is not easily defined. This confuses and repels potential users. Well, it confuses the local provider too, and you will probably have to deal with them on their terms. ISDN is by definition an *interface*, not a service and not really a protocol either. However as ISDN is presented and delivered to the user, it is effective to consider that ISDN is a bundled package of digital services and supplemental delivery services provided on the public switched telephone network (PSTN). The reality of ISDN is that it is not a single service, product, or delivery package. That can be confusing. However, ISDN is standardized as a single 56/64-kbps switched digital delivery *bearer* channel. It is bundled with at least one control *data* channel for *out-of-band signaling*. A *channel* is a separate pathway which can support at least one single two-way conversation. When at least one channel is bundled as a service, and there is usually more than one, the bundle is called a *circuit*. A circuit maps to a pair or several pairs of copper wire or a fiber optic connection. The out-of-band signaling means that no part of the 56- or 64-kbps bandwidth is used for management and control of the call. The bearer channel is also called the *B* channel and the data channel is called the *D* channel. Figure 2.15 illustrates this.

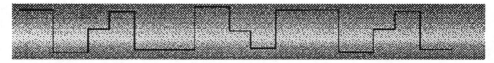

Switched digital "bearer" channel

Out-of-band call control channel
(call overhead only)

Figure 2.15 ISDN is a switched digital circuit with at least one data channel and a call control channel dedicated for management overhead.

Although most ISDN bearer channels will support 64 kbps, the actual central office switched infrastructures *at both ends* of a connection (and any part of the circuit between the ends) might limit that bearer channel to a subrate 56 kbps. That limitation is driven by whether the switches at the central offices provides out-of-band call management. Most network data and telecommunication protocols, this includes FDDI, Token Ring, Ethernet, ATM, frame relay, and even voice phone calls, use part of the delivery bandwidth for call control. This is called *in-band signaling*. Out-of-band signaling is analogous to T-1 protocol support for Signaling System Number 7 (SS7) in the United States. Out-of-band signaling and call management provide a clear channel (that is, full 64 kbps digital data capacity per channel). The difference between 64 and 56 kbps is the in-band signaling. In-band signaling technically uses a data framing protocol called *Zero Code Suppression,* or *raw-bit ones,* for signaling and timing. By the way, you can achieve the 64-kbps speed only with synchronous operation. Bandwidth is limited to

57.6 kbps with asynchronous connections. Similarly, when two lines are bonded, asynchronous connections top out at 115.2 kbps.

That bandwidth is a function of ISDN switching capabilities at the central offices and long-distance provider. Whatever speed is available (that is 56, 57.6, or 64 kbps), all can interoperate with a switched digital 56-kbps line or a leased 56-kbps line. Line speed is usually automatically adjustable when an ISDN circuit is connected. Figure 2.16 shows the many ISDN service packages. Each pipe in the illustration corresponds to a channel in the ISDN circuit. Note the 16- and 64-kbps D channels and 64-kbps B channels. These are combined in various service packages for 0B+D, D+D, 2B+D, 6B (ISDN-H), and 23B+D circuits. This chapter will define these services in greater detail.

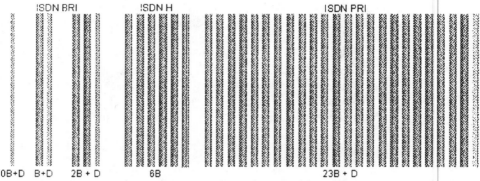

Figure 2.16 ISDN is available as BRI, H, Primary Rate Interface (PRI) formats.

ISDN is a set of digital transmission protocols defined by ITU/TSS (Telecommunications Standardization Sector parent, the International Telecommunications Union). The protocols are accepted by virtually all the world's communications carriers as standard. The characteristics that distinguish ISDN are:

- ISDN builds on groups of standard transmission channels.
- ISDN handles all types of information.
- ISDN handles many devices and telephone numbers, on the same line.
- ISDN offers variable transmission speeds.
- ISDN uses switched digital connections.

Bearer channels (or B channels) transmit user information at relatively high speeds, while separate *data channels* (or D channels) carry call setup, signaling, and other information. Up to eight separate telephones, fax machines, and computers can be linked to a single ISDN channel and have up to 64 "call appearances" of the same or different telephone numbers. Two or more channels can be combined into a single larger transmission "pipe." Channels can be assembled as needed for a specific application (a large videoconference, for example) and then broken down and reassembled into different groups for different applications (normal voice or data transmissions). Combining B channels is called *inverse multiplexing*, or *bonding*. Perhaps the most important single feature of ISDN, however, is that it offers inexpensive *dialed* digital access to the worldwide telecommunications network. Representative standards describing actual ISDN implementations include Euro-ISDN (defined by European Telecommunication

Standards Institute [ETSI], 1TR6, VN3 and VN4, ISDN 30, and DASS-2). These are all various service bundles of bearer and data channels into deliverable packaging.

ISDN Basics

ISDN is delivered from a digital switch through two types of user interfaces: the *Basic Rate Interface* (BRI) and the *Primary Rate Interface* (PRI). Each consists of a number of *bearer channels*, or *B channels*, coupled to one *data channel*, or *D channel*. As defined for all implementations of ISDN in disparate signal delivery methods, channel packaging, or usage, the B channels are subrate 56-kbps (with 8 kbps allocated for call control overhead) or 64-kbps clear-channel connections. Any bearer channel can be used for dial-up voice and data connections. The D channel is defined as a packet-switched setup and signaling connection shared by all users of ISDN.

It should be noted that the current implementation of ISDN in the United States varies from the ITU/TSS standards. The primary difference is in the transmission of the D channel call-signaling information. When ISDN connects to a switched 56 kbps line, the bandwidth is automatically reduced to that rate as well. SS7 is functional in most urban areas in the United States. Once it has been deployed elsewhere, and connected, or *internetworked*, to all areas of ISDN, all ISDN providers will match ITU/TSS recommendations with an *out-of-band* D channel network and full 64-kbps clear-channel connections.

One of the other ISDN features is that instead of the ring signal to activate your handset for an incoming call, ISDN sends a digital packet that tells the name and number of the calling party and the type of call (voice, packet data, stream data, bonded channels). ISDN lines do not "ring" without some conversion of this digital packet into the familiar sound. Your equipment can also analyze this header information and make an intelligent decision of how to route it. For example, a phone (with voice-only capacity) would completely ignore a packet data or data stream call while a data device would respond to it by accepting the call, rejecting the call, routing it to a particular party or end-user equipment, or establishing a special linkage to enabling software application.

If you have several ISDN directory number *aliases* provisioned to your line, you can program certain phones to answer calls for certain numbers only. Data calls also contain baud rate and protocol information within the setup signal so that the connection is virtually instantaneous.

Physical ISDN Interface and Reference Points

The ISDN interface is defined by the ITU/TSS. Equipment and wiring complying with this specification are guaranteed to be compatible with CPE equipment. There are two types of terminal equipment (TE), devices with a built-in ISDN interface and those without. Equipment with a built-in interface is known as *TE1*. This corresponds to *NT-1* equipment. Equipment without a native ISDN support is known as a *TE2*. This corresponds loosely to *NT-2* equipment. You need to know this distinction when ordering lines and particularly when specifying end-user equipment when you want to integrate existing analog-based services to ISDN. Terminal equipment comprises terminals, PCs, mainframes, hosts, servers, routers, telephones, fax machines, and videoconferencing equipment. Terminal adapters translate data from a non-ISDN TE2 device into ISDN signals. The relationship between these physical interfaces for ISDN is shown in Figure 2.17. The most practical information in this figure is the top line of the illustration showing the local loop, the demarcation point, and the customer wiring. This is what you physically need for most installations used for remote or Internet ISDN access.

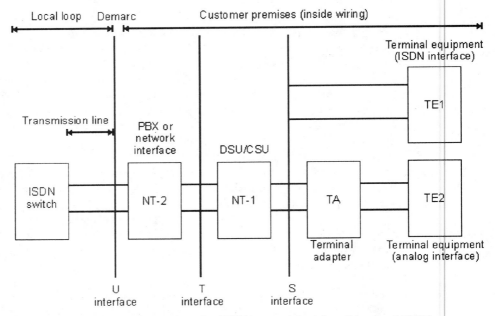

Figure 2.17 The physical interface for ISDN is consistent for all types of ISDN.

However, when you need more services, such as voice, data, or video and also connections to older PBX equipment, you need more information about the physical interfaces. Reference to the S, T, and U interfaces is omnipresent in ISDN literature and vendor specifications. The S interface is a four-wire interface that connects terminal equipment to a customer switching device, typically a PBX, for distances of up to 590 meters (m) or 1912 feet (ft). The S interface can be a passive bus to support up to eight terminal equipment devices on the same bus chain. Under this design, a B channel is dedicated to a specific piece of equipment for the duration of a call. Devices that provide on-premises switching, multiplexing, and ISDN concentration are NT-2 devices. ISDN PRI can connect directly to NT-2 devices, but ISDN BRI requires a different format—NT-1—for wiring termination.

The T interface is a four-wire interface that connects the customer site NT-2 switching equipment to the local loop termination called the *network terminator 1* (NT-1). This device physically connects the customer site to the telephone company local loop. For ISDN PRI service the NT-2 is comparable to the functions of a CSU/DSU. The NT-2 breaks out individual or bonded B channels. For ISDN BRI service, individual ISDN data service, or conversion to analog, (the NT-1 is called an NT-1) and is also required.

Because the T interface is only associated with NT-2 and TE2 equipment, most single-user applications bypass this interface completely. As such, it is often designated as the *S/T interface*.

The U interface is a two-wire local loop. This BRI standard is not supported outside of North America. It is the common interface in North America and Asia. BRI service is not very common at present in Europe where the S interface is more common for BRI service. Length limitations are presented elsewhere. The only reason you want access to the U interface is for debugging problems, special CTI applications, or compatibility with U interface equipment.

ISDN includes several physical connections and these are referred to by the ITU ISDN definition as *reference points*. The name of the S/T bus comes from the letters used in the ISDN specifications, which refer to two reference points, S and T. Point T refers to the connection between the NT-1 device and customer-supplied equipment. *Terminals* can connect directly to NT-1 at point T, or there may be a PBX or a special ISDN hub. When a PBX is present, the point S refers to the connection between the PBX and the terminal. Note that in ISDN terminology, *terminal* can mean any sort of end-user ISDN device, such as data terminals, desktop computers, network communication servers, telephone, and facsimile machine.

The T bus is a multipoint bus, sometimes called the *passive bus* because there are no repeaters on the line between the NT-1 and the devices. It can be implemented using the same cable and connectors used in 10Base-T or 100Base-T Ethernet. There may be up to eight devices on the S/T bus. The bus may be formed with splitters and T connectors. The D channel is used to control the attachment of the one to eight devices to the two B channels, but realize that although you can physically attach eight devices, no two devices can be active on the same B channel at the same time.

The major function of the NT-1 is to allow more than one device to have access to the two B channels provided by the ISDN BRI. For example, the NT-1 can connect an ISDN telephone, an ISDN fax, and an ISDN computer interface attached to the ISDN BRI circuit. Each device can listen for calls and only connect to a B channel when it identifies a message requesting a service it can provide based on the directory number.

The NT-1 only implements part of the channel-sharing scheme; the other devices participate as well, and the communication protocol used by the NT-1 and the other devices after the NT-1 are an integral part of the channel-sharing scheme. The NT-1 also performs other functions; it translates the bit-encoding scheme used on the lines between it and the telephone company (which is the U loop or local loop) to the encoding used between it and the devices. These schemes are different because the device to NT-1 encoding was designed to enable channel sharing, whereas the NT-1 to central office switch encoding was designed to allow transmission across the long local loop distances.

In this configuration, the wires at points S and T are point-to-point links. Electrically, the S and T points are the same, which explains why the term *S/T bus* is always used in this format. An NT-1 (network terminator 1) is a device which provides an interface between the two-wire twisted pairs used by telephone companies in their ISDN Basic Rate (BRI) network and an end-user's four-wire terminal equipment. The NT-1 also provides power for the terminal equipment.

Terminal adapters are available that will interface non-ISDN TE2 terminal equipment to the S/T interface. At least one RBOC provides a modem pool to allow for interchange of data with POTS subscribers. Bellcore may approve a standard to allow an analog *pair* to interface to handsets from an NT-1; the BitSurfr already does this trick. If you want to connect two active analog handsets, the interface device must support NT-2 and supplemental NT-1. To add confusion to the hardware acquisition process, some vendors do not recognize the NT-2 designation and simply call this *Advanced NT-1*.

BRI

The *Basic Rate Interface* (BRI) is defined as two 64-kbps bearer (B) channels, and one 16-kbps data (D) channel that carries both call setup and user packet data across the network. The BRI interface is also referred to as a *2B+D* connection. The first channel is labeled B1; the second is B2. You might think of the bandwidth at 144 kbps (192 kbps with the 40 kbps in out-of-band overhead) as a *fractional T-1*. Fractional T-1 is a service that is much more expensive when not supplied in the form of ISDN. The ISDN local loop consists of two wires to a U interface or four wires to an S/T interface. In Europe, the service provider owns the NT-1 and provides an S/T interface, special provisions, and the E-1 ISDN PRI service for NT-2 services.

The terminal adapter or other ISDN equipment must be compatible with the S/T interface. Livingston Enterprises, a vendor of ISDN terminal adapters illustrates the functional and electrical differences between the U and S/T interfaces with the schematic in Figure 2.18.

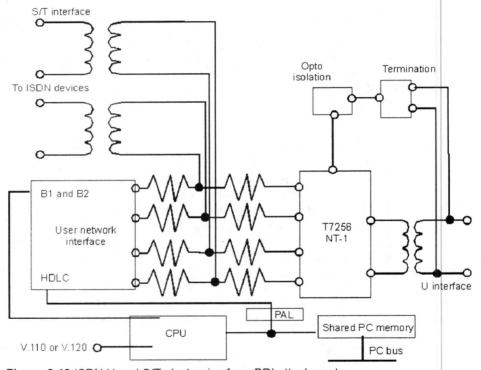

Figure 2.18 ISDN U and S/T electronics for a BRI attachment.

As noted earlier, BRIs can carry a wide and flexible range of communications. A single BRI, for example, can carry two simultaneous voice or data conversations (to the same or different locations), each channel call, or "switch," to a different phone number. In either example, the D channel can also be used for packet communications to a third location, also simultaneously by dialing a third phone number. Part of the D channel, 16 kbps, is used for call setup, and control is not accessible to the user. This leaves only 9.6 kbps for actual use data. Some of the literature refers to this 7.4-kbps overhead as the *E* channel. This is mentioned in the CCITT I.412 standard of the famous Telecommunications Red Book (1984) but has since been dropped from the Telecommunications Blue Book (1988). Unfortunately, vendors do not always have the latest documentation and create their own based on old standards. You occasionally see this outmoded reference.

The two B channels can also be bonded for transmitting data at uncompressed speeds of up to 112 kbps currently and 128 kbps with end-to-end clear channel. (Recall you also get slightly less with asynchronous ISDN line provisioning.) Computers connect to ISDN lines through standardized NT-1 interfaces. At the simplest level, ISDN BRI is a bundle of two phone lines that you can dial just like an ordinary phone. The ISDN BRI service is shown in Figure 2.19.

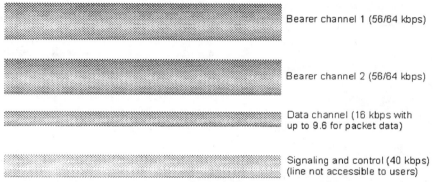

Bearer channel 1 (56/64 kbps)

Bearer channel 2 (56/64 kbps)

Data channel (16 kbps with up to 9.6 for packet data)

Signaling and control (40 kbps) (line not accessible to users)

Figure 2.19 ISDN BRI is three channels. The fourth "channel" shown here is the out-of-band call control channel.

Note that Basic Rate ISDN is delivered on one pair, so the three usable channels must be demultiplexed—that is, split from the combined signals into separate channels—with a *digital service unit/channel service unit* (DSU/CSU). Typically, this hardware device is the NT-1 or an ISDN terminal adapter with an integrated NT-1.

Each B channel is compatible with a switched 56-kbps digital line, and the entire ISDN bundle is generally less expensive than even one switched 56-kbps line. ISDN and switched 56-kbps will interoperate, and ISDN and switched 64-kbps (same thing except you get clear channel because the local service provider uses out-of-band management) also will interoperate with ISDN, providing clear-channel ISDN service to the switched 64-kbps termination. When you dial CompuServe and get 56 or 64 kbps you are really connecting into a switched digital service. Eventually CompuServe and other BBS providers will include true bonded ISDN connectivity at speeds of 112 and 128 kbps and maybe even higher as they adjust to support Internet access in addition to the core forum and mail services. This entails support of a multiple-channel ISDN bonding protocol that combines two or more B channels into a single communication pipe with the bandwidth of each B channel aggregated together. These protocols are still being developed and defined, although the most common ones in use include Multilink Point-to-Point Protocol (MP or MLPPP), Bandwidth Allocation Control Protocol (BACP), or Multilink Point-to-Point Protocol Plus (MPP or MP+). All ISDN services are referenced by a *service profile identification*, also known as the *SPID*. There is usually a SPID assigned to each and every channel. It is a reference number to the overall ISDN circuit (consisting of one or more channels), a pointer to the circuit switch configuration and the overall ISDN service provisioning. Chapter 3 details the many SPID formats in excruciating detail. Do worry because Chapter 4 shows you how to configure them. For practical purposes, think of SPIDs as your ISDN phone numbers.

ISDN BRI provisioning supports up to eight linked devices for each U or S/T reference point and up to a maximum of 64 individual directory numbers. The directory numbers are not the SPID, but rather regular logical and physical phone numbers. This is a technology defined in National ISDN level 3 that supports more application-specific features and services to provide end-users with the ability to select any increment in service from 64 kbps to 1.534 Mbps and to Euro-ISDN E-1 bandwidths.

While each ISDN BRI line can support up to 64 aliases, each B channel can daisy-chain as many as *eight separate devices* (telephones, computers, fax machines, and more) with each given as many separate telephone numbers as needed. This means it is no longer necessary to have multiple telephone lines to handle multiple telephone devices, multiple telephone numbers, or multiple telephone calls. However, you are limited to one conversation per channel at a time and to only one analog connection with NT-1. ISDN telephones can, of course, call to and receive calls from ordinary telephones everywhere, since the digital and analog systems are fully interconnected. Ordinary telephones can plug into most ISDN terminal adapters or NT-1 devices after the S/T reference point. You might also note that digital ISDN connections are absolutely quiet all the time. There are no static and background conversations as are associated with analog voice circuits.

A special unbundling of ISDN BRI services without the bearer channels and only the D channel is useful for data packet service. It is a hybrid carrier for X.25. This service is called *0B+D* because there are no bearer channels and only the single D channel. Although you might perceive that the D channel is only for call control and signaling information, this channel is really 64 kbps, of which 40 kbps is allocated for management (that users never see) on the so-called *E channel* no longer defined by current literature. As previously stated, after all the overhead is deducted, this leaves 9.6 kbps available for packet data services as a dedicated line or on-demand by dial-up call. The D channel can be connected to standard public X.25 packet service clouds at the central office to provide a cheaper alternative (about $30 per month or less) to dedicated or lease lines. The standard protocol for the D channel is *LAPD*, or *link access protocol for the D channel*. Figure 2.20 illustrates the typical configuration for a "dedicated" ISDN service to support bank ATM machines or credit card processing.

Although 16 kbps may not seem like a lot of bandwidth for LAN data transfers and remote site connectivity, realize that only about 9.6 kbps is actually available for data. The rest is used for call signaling. Although many ISDN/X.25 equipment vendors talk about 16-kbps capabilities on the D channel for data, the actual useful data stream is at most 9.6 kbps. A D channel can be configured with ISDN PRI to yield a full 64-kbps data stream, or many B channels (with ISDN BRI or PRI) can be bonded together for X.25 or various committed information rates for frame relay service. The capacity is a function of the central office switch and the ISDN provisioning.

Nevertheless, 9.6 kbps is an extraordinarily luxurious connection for automatic teller machines, point-of-sale (POS) terminals, lottery terminals, cash registers, credit card scanners, and check verifiers. These devices typically run on analog multidrop lines at speeds of 2.4 kbps or slower, and ISDN 0B+D can represent a significant improvement that can really improve overall performance.

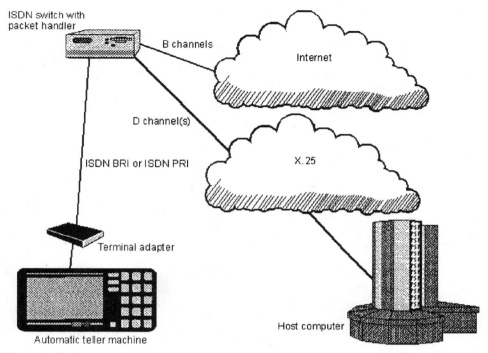

Figure 2.20 Dedicated hardware and connections for ATM and credit card processing using the ISDN D channel for packet data transmission.

Most router vendors include options for X.25 support, but if you need a single solution box for X.25 connectivity, look at the Microtronix ISDN X.25 Transactor because it is a stand-alone solution. In fact, most POS and credit card terminals operate at 300 or 2400 bps. 0B+D is attractive for retailers, banks, and order-processing organizations because the cost is less than for standard POTS connections. Common uses for ISDN D channel and X.25 services are illustrated in Figure 2.21.

You might also check the service costs for packet bandwidth greater than just 9.6 kbps. Typically, a DS-0 circuit will provide anywhere from 300 bps to 56 kbps and, with central office switching hardware will support ISDN for virtually the same costs as 0B+D. If a retailer needs this connectivity and uses telephones also, 2B+D, the standard ISDN BRI line, actually makes more sense. The BRI bundle can be split at the NT-1 for analog telephone service on the first B channel, a fax or modem line (and even an integrated CTI solution) on the second line, and packet data services on the D channel for a quarter of the cost of two separate analog phone lines and a dedicated digital line.

The incremental cost for converting an analog line to an ISDN line averages $5 to $10 in the United States, according to Jim Love at the Consumer Project on Technology. This means the cost for each of these three service lines is less than a comparable analog line. Most service providers are charging rates roughly equivalent to comparable analog services, but less than one tenth of a single comparable digital line.

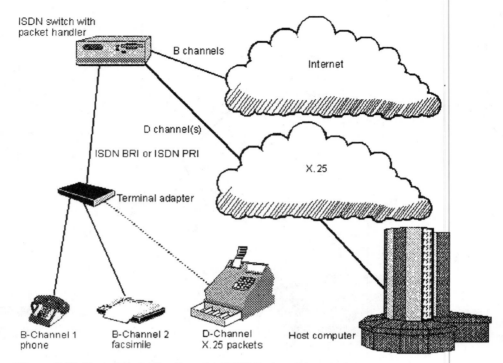

Figure 2.21 Common applications for ISDN D channel packet data transmission.

PRI

The *Primary Rate Interface (PRI)* in the United States is a T-1 circuit that transports the ISDN protocol. The circuit consists of 23 separate 64-kbps bearer channels and one 64-kbps data channel; it is often called a *23B+D* connection. Channels are labeled B1, B2, through B23. B24 is available under special cases when the D channel call setup and management is provided on another circuit. Think of this circuit as the next size up from ISDN BRI. With a total bandwidth of 1.534 Mbps, it is designed for transmission through the standard North American and Japanese T-1 trunk. In Europe and the Pacific Rim, because transmission standards differ slightly, the Primary Rate Interface is supplied through a standard 2.048-Mbps E-1 channel, and consists of either thirty or thirty-one 64-kbps B channels, and one 64-kbps D channel, thus *30B+D* or in somewhat rare instances, *31B+D*. By the way, this D channel can support X.25 bandwidth to the full 64 kbps. The official ITU/TSS designation for North American ISDN PRI lines is *H11* at 1.536 Mbps. In Europe it is *H12* at 1.92 Mbps and *H13* at 2.048 Mbps. Figure 2.22 illustrates these differences.

Each pipe in Figure 2.22 corresponds to a channel in the ISDN PRI circuit. Note the 64-kbps D and B channels. The single D channel is on the left of each service bundle. These channels are combined in various service packages for 23B+D, 24B, 29B+D, 30B circuits. Although the specifics of ISDN implementation are still slightly different from nation to nation, interconnections between any two systems in the world are now not only possible, but increasingly practical. Recall that each B channel is a switched digital line that interconnects at either 56 or 64 kbps depending on clear-channel capacity or provides an analog capacity that can be called from or can call to any other analog circuit in the world. ISDN is fully interoperable with both POTS and dedicated circuits.

Figure 2.22 The Primary Rate ISDN (PRI) format differs around the world.

PRIs are dedicated trunks (carried of course by the T-1 or E-1 circuit) that connect medium and large locations to a telephone company central office. Virtually all modern telephone and computing systems can be connected to ISDN through a PRI, including PBXs, mainframe and distributed systems, LANs and WANs, multiplexers and ISDN controllers, and videoconferencing units. PRIs are designed to facilitate the flexible use of these systems by allocating multiple channels to larger units as needed, while supporting individual BRIs for both internal and external voice and data communications.

Each ISDN PRI line can support up to 1536 aliases. Like ISDN BRI, each B channel on ISDN PRI can daisy-chain as many as *eight separate devices* (telephones, computers, fax machines, and more) and each can be given as many separate telephone numbers as needed. The primary interface is the NT-2 and you will require at least TE2 equipment. You are still limited to one conversation per channel at a time without additional sophisticated multiplexing equipment. Once the channels have been split from the T-1 or E-1 circuit, ISDN telephones can, of course, call to and receive calls from ordinary telephones everywhere, since the digital and analog systems are fully interconnected.

You can link a T-1 into a PBX or portable ISDN hub in areas where the Primary Rate Interface is not available. Depending on the technology of a PBX and its currency, you can connect the PRI directly into the PBX or else separate the channels and route these separately into it. The required PBX is similar to existing business telephone switches used today but must be ISDN-compatible and requires additional ISDN line cards to support the BRI lines to the end terminals. The PBX or ISDN hub will then provide BRI or single B channel services. The PRI is usually linked into an ISDN-enabled PBX or portable ISDN hub, and the PBX or hub then provides BRI to multiple units. In this method, the PRI can provide ISDN access to the public switched network or directly to the Interexchange Carrier (IEC) for multiple terminals. A portable ISDN hub is essentially a smaller, less expensive PBX that can provide BRI lines to end points with a PRI, T-1, or E-1 interface to the network. ISDN can use the same physical wires as POTS but requires digital switching equipment in the telephone company's central offices.

One of the limitations of T-1, and ISDN in particular, is that the local loop must be less than 3.7 km (2.3 mi) without local loop repeaters. Local loop repeaters are not the same as loading coils used for analog T-1 lines. Repeaters are typically expensive to install and maintain. Since most local access providers used to charge installation rates based upon distance from the central office for T-1, getting loading coils for T-1 wasn't a problem to support sites very distant from the central office. However, because ISDN PRI tariffs are computed differently, extreme distances and fixed installation costs provide limited incentive for the local provider to reach distant users. Central site hardware and several new transmission signal methods including high bit-rate digital subscriber loop (HDSL) makes it possible to install local ISDN loops with repeaters and make it attractive for the telcos to wire virtually every site.

ISDN PRI is scalable beyond 1.534 and 2.048 Mbps. That is by no means the top end in bandwidth for ISDN. In fact, a single 64-kbps D channel can carry, negotiate capabilities, provide call setup, and enable B channels for up to 19 separate ISDN PRI lines. This is very efficient because a single control line can manage call control for up to 455 B channels. The ITU/TSS specifications name 569 B channels as the functional limit. That aggregates to 36.416 Mbps. This bandwidth enhancement for ISDN is illustrated in Figure 2.23.

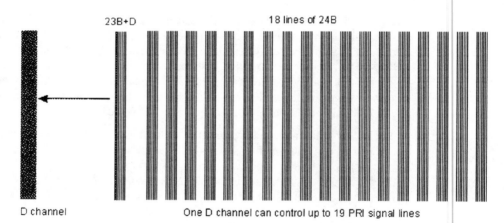

Figure 2.23 A single 64-kbps D channel can manage the 23 bearer channels and 432 other bearer channels in 18 other Primary Rate ISDN circuits configured as 24 B channels. It is usually advisable to provide some D channel backup.

Realize that any failure of that single managing D channel or failure of the PRI circuit with that D channel will disable all other subsequent PRI circuits. In general, the operational philosophy is to allocate a backup D channel for each eight PRI circuits. Switchover to the backup D channel is handled by the central office switch and the customer premise equipment. Backup D channel switchover takes approximately 1 to 4.5 s at the central office in the event of a primary D channel or circuit failure. This assumes that the failure was not caused by a power failure, central office switching failure, or other catastrophic events at the site.

One interesting financial consideration is the break-even point for multiple ISDN BRI and a single ISDN PRI connection. In general, the break-even point occurs with eight B channels or four ISDN BRI lines in *most* places. In Europe, ISDN BRI is difficult to find and in some U.S. regions ISDN PRI costs the same as T-1, about $1000 per month. Hence the break-even point occurs with 20 B channels. If you already have

ISDN BRI in place and outgrow it, the break-even point is higher because you must factor in the costs of a new NT-2 (and possibly integral NT-1) support and bus system for the ISDN PRI line. The BRI NT-1 and terminal adapters are not generally of use with the new service unless the PRI hub can break out single B channels for user terminal adapters and NT-1 units.

Protocol Formats

ISDN PRI is carried within the common T-1 circuit using a protocol based either on zero code suppression with raw-bit ones or on SS7. The zero code suppression framing overhead reduces bandwidth to 56 kbps per channel while SS7 yields 64 kbps per channel. The SS7 protocol alternates streams of 3-volt pulses over 100-ohm twisted-pair wire. Each pulse is the opposite of the polarity in order to maintain an average DC-voltage level close to 0. In order to keep the transmissions phased locked, the transmitters must send a minimum number of pulses per second. One way to do this is to use a bipolar eight-zero substitution (B8ZS) code. This is a stream of four pulses into each string of eight zeros. However, the transmitter keeps the polarity the same in a technique called *bipolar violation* so the receiver knows that this is a code word. E-1 is similar to T-1 but it uses a high-density bipolar three coding (HDB3) whenever there are four consecutive zeroes. The concept of a digital frame was designed by AT&T. The extended superframe format used for most AT&T-compatible services is shown in Figure 2.24.

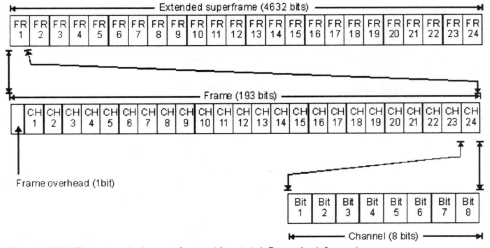

Figure 2.24 T-1 extended superframe (Accent 1.5 service) format.

The difference between T-1 and E-1 is that the E-1 frame has 32 bits per frame and a higher signaling rate. E-1 also reserves a byte of each frame for overhead and the seventeenth byte of frame is reserved for signaling the *hook status* whereas T-1 uses every sixth frame to transmit the hook status. Since this is unacceptable for ISDN data streams. ISDN uses a *call-control message* sent over the D channel. This message can contain:

- Call requests
- Dialed digits

- Party ringing
- Party answered
- Party hung up

The call control and call management is of course routed over the D channel. The primary protocol used to frame and transmit messages on the D channel is a High-Level Data Link Control- (HDLC-) based packet protocol defined by Q.921 and Q.931. Specifically, Q.921 provides flow control, error detection, and messages. Q.931 establishes connections, maintains connections, and terminates them. The calling party can specify the type and bandwidth of service for the call; this information is transported in HDLC frames defined by Q.931. In addition Q.321 provides maintenance and diagnostic messages.

The lower level of protocols includes X.25 level 2 (LAPB), Q.921 (LAPB), V.120 (a symmetric LAPD), and frame relay. LAPB is used to connect computers together using the X.25 frame format. LAPD is applied mostly to attach telephone equipment together. Symmetric LAPD is used with ISDN terminal adapters for peer-to-peer connections so that the connecting devices do not have to negotiate who is in control of the channel and what the connection rate will be. Frame relay is used for fast packet switching.

On top of these protocols, there are ISDN services driven by other protocols. The resource reservation protocol (RSVP) dedicates a portion of the switched connection to time-sensitive data transmissions. Bandwidth allocation control protocol (BACP) is a bandwidth on demand (BOND) mechanism that allows users to aggregate and drop ISDN lines as bandwidth is needed. BACP is useful for X.25 and frame relay as well and works in conjunction with other multiple-line protocols. For instance, multilink point-to-point protocol (variously referred to as MLPPP, ML-PPP, or MP in the literature) is an enhancement of point-to-point protocol (PPP) for adding bandwidth by bonding additional lines together. This is called *channel aggregation*. MLPPP does not work over Serial Line Internet Protocol (SLIP). These protocols are important for bandwidth-strapped Internet users and connections from a public corporate web server (PCWS) to an Internet service provider (ISP). Lines can be added or "torn down" as needed to supplement an analog dial-up, switched digital, T-1, or other carrier connection. For example, if a user is utilizing a single bearer channel and tries to replicate or synchronize a database, a process which is bandwidth-intensive, BACP "bonding" dynamically enables additional bearer channels. BACP would enable the other bearer channel in BRI or as many as available and configured in PRI. The extra channels are dropped from the connection when the bandwidth demand decreases.

Common-ISDN-API, or CAPI, is the application programming interface to ISDN BRI and PRI equipment. CAPI supports facsimile, video telephony, and voice call management. This includes:

- Basic call setup and clearing
- Support for all the B channels
- Support for multiple logical connections within one physical channel
- Support for different services and protocols on the fly
- Transparent protocol interface
- Operating system independence
- Asynchronous event-driven commands

Multirate ISDN

ISDN BRI and ISDN PRI are not the only ISDN services. They are, however, the most widely implemented. A midpoint ISDN service is called *ISDN-H*, which is 384 kbps and also a fractional T-1 service. ISDN-H corresponds to six B channels. The *H* stands for *hyperchannels*. This also is the basic bearer *channel* in the Broadband ISDN (B-ISDN) *circuit* which begins at the highest transmission rate generally available in the European digital infrastructure at 34.368 Mbps. This corresponds to 16 E-1 circuits. In the United States, B-ISDN corresponds to a minimum *circuit* signaling speed of 44.736 Mbps and forms the basic connection for ATM services. There is no D channel overhead or a D channel for that matter, but there is nevertheless out-of-band signaling overhead. This service is primarily designed for videoconferencing, whiteboarding, teleteaching, or support of multisite connectivity. Since each user represents a B channel, this format represents six different participants at a time.

When ISDN-H is used for multisite videoconferencing, a controlling computer (apart from any end-user videoconferencing equipment) manages call setup, call control, and the image for each site. The outgoing video signal on each B channel is typically copied as small remote windows for each participant and directed to the other participants. The management hardware and software, like the PictureTel system, manages the call setup and session. Figure 2.25 illustrates this conference in action.

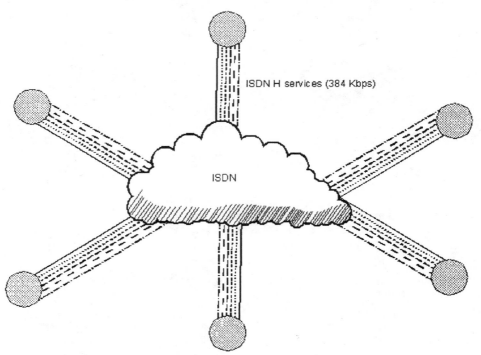

ISDN H services (384 Kbps)

ISDN

Figure 2.25 ISDN-H wired for six-way conferencing.

There are four types of hyperchannels. H0 is 6 B channels, H10 is defined as 23 B channels (North American PRI), H11 is described as 24 B channels (secondary North

American PRI with a primary D channel supplying control), and H12 in Europe for E-1 is 30 B channels.

IsoEthernet

IEEE 802.9a is called *Isochronous Ethernet,* or *IsoEthernet.* This is an enhancement of standard 10-Mbps 10Base-T Ethernet with four ISDN PRI channels, sufficient for 96 B channels. Bandwidth is 10 Mbps plus the 6.144 Mbps for the ISDN B channels. By the way, just to give you a sense of the ongoing integration of digital services with analog telephony, 6.144 Mbps is the exact bandwidth to be implemented with ADSL and HDSL services for broadband home connectivity. IsoEthernet is distributed over standard 10Base-T lines (four pairs) but requires special hubs and network interface controllers (NICs) to break out the ISDN B channels. The PRI channels can be aggregated to carry voice, data, or video. Figure 2.26 illustrates this special LAN and videoconferencing solution. You might also refer to my book *Web Video Toolkit* for additional implementational details on LAN and Internet videoconferencing.

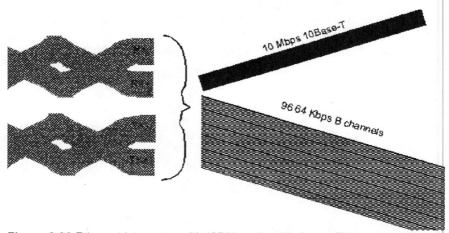

Figure 2.26 Ethernet integrates with ISDN easily with the IsoEthernet variant.

It is not readily known that the combined IsoEthernet B channels are called a *C* channel. There are also two more channels. There is a standard 64-kbps D channel for overhead and call setup and a separate 96-kbps *M* channel for network maintenance. In addition, the standard 10Base-T connection is called the *P* channel and provides Ethernet's first-come, first-served 10-Mbps service. It is also possible to configure the entire 15.872 Mbps available on the IsoEthernet connection into 248 ISDN B channels when the NICs and hubs support that part of the IsoEthernet specification.

Already, there is talk about *Fast IsoEthernet* as a heterogeneous infrastructure for Fast Ethernet and integrated videoconferencing support. The Fast IsoEthernet would combine the 96 B channels with 100 Mbps 100Base-T Ethernet.

ISDN Standards

The formal standards defining ISDN and ISDN-encapsulated services are listed on the next page in Figure 2.27. Refer to Appendix A for web addresses for the ITU web site and other Internet sources for the actual ITU standards.

G.711 Pulse Code Modulation (PCM) of Voice Frequencies

G.722 7-kHz Audio Coding Within 64 kbps

G.728 Coding of Speech at 16 kbps Using Low-Delay Code Excited Linear Prediction (LD-CELP)

H.221 Frame Structure for a 64 to 1920 kbps Channel in Audiovisual Teleservices

H.230 Frame Synchronous Control and Indication Signals for Audiovisual Systems

H.242 System for Establishing Communications Between Audiovisual Terminals Using Digital Channels up to 2 Mbit/s

H.261 Video Codec for Audiovisual Services at p x 64 kbits/s

H.243 Basic MCU Procedures for Establishing Communications Between Three or More Audiovisual Terminals Using Digital Channels Up to 2 Mbit/s

H.320 Narrow-band Visual Telephone Systems and Terminal Equipment

I.2xy ISDN Frame Mode Bearer Services

I.310 ISDN—Network Functional Principles

I.320 ISDN protocol reference model

I.324 ISDN Network Architecture

I.325 Reference configs for ISDN connection types

I.330 ISDN numbering and addressing principles

I.331 Numbering plan for ISDN (and several more in I.33x relating to numbering and addressing and routing)

I.340 ISDN connection types

I.350/351/352 refer to performance objectives

I.410 User network interfaces

I.412 User network interfaces

I.420 User network interfaces

I.421 User network interfaces

I.430-1431 Layer 1 specs

I.440 Layer 2 specs (Q.921)

I.441 Layer 2 specs (Q.921)

I.450 Layer 3 specs (Q.931)

I.451 Layer 3 specs (Q.931)

I.452 Layer 3 specs (Q.931)

I.450 General Overview

I.451-1452Basic ISDN call control and Extensions

I.460 Multiplexing and rate adaptation

I.461 Multiplexing and rate adaptation

I.462 Multiplexing and rate adaptation

I.463 Multiplexing and rate adaptation

I.464 Multiplexing and rate adaptation

I.465 Multiplexing and rate adaptation

I.470 Relationship of terminal functions to ISDN

Q.921 ISDN User-Network Interface Data Link Layer Specifications

Q.931 ISDN User-Network Interface Layer 3 Specification for Call Control

Q.930 General Overview

Q.931 Basic ISDN call control

Q.932 Generic procedures for the control of ISDN supplementary services

Q.933 Frame Mode Call Control

Q.2931 (ex-Q.93B): B-ISDN Call control

V.110 Support of DTE's with V Series Type Interfaces by an ISDN Terminal rate adaptation by bit stuffing

V.120 Support by an ISDN of Data Terminal Equipment with V series Type Interfaces with Provision for Statistical Multiplexing (that is, "bonding")

Figure 2.27 The formal standards defining ISDN and ISDN-encapsulated services.

ISDN Configurations

Prior sections defined ISDN in terms of the basic wiring and logical partitioning of the service in a mix of B and D channels. While that may seem interesting, the logical mapping of actual channels has more relevance to this book, most users, and their applications. An important component in ISDN is the complex mapping of channels to services. For example, while 2B+D could be a bonded 128-kbps service, it could also be a daisy chain of directory numbers for digital service, X.25, fax machines, desktop phones, and analog bulletin board service (BBS) connections. This is actual useful information for making a SOHO for one person look like a multiperson office, routing designated calls into a Caller ID-enabled PC-based application, or enabling more complex CTI functions. This hierarchy is best illustrated for ISDN BRI in Figure 2.28.

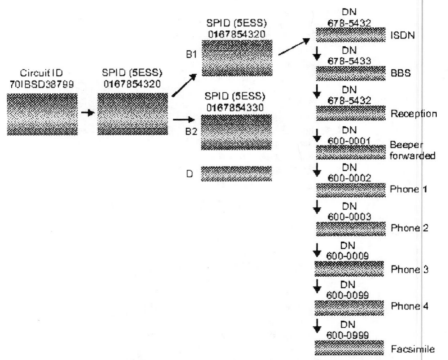

Figure 2.28 The logical mapping of physical ISDN lines to directory numbers.

You can map from two directory (that is, phone) numbers which are based on the SPID to as many as a full directory exchange to each circuit. While the service provider may provide from 10 to 20 directory numbers, so-called because these are publishable phone numbers, to your ISDN circuit as part of the free provisioning, you can get an entire exchange such as XXX-0000 through XXX-9999. However, how you use these numbers is a function of the provisioning and the limitations of the ISDN terminal adapter. There also may be a price to pay for the right to have the exchange. Although the North American Number Plan (NANP) has relieved some of the pressures on telephone numbers and exchanges, the cost and management of the numbers is not free. In general, you can use no more than the two BRI B channels for digital or analog service at the same time even if you map an entire directory exchange to that single ISDN circuit. Some vendor products will not support more than a single active addressable analog number at a time. This means that the first B channel is in use for data, and the second channel can be either an active phone call, an incoming or outgoing fax, or an analog connection to the BBS. If both B channels are bonded, all the lines are "busy."

You may ask why then allocate an entire exchange sequence to a limited BRI line because it seems unreasonable to assign 1000 phone numbers to a circuit that at most can support only two (or three with a packet service) active channels at any one time. The benefit is not limited to making an organization look bigger than it really is, but is also important for large organizations that really have many people.

The primary benefit from the overabundance of directory numbers is for inbound call routing. You can capture the automatic number identification (ANI), better known as Caller-ID (CID), which is built into ISDN and is delivered as part of the call setup packets or is delivered between the first and second "ring" as a mail box routing feature. Voice messages can be delivered based on that ANI number, faxes can be routed through the network based on this number, or this number can be used as a key into a lookup table for enabling secured call-back services. Although you are limited to one, two, or three active lines at any time with one BRI circuit, ISDN is very flexible and you can overcome this obvious limitation.

For example, some of these calls last no more than 1 s, or the general load can sustain the 10 min incoming fax without too many busy signals. In addition when loads are higher, multiple BRI lines can be conjoined with a hunt sequence to handle the overflow. Chapter 4 discusses "hunt groups" in detail and the benefits of this. Chapter 6 discusses call-back and security services, while Chapter 7 details how to implement small office telephony services.

This same concept is extensible to ISDN PRI and multiple PRI circuits as well. You can use all of the PRI or multiple ISDN PRI circuits at the same time and each B or D channel is just another telephone line. With PRI, each circuit can sustain at least 23 digital or analog connections at one time and the benefits of mapped directory numbers become very valuable. If you install special matching PBX systems at corporate sites with signal compression and conversation management software, you can multiplex voice channels into 4.8 to 9.6 kbps and thus get 6 to 14 simultaneous conversations on each B channel. In other words, for the cost of a single phone line, you can get many more. Outside calls still map to a single directory number and must be brought into the PBX on a single B channel, so the telephony integration is seamless. The D channel controls directory number management for the internal multiplexed voice connections; that is the benefit of the out-of-band management on the D channel.

Electrical Characteristics

ISDN basically has no voltage or very low current on the line. This is true for BRI and PRI. The ISDN local loop is mostly an "inert" connection between the central office switch and the customer premise equipment. Electrical characteristics vary based on whether the line is the local loop, whether you are past the demarcation point, and whether you are exploring the U or S/T reference point. Once the U is converted into the S/T, the NT-1 can support wiring to terminals at a distance of up to 250 m (817.5 ft). This includes data terminal adapters and ISDN phones. This is important information because you cannot split an ISDN terminal adapter from an attached fax machine by more than that distance. Although 250 m (817 ft) is an extraordinary distance inside most office spaces, it is possible to create a daisy chain longer than that. These limitations are illustrated in Figure 2.29.

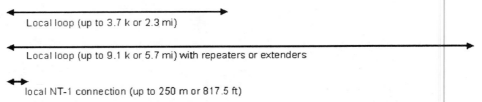

Local loop (up to 3.7 k or 2.3 mi)

Local loop (up to 9.1 k or 5.7 mi) with repeaters or extenders

local NT-1 connection (up to 250 m or 817.5 ft)

Figure 2.29 ISDN wire path limitations.

Typically, ISDN includes a low voltage carrier signal, but this is not guaranteed. ISDN requires an external energy source, that is AC power to drive the local loop. This is illustrated by Figure 2.30.

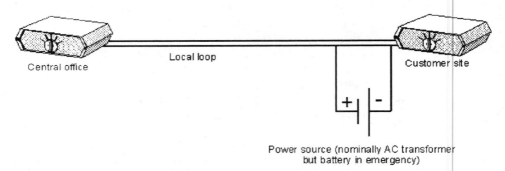

Central office

Local loop

Customer site

Power source (nominally AC transformer but battery in emergency)

Figure 2.30 ISDN is basically an inert connection requiring external power.

Analog telephone and modem lines are typically ±48 VDC. This can fluctuate by as much as 100 percent. Analog equipment derives power directly from the phone lines. Contrast this with digital ISDN services requiring an external minimum of 34 VDC and maximum of 56.5 VDC at a maximum current of 1.3 amps. The maximum current for an external power source must be less than 17.5 amps. ISDN has a carrier signal that is called a heartbeat and sounds like a "pop" if you attach an analog headset to the ISDN line. If you attach an analog handset (that is, a POTS phone) directly to the ISDN line, you are likely to confuse the central office switch so that it disconnects or disables the line. If you disable the central office switch in this manner, you may have to call in a service report so that it can be reactivated, or it may be reactivated after the next day's automatic central office switch reset procedure.

I plugged an expendable handset into a BRI and a PRI line to experiment. The ISDN BRI sound was a loud static hiss for a few to 20 s. After that there was an instant of silence, then a stable tone. This sound was not a POTS dial tone, but was similar. This stopped after several more seconds. The ISDN PRI sound was a loud static hiss that became silence after several seconds. However, ISDN gets power from a separate AC supply. This makes a phone system with emergency 911 access susceptible to power outages; some utility commissions do not allow 911 support over ISDN for this reason.

This limits the value of ISDN as a primary phone service since any power fluctuations, brownouts, or blackouts disable the 911 (emergency call) feature of the North America phone system. In Florida, the lightning capital of the world, fires at substations and problems with feeder lines are very common during thunderstorms. However, a backup power supply—some are even built into ISDN PBX systems or supplied as an external power conditioner and battery backup—is critical for primary phone systems and mission-critical ISDN data connections.

You may need to supplement the ISDN network interface with a backup device to provide -48 VDC to the U or S/T interface. In addition, realize that ISDN lines and equipment are susceptible to power surges, lightning, polarity reversal, and other wiring problems and events. In general, you will want tools to test for power surges, polarity, power conditions, and ISDN line conditions. American Power Conversion was one of the first vendors to provide a device dedicated to protecting ISDN equipment from power spikes and surges and to provide supplemental power backup in the event of a power failure. Motorola is another. An ISDN UPS device is shown in Figure 2.31.

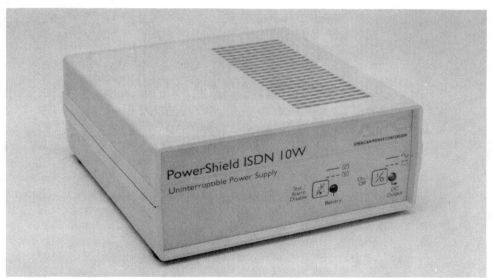

Figure 2.31 A UPS designed for ISDN (courtesy of American Power Conversion).

It is important to recognize that this device does not protect the primary line interface but does protect ISDN terminal devices. In addition, it also provides about 4 hours of backup power (at 10 watts) to keep ISDN working in the event of an AC power failure. Note that the ISDN terminal adapter requires power for data connectivity and standard POTS service in the event of a power outage, whereas the NT-1 and NT-2 require the power if the phone system is entirely based on ISDN. Most ISDN-based PBXs include an integrated power backup system. The equipment connects as shown in Figure 2.32.

This power backup device is designed only for older and two-step ISDN installations. The surge filter for the incoming line could also be a standard line filter, and backup power for integrated ISDN terminal adapters could be provided with a standard UPS battery pack. Since internal ISDN terminal adapters draw power from a PC, you really only need to provide backup power to the PC itself.

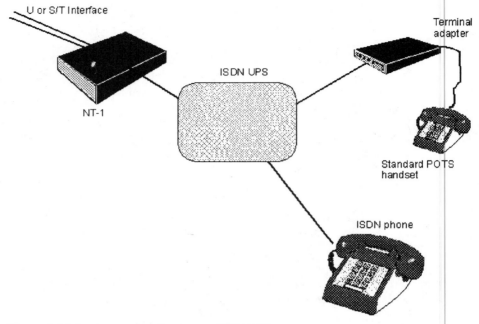

Figure 2.32 Backup connection for an ISDN UPS.

External ISDN adapters typically include an anchorlike supplemental power transformer; plug this into a standard UPS battery pack for supplemental power. In fact, some of the mass-marketed PC UPS devices may be cheaper and suitable for ISDN power; you just need to match wattage. Since most ISDN devices draw from 10 to 40 watts, a 640-watt UPS will supply at least 16 hours of backup to as much as 64 hours. ISDN services provided through a T-1- or E-1-rate facility typically are supplied with emergency power as part of the PBX backup system. This is illustrated in Figure 2.33.

By the way, there are functionally two types of UPS systems. The first filters, tempers, and is always on line. These are typically more expensive devices. The other type of backup device is a switching device with batteries always being charged. Most switching UPS units include lead-acid or gelled electrolytic batteries that degrade over time. The useful life time is about 2 years. Test the batteries and their current efficiency level, which defines how much strength they have and how long they yield battery power in a real outage, on a frequent basis. You could discover that the UPS hardware itself may work in the event of a power dip or outright outage while its batteries might have no cranking power. The network and any devices attached to them will fail. If you are investing time and money into power backup, make the extra effort and pick more capable UPSs and test them periodically.

Hardware

ISDN hardware comes in three main types. There is the external "modem" with its boat-anchor power supply (AC-to-DC transformer and rectifier), the internal adapter card,

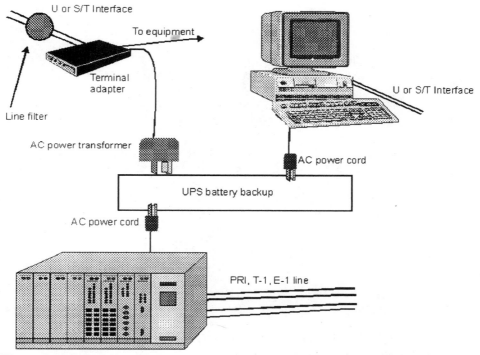

Figure 2.33 Connection for an internal, external, and multichannel ISDN UPS.

and PBX-style modules. External models are simple to configure, slightly more expensive than internal adapters, have nice lights to provide progress and debugging information, but are inherently limited by the speed of the serial port. Serial port speed is both a hardware and software issue. Standard serial ports on PCs typically top out at 112,000 bps, less than the full bonded speed of two B channels on BRI. PCs supporting the IEEE P1394 "Firewire" high-speed serial connection can easily drive the full 128-kbps capacity of a bonded two-channel connection. Most other platforms can sustain that speed as well at the hardware level.

However, the software typically cannot sustain that type of load. Realize that multitasking operating systems such as Unix, Windows NT, NetWare, or Windows 95 may not have the continuous horsepower needed to drive the serial ports at those speeds. Windows 3.x and Windows 95 must still handle I/O through the DOS INT 12 interrupt. Since an interrupt must be used for each 16 bits (the maximum buffer on the Universal Asynchronous Receiver/Transmitter [UART] 16650 chip), throughput saturates at 40 kbps, far less than even one B channel. Even the newest UARTs provide only 8 bytes of buffer. In addition, the standard serial port typically tops out at 40 kbps because of DOS interrupts; this is true for Windows 3.x, Windows 95, and even Windows NT because of the vaguely similar operating system architectures. Windows NT has a better performance and sustains data throughput up to the capacity of the serial port. Solaris, and Unix in general, is optimized for internal I/O and does not support parallel or serial I/O at speeds above 19.2 kbps unless the operating system or communication server and software have been optimized for high-speed I/O. This places a severe crimp in most data streams. This has a bearing on most Internet service providers (ISPs), which typically use SPARC stations and Solaris, because they are unlikely to provide transmission rates greater than analog rates. The architecture basically is to have a network server, a communication server with banks of Telebit serial I/O ports, and stacks of external modems or ISDN terminal adapters on hunting lines.

Internal ISDN adapters bypass the serial port bottlenecks, particularly when they deviate from the serial UART design and instead are designed as network adapters. This will not solve the software bottleneck of the DOS INT 12 interrupt. In addition, the adapter (whether ISA, EISA, or PCI) requires processing power from the host CPU, leading to an extra 10 to 25 percent CPU load. This is not material for when surfing the Internet, but it is relevant when downloading, decoding, and viewing large image files. It is material when the communication server is also a database server and the CPU load is high just managing the data operations.

What ISDN Really Means

While ISDN is the acronym for Integrated Services Digital Network, as you know by now, ISDN really represents the scalable delivery of all types of communications over a switched (standard dial-up) connection. ISDN does not mean you will get faster and better Internet service or be able to download files or browse web pages 5 or 10 times faster than with an analog modem. Although bandwidth is a significant performance limitation for most Internet access, the growth in the Internet user population and the load on the Internet backbone and service providers suggests serious CPU or bandwidth problems at that end. As a result, ISDN may not provide the type of throughput you actually expect.

Also realize that BRI ISDN services are either 56 or 64 kbps, or bonded to 112 or 128 kbps (and in between). While this is only about half the sustained load seen with a typical Token-Ring or Ethernet client connection, your PC may be unable to sustain any of those data transmission rates. This limitation is a function of Windows drivers, disk I/O speeds, and also serial port limitations for external ISDN modems. Many PCs have serial-port UARTs with only 1 byte of buffer space. Even the newer enhanced parallel port (EPP) does nothing to increase the performance of *serial* I/O. Since data compression even with ISDN can yield increases in data throughput from 2 to 7 times, realize that throughputs up to 1 Mbps are simply not sustainable without careful ISDN terminal adapter configuration.

Utility for ISDN

ISDN represents the future of telephony in terms of its paradigm shift from the analog communication system to the digital. This is represented by these features, functions, and services:

- Fast call set up and tear down
- Scalable bandwidth from 0B+D to 23B+D or 30B+D (in Europe)
- Mixed analog and digital support
- Switched packet traffic
- Switched calls
- Packet transport
- Compression on top
- Minimized wiring complexity
- Minimized hardware (i.e., modem pool device count)

Security with ISDN

ISDN is no more and no less secure than any analog phone connection. However, since it does not provide dial tone or an automatic call answer response, it does provide some additional anonymity that a modem-terminated line on a BSS or dial-in communication server cannot. However, the advent of National Caller ID, the widespread availablity of devices to capture automatic number identification (ANI) information, and increased sophistication of hackers and disgruntled former or current employees should not be downplayed. You may deter children, or people without ISDN equipment for some while, but with the growing importance and prevalence of ISDN, it will not provide protection for very long. Also, realize that if you install ISDN NT-2 equipment with terminal adapters supporting both ISDN and analog V.32 and V.35 modem services, you have a responsive modem on the line. It doesn't matter whether that line is digital or analog, ISDN or POTS. Hence, security should not be overlooked at any stage of ISDN implementation.

You can reject all masked calls (those suppressing the Caller ID header with *67 as part of the dialing string) out of hand. However, because ISDN provides more bandwidth than analog connections, there is more overhead for encryption and other security measures. In addition, you can use the ANI, calling party, or Caller ID information passed with every accepted call as an additional verification or as a callback number.

Geographical Availability

ISDN service is available throughout most of the European Union, North America, excluding Mexico, and industrialized areas of Japan, Taiwan, Singapore, Hong Kong, Australia, and New Zealand. Most of the Regional Bell Operation Companies, Canadian service providers, and Sprint, Stentor, AT&T, MCI, GTE, BT, and NTT are some of the local and long-distance (Interexchange Carriers) providing ISDN. However, BRI, PRI, and ISDN-H may not be available within your area. In such cases, you may be able to make arrangements with the local provider for special connection to another nearby provider, installed switched digital services, or an installed dedicated line routed directly to an Interexchange Carrier. Just because a local office says that they do not support ISDN does not mean you cannot get the service or apply ISDN connectivity to all remote sites. The next sections list some of the states and regions covered.

North America

These are the carriers for North America and the regional operating companies for the United States:

```
Canada
    Stentor
Mexico
    AT&T
    TelMex
Ameritech
    Illinois
    Indiana
    Michigan
    Ohio
    Wisconsin
```

Bell Atlantic
   Delaware
   Maryland
   New Jersey
   Virginia
   Washington, D.C.
   West Virginia
BellSouth
   Alabama
   Florida
   Georgia
   Kentucky
   Louisiana
   Mississippi
   North Carolina
   South Carolina
   Tennessee
Cincinnati Bell
   Ohio
   Kentucky
NYNEX
   Massachusetts
   Maine
   New Hampshire
   New York
   Rhode Island
   Vermont
Pacific Bell
   California
Southern New England Telephone
   Connecticut
Southwestern Bell Telephone
   Arkansas
   Kansas
   Missouri
   Oklahoma
   Texas
US West Communications
   Arizona
   Colorado
   Idaho
   Iowa
   Minnesota
   Montana
   North Dakota
   Nebraska
   New Mexico
   Oregon
   South Dakota
   Utah
   Washington
   Wyoming

Here are some contact numbers for providers with ISDN services:

Ameritech: 800-832-6328

Bellcore National ISDN information clearing house hotline: 800 992-4736

Bellcore's "ISDN Deployment Data," Special Report (SR) 2102.
Bellcore document ordering: U.S.: 1-800-521-2673; other: 1-908-699-5800

Bell Atlantic: 1-800-570-ISDN (pricing, availability, tariffs, applications or ordering)
  or +1-301-236-8163 (access to above for international callers)

BellSouth: 1-800-858-9413

Cincinnati Bell: 513-566-DATA

Pacific Bell:
  800-995-0346—ISDN Availability Hotline (automated audio response)
  800-662-0735—ISDN Telemarketing (ordering information)
  800-4PB-ISDN—ISDN service center
  Also, try the gopher servers at gopher.pacbell.com or gw.pacbell.com.

GTE: Menu-driven information service at (800) 4GTE-SW5.
  Florida, North Carolina, Virginia, and Kentucky: 1-800-483-5200
  Illinois, Indiana, Ohio, and Penn. 1-800-483-5600
  Oregon and Washington 1-800-483-5100
  California 1-800-483-5000
  Hawaii 1-800-643-4411
  Texas 1-800-483-5400

Nevada Bell: 702-688-7124 (contact Lyle Walters)

NYNEX: 1-800-438-4736, 800-GET-ISDN, 800-698-0817, or 212-626-7297.

Rochester Tele: 716-777-1234

Southwestern Bell (Texas):
  Austin 512-870-4064
  Dallas 214-268-1405
  Houston 713-567-4300
  San Antonio 512-351-8050
Southwestern Bell (Missouri):
  St. Louis 800-SWB-ISDN (800-792-4736)

US West: 303-896-8370 (contact Julia Evans)

Elsewhere

Since the issue in not just connecting corporate offices to ISDN, but the remote sites as well, you often need to know whether service is available at the remote sites. Figure 2.34 lists ISDN availability by country at the time of writing:

Country	ISDN availability (1995)
Argentina	Not available
Australia	Some availability
Austria	General availability
Belgium	General availability
Belorussia	Not available
Brazil	Not available
Canada	General availability
Chile	Not available
China	Not available
Czechoslovakia	Not available
Denmark	General availability
Egypt	Not available
Finland	Very available, low prices
France	General availability
Germany	Very available, low prices
Greece	Not available
Hong Kong	General availability
Hungary	General availability
India	Not available
Indonesia	General availability
Ireland	General availability
Israel	General availability
Italy	General availability
Japan	General availability
Kuwait	General availability
Luxembourg	General availability
Malaysia	General availability
Mexico	Not available
Mongolia	Not available
Netherlands	General availability
New Zealand	General availability
Nigeria	Not available
Norway	General availability
Philippines	General availability
Poland	General availability
Portugal	General availability
Romania	Not available
Russia	Not available
Singapore	General availability
South Africa	General availability
South Korea	General availability
Spain	General availability
Sweden	General availability
Switzerland	General availability
Taiwan	General availability
Thailand	General availability
Tibet	Not available
Turkey	Not available
Ukraine	Not available
United Arab Emirates	Basic availability
United Kingdom	Generally available
United States	Generally available
Yugoslavia	Not available
Zaire	Not available
Zimbabwe	Not available

Figure 2.34 ISDN availability by country.

Figure 2.35 provides contact numbers for international telecom providers.

Country	Provider	Contact	Telephone	Fax
Australia	Australia Telecom		008 077 222	(07) 220 0080
Austria	PTT Austria	Michael Schneider	+43 1 317 30 39	+43 1 31 3.13.66.63
Belgium	BELGACOM	Egied Dekoster	+32 2/213.46.49	+32 2/921.02.13
Denmark	Tele Danmark	Soren Christensen	+45 3399 6940	+45 3314 5625
Finland	Telecom Finland	Terho Salo	+358 31 243 22 67	+358 31 243 23 83
	The ATC Finland	Matti Tammisalo	+358 0 606 35 08	+358 0 606 33 22
France	France Telecom	Pascal Meriaux	+331 44 44 53 59	+331 44 44 75 50
Germany	DBP Telekom	Volker Fink	+49 6151 83 30 67	+49 6151 83 50 68
Greece	OTE	Mrs Vas. Danelli	+30 1 611 89 96	+30 1 805 20 64
Ireland	Telecom Eireann	John Lawler	+353 1 790 10 00	+353 1 677 49 41
Italy	Iritel	Rocco Gentile	+39 65 494 52 56	+39 65 94 20 54
	Itacable	Rolando Bottoni	+39 65 734 45 23	+39 65 7 34 48 05
	SIP	Bernardino de Rito	+39 6 36 88 40 38	+39 6 36 44 88
Luxembourg	EPTL	Hubert Schumacher	+352 49 91 56 56	+352 49 12 21
Netherlands	PTT Telecom	Ms Corinne der Kinderen	+31 70 34 39 747	+31 70 34 32 473
Norway	Norwegian Telecom	Odd Egil Asen	+47 22 77 71 22	+47 22 2 0 78 00
Portugal	TLP	Antero Aguilar	+351 1 147 797	+351 1 544 796
	Telecom Portugal	Jose Brito	+351 1 35 04 710	+351.1 35 04 197
Spain	Telefonica Espana	Fernando Moratinos	+34 1 584 96 81	+341 584 95 58
Sweden	Telia	Peter Ostergren	+46 8 713 17 99	+46 8 713 73 62
Switzerland	PTT Telecom	Jean-Yves Guillet	+41 31 62 72 27	+41 31 6 2 85 26
United Kingdom	British Telecom	J.M. Pickard	+44 171 356 8952	+44 171 796 9120
	Mercury	Clive Curtis	+44 171 528 2635	+44 171 528 2066

Figure 2.35 Contact numbers for telecom providers.

NTT in Japan even provides ISDN pay phones. This increases the utility and mobile utility for ISDN, but its availability is particularly interesting in light of Japan's minimal use of PCs and the Internet. However, it does suggest that a powerful infrastructure is in place for more advanced communication services than are generally available in Europe, North America, or elsewhere in Asia. NTT's reliance on digital telephony may also reflect the intense population density, the Japanese desire for telephones and cellular phones, and an earlier need for the higher transmission capacities made possible with digital transmission technologies.

Geographical Interoperability

ISDN BRI, PRI, ISDN-H and switched digital services typically interoperate. This presupposes some measure of ISDN terminal adapter rate adaptation to resolve 56- and 64-kbps channel transmission discrepancies. D channel or packet data support is a function of the central, routing, or switch conversion utility at a central office. However, realize that since the costs for national long-distance calls hover around $0.10/min, it is conceivable you can route ISDN connections to other remote sites where the desired services or connectivity is available. The same is also true for worldwide calls, although with international call-back costs at $0.50/min and this may be prohibitive. Nevertheless, you can usually provide full geographic interoperability with some imagination. In addition, ISDN is interoperable with the analog infrastructure for both incoming and outgoing analog phone calls, although not all special digital or supplemental analog call services are supported. This may include conferencing, call forwarding, call routing, and call-back services.

Applications

In effect, ISDN offers the benefits of a private virtual network yet uses standard telephone lines and the public telephone network. Around the world these ISDN advantages are being put to use in a growing number of applications. Some examples include:

- A credit card service company for transaction submittal
- Oil companies for POS pump sales
- HMO to authorize medical insurance benefits
- State agencies to reduce fraud
- Banks to interconnect ATM machines
- Insurance agents to retrieve mortgage, lien, and coverage records
- Real estate agents to retrieve mortgage, lien, and ownership records
- State lotteries to lower costs and improve runaround time
- Restricted entry to buildings and telemetry
- Student ID cards for meal payments, library check outs, dormitory entry, and test authorization

The credit card service company has linked thousands of point-of-sale card readers to its computerized database over a mix of ISDN, switched 56-kbps, dedicated, and leased lines for credit card authorization and transaction processing. Times have been reduced from an average of more than 30 to less than 2 with the transition from analog dialing to digital. Costs have been reduced from an average of about $.05 to less than a penny per transaction. Several major oil companies are testing ISDN telephone lines to link gas pumps, cash registers, and even vending machines into nationwide data networks. Central computers authorize credit purchases, control inventory, and schedule "just-in-time" replenishment of everything from gasoline and oil to candy and potato chips. Health maintenance organizations use point-of-sale card readers and ISDN telephone lines to authorize medical insurance benefits and issue payment requests to a range of health-care insurers. Many state agencies are using "food stamp credit cards" to reduce or eliminate the fraud-prone use and collection of paper food stamps. A magnetic card is inserted into the standard card-swipe terminal now appearing in many supermarkets. It quickly authorizes a purchase, and automatically deducts the amount from the card holder's account. A growing number of banks are now linking remote automatic teller machines to a central computer through ISDN and packet switching.

D channel connections eliminate the need for dedicated lines to an ATM-enabled central office and make it economical to serve many more locations. In several tests, insurance and real estate agents are now able to access records from a range of computers. Insurance agents, for example, can retrieve policy information from several participating companies, and real estate agents can often tap into home-for-sale listings from neighboring towns and cities. Several state lottery agencies are experimenting with ISDN-based approaches to playing state lottery and numbers games. The attraction of ISDN is that it uses existing telephone lines, and thus reduces the current dependence on dedicated connections to these statewide systems. It would make lotto terminals much more widely available, almost at any public location that had phone service. Although I can raise some valid social concerns over the wide-spread distribution of automated lotto and gambling machines, the point-of-sales technology is valid for a whole host of vending machines. In fact, when communication becomes efficient and cheap, services can be censored based on age, credit rating, past sales history, pass card, or other tracked call source. call destination, calling type, or service type parameters.

At many corporate and government locations, low-cost D channel connections are being used for entry to buildings, laboratories, warehouses, and other restricted areas, as well as for security and other forms of telemetry, such as tracking user whereabouts within a building or campus facility. Several other companies are also using ISDN for physical security in warehouses, receiving docks, and other similarly vulnerable sites. Both B and D channels can be used as conduits for remote TV cameras and to monitor locks, alarms, and strategically placed sound, movement, heat, and other sensors.

Most transfers of large files today take place through dedicated broadband digital connections or on magnetic media physically transported by messengers and overnight couriers. The flexibility of ISDN PRI connections, in contrast, expands a system manager's options. It enables the same dialed B channels used for individual connections during the day to be bonded into higher-speed links for after-hours file transfers to and from multiple points.

Many colleges and universities that installed ISDN for voice and high-speed data connections have found they can economically use student ID cards for meal payments, library check outs, classroom and dormitory entry, test authorization, and other applications. D channel card readers are easily linked to an appropriate computer through the nearest ISDN telephone connection. Note that in many of these applications, the use of a dialed modem on an ordinary telephone line is simply not practical. Call-by-call connection times are far longer than a customer would wait.

Conclusion

This chapter presented the details, definitions, standards, and usage for the Integrated Switched Digital Network. Primarily, ISDN represents the shift from analog frequency-division multiplexing and mechanical circuit switching to digital time-division multiplexing with packet multiplexing and fast-packet electronic switching. It is a technology that is available in many places and provides the extraordinary benefit of integrating voice, data, video, streaming video, local area networks, and wide area services into a single manageable infrastructure. The next chapter addresses what has been fundamentally the most difficult part of ISDN installation, that of ordering and provisioning service to match usage requirements and equipment.

3

Deciphering ISDN Options

Introduction

ISDN is really nothing more than a bundle of switched digital service lines. However, a typical configuration and services order list from the service provider runs from 30 to 60 pages, with about 10 options per page. This is typically too much information for the average user, even a sophisticated corporate user, and it often leads to provisioning, configuration, and operational errors. The problems are expensive for service providers, equipment vendors, and ISDN users alike. To spur implementation and the utility for ISDN, the North American ISDN Users Forum (NIUF), a group funded by the U.S. Department of Commerce, has created a set of ISDN ordering codes and a separate set of application-specific ordering codes. In addition, the telecommunication carriers including AT&T, Ameritech, Bell South, NYNEX, and Pacific Bell have begun an initiative with ISDN equipment manufacturers that include Boca Research, IBM, Intel, Motorola, and US Robotics, which collectively is called the ISDN Consortium. The ISDN Consortium hopes to simplify ISDN service orders and implementation. Their initial goal has been to develop standard ordering codes to represent a typical *working* ISDN service for each piece of vendor equipment. The ordering codes are supposed to simplify ISDN installation. The secondary goal is to develop hardware and software that will make the most of ISDN services and interoperate fully.

Unfortunately, ISDN features and functions are complex enough without the NIUF creating new matrices of services that are at least as complex as the original service and one order of normality removed from the original ISDN service order and ISDN features list from a service carrier. This confusion is only complicated by the ISDN Consortium creating ordering codes that compete with NIUF. Basically, you have the service provider's features list, the NIUF ordering codes, the NIUF application-specific ordering codes, and the ISDN Consortium standard ordering codes. That is plainly too much and confusing because a government-sponsored forum has created a new list that is now challenged by the service providers and equipment. I believe the ISDN Consortium list will win, if only because it is being driven by these significant players. At least, with only four options, that set is also simpler. I have included all these ISDN facts to live up to the title of this book. However, I find this information impractical, confusing, and impossible to simplify. If anything, these descriptions present some of the possible and financially feasible options for using ISDN for telemetry, distance learning, transaction processing, and integrated telephony environments. Take some solace in that I think ISDN is really very simple to order, install, wire, get provisioned, and make work. While there might have been an extraordinary past period when service providers did not know that they even provided ISDN and did not know what it was, that time is gone.

The competing attempts by various groups to create ordering codes that can cover all possible situations does not simplify the process; it only creates new layers to the ISDN onion. As such, the information in this chapter is best used to explore the full features of ISDN, and how ISDN might actually be used. Then it should be politely ignored. The local service provider or hardware vendor will know what works best.

Typical Services

ISDN is an *intra*LATA service providing the simultaneous transmission of voice, data, and packet services on the same exchange access line. LATA is an acronym for either Local Area Telephone Access or Local Access And Transport Area. ISDN service is supported between *inter*LATA connections through a long-distance service provider. The distinction is that the interLATA service is not ISDN, but a pipe used to connect two end-point services. Services and options are a function of the carrier, the carrier's switch capabilities, and the utility's tariff arrangements with the local utility commission. Some services are built into the basic ISDN service. These include:

- Calling number delivery
- Called number delivery
- Called hold

ISDN service is fundamentally some number of data or bearer channels which are switched digital bi-directional synchronous circuits for interoffice transmission of voice and/or data. Each switched B channel is charged at a flat or measured rate; providers do not mix rates on the same *premises*. ISDN inter-office or interLATA interconnection of 64 kbps between compatible central office switches is called *clear-channel capacity* (CCC). It may also be called *clear 64-kbps capability* when you talk to telco sales representatives. They may not know about T-1 services, SS7, zero-bit rate return protocols, and other esoteric technical details. They are likely to know about CCC or clear channel. Subrate interconnection between ISDN central offices may be at rates of 56 kbps, or subrate interconnection to voice, switched digital, or other analog services may be provided at 56 kbps. If you want to run data with at a voice-tariff rate, the maximum is 56 kbps. Each bearer channel includes a single directory number (DN) per bearer channel.

The process for ISDN configuration is called *provisioning*. That is the official terminology. Each B channel can be designated at the switch as one of these choices:

- Circuit-switched voice
- Circuit-switched data
- Circuit-switched voice/data
- Packet-switched data
- On-demand high-speed packet-switched data

You generally cannot make a data call over a voice channel or a voice call over a data channel. Digiboard and other vendors provide special hardware to enable this feature and can even compress multiple voice conversations to fit on a single B channel. With Basic Rate Interface (BRI) ISDN, the service can include two-way switched virtual circuits, an incoming switched virtual circuit, an outgoing switched virtual circuit, or combinations of those. A switched virtual circuit is analogous to a plain old telephone service (POTS) line. The limitations for two-way, incoming, or outgoing calls are simply a matter of provisioning. Each BRI can have one to eight user profile arrangements defining the service. Basic channel service options include flow control, class negotiation, agency selection, interexchange packet, fast call setup, and reverse sharing options. These options are defined below:

- Flow control parameter negotiation: negotiates flow control parameters (such as maximum packet size, window size, and protocols) on a per call basis for each direction of data.
- Throughput class negotiation: allows the calling (originating) station to request specific throughput classes in the call request packet for each direction of data.
- Interexchange packet preselect: sets the ISDN interLATA carrier at time of subscription.
- Recognized private operation agency (RPOA) selection: sets the ISDN interLATA carrier for packet switching on a per call basis.
- Fast select option: allows the user data to be passed in the call setup packets.
- Reverse charging option: permits data communication equipment to transmit incoming calls requesting reverse charging. All parties must subscribe to X.25 destination lines.

Group hunting is also supported with ISDN. This is an important feature when establishing inbound communication services. In a data environment, one busy line means a caller cannot get through on a number because the line is busy. However, with group hunting, the switch looks for a free line and switches the call through on that one. The central office switch pushes the caller to the second ISDN number in the sequence. Group hunting is illustrated below. Since the service profile identification (SPID), which is like a phone number, for the first B channel is typically 1 or 01 and the second B channel is 2 or 02, group hunting is automatic with BRI but must be provisioned with primary rate interface (PRI). The SPID sequences do not have to be sequential, consecutive, or orderly. Neither do the directory number sequences. It is merely a logical mapping of available channels so that there is a sequence to them. The following features are available only with a custom ISDN or National ISDN service:

- Call forwarding variable.
- Call forwarding variable—feature button.
- Call forwarding—busy line.
- Call forwarding—do not answer.
- Call pickup.
- Conference; drop, hold, transfer: allows the EKTS user to add, drop, transfer a third party to a conference call. Transfer is available only for voice calls.
- Multiline hunt group: provides a predetermined search for idle directory numbers.
- Speed calling: allows user to assign up to 30 numbers to a two-digit code.
- Calling/called number delivery: the same as Caller ID.

- Calling number identification services—National ISDN: provides redirected caller ID information in addition to calling/called number delivery.
- Visual message waiting indicator: provides user with an indication of a message service that at least one message is waiting.
- Audible message waiting indicator: provides user with an indication of a message service that at least one message is waiting.
- Additional call appearance: allows a terminal to have more than one directory number assignment.
- Call tracing: provides user with ability to perform an automatic trace of last call received.
- Call return: provides user with ability to place a call to the number of the most recent call (attempted or completed).
- Preferred call forwarding: allows user to transfer selected calls (from a screening list) to a forwarded telephone number.
- Call block: allows user to block selected calls (from a screening list).
- Call selector: provides user with an indication of selected calls (from a screening list).
- Repeat dialing: when active automatically, redials the last number attempted or completed.
- Automatic line/direct connect: automatic directory number reprogramming when internal station or operator is unavailable.
- Make set busy: makes channel appear to be busy.
- Selective call acceptance: allows user to select calls (from a screening list) for acceptance.
- Station restriction: various station restrictions.

At the current time, there are many possible prefixes and suffixes for the terminal identifications (TIDs) and SPIDs, the switch type, and the terminal end point. At least the SPID problems are supposed to disappear because the National ISDN Council, a group of U.S. regional telephone operating companies, have agreed to a uniform 14-digit format. In addition, they will automate the SPID capabilities in network switches by 1998 so that network devices can configure themselves. Now, you will need to know bearer channel capabilities to avoid bearer provisioning errors, and the table in Figure 3.1 lists the most common switch and switch provisioning settings.

ISDN switch types	*Provisioning*
AT&T 5ESS	
	NI-1
	Custom
Northern Telecom (Nortel) DMS 100	
	NI-1
	Custom
Siemens	EWSD

Figure 3.1 Common ISDN switch types and basic provisioning configuration.

Although these switches are from different manufacturers and basic configuration is different for each, all support a common subset of services and features designated as *National ISDN 1* (NI-1) and generally implemented in most North America local service provider central offices. The actual switch provisioning settings under National ISDN 1 are listed below:

- Number of channels for CSV
- Number of channels for CSD
- Terminal type A or D
- Display
- Prefer ringing or idle
- Autohold
- One touch dialing
- Electronic key telephone system (EKTS)
- Multipoint
- Functional signaling
- Private virtual circuit (PVC) protocol version
- Maximum number of programmable keys
- Release key
- Ringing indicator

These options and services are defined in the remainder of this chapter. This is useful and important information. Use it to explore the ISDN services that might be useful to you or your organization for provisioning and configuring ISDN service so that it actually works and also for debugging the rare problems with ISDN services. Although these 14 ISDN service options and features seem simple enough, there is actually quite a bit of refinement and detail to them. When these items are expanded and defined in a formal tariff statement and hence a carrier's formal ordering and provisioning paperwork, you end up with 30 to 60 pages of confusing features, functions, and services.

Voice Options

ISDN supports voice communications when encapsulated and encoded inside data or when routed through customer premise equipment (CPE). Typically, the CPE supporting voice is called an *electronic key telephone service (EKTS)*, and it provides user flexibility for allocating data and voice channels from the BRI or PRI service. Voice options with a customer EKTS system include:

- Share primary directory number: the basic phone number.
- Secondary-only directory number: for identifying one or more terminals.
- Share secondary-only directory number—first appearance: first appearing identifier for identifying one or more terminals, but not the primary directory number.
- Share-Non-ISDN directory number: provides call coverage for an analog (POTS) number.
- Privary release: disables bridging (call forwarding). Privary (that is, *privacy*) release can be disabled on a call-by-call basis.
- Manual exclusion: opposite of privacy release. enables bridging (call forwarding). Privary release can be enabled on a call-by-call basis.
- EKTS intercom dialing—dial: provides user access to other terminals in an EKTS group with one- or two-digit dialing.
- EKTS intercom calling—automatic: one-button call to another terminal.
- EKTS intercom calling—call appearance: provisioning option for call appearance.
- Conference; drop, hold, transfer: allows the EKTS user to add, drop, transfer a third party to a conference call. Transfer is available only for voice calls.

Data Options

Packet switching options are available with both a B channel or D channel:

- X.25 hunting: multiple directory number assignments to X.25 terminals.
- International close user groups (ICUG): the optional formation of subgroups or subnetworks with a close membership for incoming calls, outgoing calls, or calls from outside the ICUG.

Recognize that it is insufficient to provision an ISDN channel or entire circuit for data switching unless you have a comparable or compatible service on the order side of the central office switch.

Order Code Simplification

As previously stated, at least two ISDN-related groups and consortiums and even some big ISDN vendors have provided distinct definitions for the service providers. Generally, you will want to pick the hardware, make sure that it works with your central office equipment, get that equipment in place, and then provide the telephone company with a list of the exact options desired. It is better to get these option correct first, rather than later, or you could find additional service fees on your monthly statements.

An ad hoc Group on the Simplification of ISDN Ordering, Provisioning, and Installation has been formed under the auspices of the Issues Family of the NIUF to work toward industry agreement on the best solution to the simplification issue.

The simplification of the ISDN ordering, provisioning, and installation process is a critical National ISDN issue, according to the NIUF. I think it will have significant bearing only on the NI-1, NI-2, and NI-3 initiatives and extending the future compatibility of ISDN with broadband ISDN (B-ISDN), asynchronous mode transfer (ATM), switched multimegabit data service (SMDS), and synchronous optical network (SONET). I do not think that these ordering codes will last long because there are just too many with too subtle differences. I still believe the ISDN Consortium ordering initiative will be more practical and hence important to the average end-user. Currently, the ISDN service negotiation procedure involves the negotiation of hundreds of subscription parameters necessary to set up ISDN service. This complexity causes customer confusion, as well as telephone company administrative and CPE installation problems. The simplification of this process with a limited number of codes will make ISDN easier to order for customers and will also ease ISDN provisioning and installation hurdles for videoconferencing, complex computer telephone integration (CTI), and network interconnections. I am going to hammer home this point with a discussion of the complexities and subtleties of the NIUF ordering codes and then shows you the ISDN Consortium's four negotiated codes. When you give more than two choices to an infant, the infant gets confused and upset. When you give more than a few ISDN choices to adults, you can confuse and upset them, too. I know. Network management, MIS, internal or client consulting are pressured environments that need clear and concise answers. If my practiced and practical ISDN philosophy comes across—and I hope it does—ISDN is simple and easy to install and make work. You can get most problems resolved in 1 or 2 hours with service provider or vendor technical support, and even the most tenacious problems can be solved with only a couple of hours spread out over a week. Obviously, there will always be leading-edge integration projects that stress and confuse the development effort, and require a complete understanding of the ISDN features, functions, and services, as well as how these impinge on provisioning and equipment selection. That is the exception, not what will be normal and simple for most sites and users.

This simplification issue was introduced into the NIUF in June 1991; at that time, the NIUF identified it as a red flag issue, needing priority attention. The ad hoc Group on the Simplification of ISDN Ordering, Provisioning, and Installation was formed to address this issue. This group has had broad representation from all parts of the telecommunications industry. The main focus at this time is:

- A generic SPID format (as it varies by provider and their switch types)
- Assignment of feature keys to default terminal buttons
- CPE process for automatic SPID selection

A generic SPID format would provide a uniform nationwide (United States) 14-digit sequence to include the subscribing user directory number, sharing identifier, and terminal identifier fields. The assignment of feature keys would address the problem of a noninitialized CPE (that is, an ISDN terminal or terminal adapter that knows nothing about the channel or circuit) because a user or network administrator did not enter a valid SPID. This would allow noninitialized terminals to access additional terminal features such as conference calls, transfer, and message waiting indicators. Automated SPID selection automates the CPE initialization process by having the central office switch send the SPID to the terminal rather than having the user or network administrator enter it. The terminal initiates the automated SPID selection process by sending an initialization request to a switch with a fixed (default SPID). The switch determines the service profiles associated with the terminal's interface and any explicit provisioning and sends the corresponding SPID to the terminal. Specifically, the identification and description of key application scenarios are:

- The development of a limited number of North American-accepted ISDN Parameter Groups (IPGs), which contain ISDN subscription parameters and their settings. A selection of one or more of these IPGs, as appropriate to support an end-user's application(s), can replace negotiation of each parameter on a parameter-by-parameter basis.
- The development of a process for CPE button assignment.
- Promotion of the implementation of IPGs in the telephone company service negotiation and provisioning process.
- Promotion of industry acceptance customer awareness concerning IPGs.

As a result, the basic (currently 20) NIUF capability codes for ordering ISDN are as follows:

Capability Package A (0B+D) includes basic D channel packet. No voice capabilities are provided.
Capability Package B (1B) includes circuit-switched data on one B channel. Data capabilities include Calling Number Identification. No voice capabilities are provided.
Capability Package C (1B) includes alternate voice/circuit-switched data on one B channel. Data and voice capabilities include Calling Number Identification.
Capability Package D (1B+D) includes voice on one B channel and basic D channel packet. Only basic voice capabilities are provided, with no features.

Capability Package E (1B+D) includes voice on one B channel and basic D channel packet. This package provides non-EKTS voice features, including Flexible Calling, Additional Call Offering, and Calling Number Identification.

Capability Package F (1B+D) is equivalent to Capability Package E, with the change that Call Appearance Call Handling (CACH) EKTS service is used for the voice service. (Note that Additional Call Offering functionality is incorporated in the EKTS service.)

Capability Package G (2B) includes voice on one B channel and circuit-switched data on the other B channel. This package provides non-EKTS voice features, including Flexible Calling, Additional Call Offering, and Calling Number Identification. Data capabilities include Calling Number Identification.

Capability Package H (2B) is equivalent to Capability Package G, with the change that CACH EKTS service is used for the voice service. (Note that Additional Call Offering functionality is incorporated in the EKTS service.)

Capability Package I (2B) includes circuit-switched data on two B channels. Data capabilities include Calling Number Identification. No voice capabilities are provided.

Capability Package J (2B) includes alternate voice/circuit-switched data on one B channel and circuit-switched data on the other B channel. Only basic voice capabilities are provided, with no features except Calling Number Identification. Data capabilities include Calling Number Identification.

Capability Package K (2B) includes alternate voice/circuit-switched data on one B channel and circuit-switched data on the other B channel. This package provides non-EKTS voice features, including Flexible Calling, Additional Call Offering, and Calling Number Identification. Data capabilities include Calling Number Identification.

Capability Package L (2B) is equivalent to Capability Package K, with the change that CACH EKTS service is used for the voice service. (Note that Additional Call Offering functionality is incorporated in the EKTS service.)

Capability Package M (2B) includes alternate voice/circuit-switched data on two B channels. Data and voice capabilities include Calling Number Identification.

Capability Package N (2B+D) includes alternate voice/circuit-switched data on one B channel, circuit-switched data on the other B channel, and basic D channel packet. This package provides non-EKTS voice features, including Flexible Calling, Additional Call Offering, and Calling Number Identification. Data capabilities include Calling Number Identification.

Capability Package O (2B+D) is equivalent to Capability Package N, with the change that CACH EKTS service is used for the voice service. (Note that Additional Call Offering functionality is incorporated in the EKTS service.)

Capability Package P (2B+D) includes alternate voice/circuit-switched data on two B channels and basic D channel packet. This package provides non-EKTS voice features, including Flexible Calling, Additional Call Offering, and Calling Number Identification. Data capabilities include Calling Number Identification.

Capability Package Q (2B+D) is equivalent to Capability Package P with the change that CACH EKTS service is used for the voice service. (Note that Additional Call Offering functionality is incorporated in the EKTS service.)

Capability Packages R and S are closely related to Capability Packages I and M; the main difference is that Capability Packages R and S consistently assign two directory numbers to the interface, whereas the Capability Packages I and M assign one or two directory numbers based on the central office switch type.

These new Capability Packages grew out of ISDN Solutions '94, sponsored by the Corporation for Open Systems, International (COS). As part of ISDN Solutions '94, industry participants began the implementation of the Capability Packages developed by the NIUF. This implementation, in the form of ISDN Ordering Codes (IOCs), is seen as essential for delivering ISDN products and services to the mass market. Capability Packages R and S are identical to two nongeneric IOCs, Generic Data I and Generic Data M, developed by three CPE suppliers during ISDN Solutions '94. These new packages were seen by equipment/application developers to be more useful for many applications than Capability I and M. In addition, they are consistent with National ISDN, they are already in use by several CPE suppliers, and they further promote the goals of simplification.

> **Capability Package R** (2B) includes circuit-switched data on two B channels. Data capabilities include Calling Number Identification. No voice capabilities are provided.
>
> **Capability Package S** (2B) includes alternate voice/circuit-switched data on two B channels. Data and voice capabilities include Calling Number Identification.

Capability Package T was developed primarily to meet the needs of Transaction Services applications. Several CPE suppliers expressed the need for such a package during ISDN Solutions '94. Several ISDN Ordering Codes (IOCs) have been created which are related to this Capability Package. Capability Package T includes two voice channels and D channel packet. A typical application scenario might involve a multipoint application, with the voice channels used either for two voice telephones or for voice and fax, and with the D channel packet used for transactions. Although this application can be provided with two directory numbers on the AT&T and Siemens switches, this Capability Package contains three DNs on all switch types, to meet the requirements of some CPE and for consistency across switches.

> **Capability Package T** (2B + D) includes voice on two B channels and basic D channel packet. Only basic voice capabilities are provided, with no features.

When I look at these codes I think of buying an automobile at a high-pressure auto dealership. The salesperson is giving me choices between a stripped vehicle which doesn't have air conditioning and power windows and is clearly unsuitable and a luxury touring package with power seats, leather upholstery, tinted glass, alloy wheels, whitewall tires, floating suspension, racing stripes, surround sound music, and other doodads perhaps desirable to a retiree but not my needs or personality. This dichotomy of either basic or luxury is presented as the simple choice but is not focused to typical usage patterns. Consider more realistic packaging for the traveling salesperson, the mommy-mobile, the mid-life crisis, the basic retiree comfort package, and the "I just need a car this big with Teflon-coated child-proof seats and A/C for hot days" package. That last one is for me.

If this ordering code confusion isn't sufficiently clear to you, the next section adds in the uses for which you are likely to apply ISDN. This is like buying a car (sight unseen) based on a list of utilitarian and preferential requirements. This is similar to a description like "I need a car that carries a family, goes through anything anytime, and is easily repaired in the few cases when it needs service" that returns anything from a motorcycle sidecar to a Humvee. Perhaps it is more useful and businesslike to define ISDN needs in this way, but as you can see in the next section, the ISDN application solutions end up as general-purpose pickup trucks suitable for all situations and virtuoso for none.

The more common of these ordering codes are condensed somewhat in the table shown by Figure 3.2.

Package	Line	Interface			Features	A lá Carte	Number of DNs		
	Set	B	B	D			AT&T	NT-1	SSC
A	1			p	Basic D	-	1	1	1
B	3	c				CNI(c)	1	1	1
C	4	v/c				CNI(v,c)	1	1	1
D	6	v		p	Basic D	-	1	2	1
E	6	v		p	Flexible voice	ACO(v),CNI(v)	1	2	1
F	6	v		p	EKTS	CNI(v)	1	3	1
G	11	v	p		Flexible voice	ACO(v),CNI (c,v)	1	2	1
H	11	v	p		EKTS	CNI(v,c)	1	3	1
I	14	c	p			CNI(c)	1	2	1
J	15	c	v/c			CNI(v,c)	1	2	1
K	15	c	v/c		Flexible voice	ACO(v),CNI (c,v)	1	2	1
L	15	c	v/c		EKTS	CNI(c,v)	2	3	1
M	17	v/c	v/c			CNI(c,v)	2	2	1
N	25	v/c	v/c	p	Flexible voice	ACO(v),CNI (c,v)	1	3	1
O	25	v/c	v/c	p	EKTS	CNI(c,v)	2	4	1
P	27	v/c	v/c	p	Flexible voice	ACO(v),CNI(c,v)	2	3	1
Q	27	v/c	v/c	p	EKTS	CNI(c,v)	2	3	1

v = Circuit-switched voice   DN = Directory number
c = Circuit-swtched data   ACO = additional call offering
v/c = Alternate voice and data   CNI = Calling number identification
p = Packet-switched data

Figure 3.2 NIUF ISDN BRI capability packages.

NIUF ISDN Application Solution Sets

NIUF Solution Set Descriptions also define some other ordering packages to further simplify the process. These have nothing to do with the letter codes but represent functional applications with ISDN. Solution sets include ISDN ordering and provision codes for working at home, transaction processing, videoconferencing, work collaboration, telemetry, distance learning, and fax services. The next 26 paragraphs might not appear useful for you; however, this list provides good insight into practical and proven applications for ISDN.

Work at Home 1 is a basic work at home Solution Set. It includes a D channel packet. This Solution Set can be used to meet users data needs requiring bursty interactions, such as access to electronic mail, remote database access, and possibly LAN access. No voice capabilities are provided; voice telephone service may be provided by POTS.

Work at Home 2a is a basic work at home Solution Set. It includes voice on one B channel and D-channel packet. This Solution Set can be used for advanced call management (using non-EKTS voice features, including Flexible Calling, Additional Call Offering, and Calling Number Identification). In addition, it can be used for data needs requiring bursty interactions, such as access to electronic mail, access to remote databases, and possibly LAN access. *Work at Home 2b* is equivalent to Work at Home 2a, with the change that CACH EKTS service is used for the voice service. (Note that Additional Call Offering functionality is incorporated in the EKTS service.)

Work at Home 3a is an advanced work at home Solution Set. It includes voice on one B channel and circuit-switched data on the other B channel. This Solution Set can be used for advanced call management (using non-EKTS voice features, including Flexible Calling, Additional Call Offering, and Calling Number Identification). In addition, it can be used for data needs, such as access to a LAN or host (at speeds up to 64 kbps), and possibly access to electronic mail; data capabilities also include Calling Number Identification. *Work at Home 3b* is equivalent to Work at Home 3a with the change that CACH EKTS service is used for the voice service. (Note that Additional Call Offering functionality is incorporated in the EKTS service.)

Work at Home 4a is an advanced work at home Solution Set. It includes alternate voice/circuit-switched data on one B channel and circuit-switched data on the other B channel. This Solution Set can be used for advanced call management (using non-EKTS voice features, including Flexible Calling, Additional Call Offering, and Calling Number Identification). In addition, it can be used for data needs, such as access to a LAN or host (at speeds up to 128 kbps), and possibly access to electronic mail; data capabilities also include Calling Number Identification. *Work at Home 4b* is equivalent to Work at Home 4a with the change that CACH EKTS service is used for the voice service. (Note that Additional Call Offering functionality is incorporated in the EKTS service.)

Work at Home 5a is a multiple use work at home Solution Set. It includes alternate voice/circuit-switched data on one B channel, circuit-switched data on the other B channel, and D channel packet. This Solution Set can be used for advanced call management (using non-EKTS voice features, including Flexible Calling, Additional Call Offering, and Calling Number Identification). In addition, it can be used for data needs, such as access to a LAN or host (using circuit-switched data at speeds up to 128 kbps) and access to electronic mail (using packet-switched data); circuit-switched data capabilities also include Calling Number Identification. *Work at Home 5b* is equivalent to Work at Home 5a with the change that CACH EKTS service is used for the voice service. (Note that Additional Call Offering functionality is incorporated in the EKTS service.)

If that does not clear up the confusion surrounding ISDN, information for transaction services surely will. *Transaction Services 1* is a basic transaction services Solution Set. It includes D channel packet. This Solution Set can be used for stand-alone transaction devices, such as devices for credit card verification from a remote database, using packet service with speeds up to 9.6 kbps. No voice capabilities are provided; in this Solution Set, the transaction device is not directly associated with the business' voice service.

Transaction Services 2 is a basic transaction services Solution Set. It includes voice on one B channel and D channel packet. This Solution Set can be used for credit card verification from a remote database, using packet service with speeds up to 9.6 kbps. Only basic voice capabilities are provided, with no features.

Transaction Services 3a is an advanced transaction services Solution Set. It includes voice on one B channel and D channel packet. This Solution Set can be used for credit card verification from a remote database, using packet service with speeds up to 9.6 kbps. This Solution Set can also be used for advanced call management (using non-EKTS voice features, including Flexible Calling, Additional Call Offering, and Calling Number Identification). *Transaction Services 3b* is equivalent to Transaction Services 3a, with the change that CACH EKTS service is used for the voice service. (Note that Additional Call Offering functionality is incorporated in the EKTS service.)

There is even a special ordering code for videoconferencing. *Video Conferencing 1* is a low-bandwidth video Solution Set. It includes circuit-switched data on one B channel, for low-speed (at speeds up to 64 kbps) video. Data capabilities also include Calling Number Identification. No voice capabilities are provided.

Video Conferencing 2 is a high-bandwidth video Solution Set. It includes circuit-switched data on two B channels, for high-speed (at speeds up to 128 kbps) video. Data capabilities also include Calling Number Identification but without voice capabilities.

Video Conferencing 3 is a high-bandwidth video Solution Set with basic voice. It includes alternate voice/circuit-switched data on one B channel and circuit-switched data on the other B channel. High-speed (at speeds up to 128 kbps) video is supported. Only basic voice capabilities are provided, with no features except Calling Number Identification. Data capabilities also include Calling Number Identification.

Video Conferencing 4a is a high-bandwidth video Solution Set with advanced voice. It includes alternate voice/circuit-switched data on one B channel and circuit-switched data on the other B channel. High-speed (at speeds up to 128 kbps) video is supported. Data capabilities also include Calling Number Identification. This Solution Set can also be used for advanced call management (using non-EKTS voice features, including Flexible Calling, Additional Call Offering, and Calling Number Identification). *Video Conferencing 4b* is equivalent to Video Conferencing 4a, with the change that CACH EKTS service is used for the voice service. (Note that Additional Call Offering functionality is incorporated in the EKTS service.)

If these conventions are not sufficient for your needs, there is also a new Solution Set for desktop collaboration. *Desktop Collaboration 1a* is a basic desktop Solution Set. It includes voice on one B channel and D channel packet. This Solution Set can be used for low-speed (at speeds up to 9.6 kbps) file transfer between individuals, and possibly for access to electronic mail, LAN access, and access to remote databases. In addition, this Solution Set can be used for advanced call management (using non-EKTS voice features, such as Flexible Calling, Additional Call Offering, and Calling Number Identification). *Desktop Collaboration 1b* is equivalent to Desktop Collaboration 1a, with the change that CACH EKTS service is used for the voice service. (Note that Additional Call Offering functionality is incorporated in the EKTS service.)

Desktop Collaboration 2a is an advanced desktop Solution Set. It includes alternate voice/circuit-switched data on one B channel and circuit-switched data on the other B channel. This Solution Set can be used for high-speed (at speeds up to 128 kbps) file transfer between individuals, and possibly for video, access to electronic mail, LAN access, and access to remote databases. Data capabilities also include Calling Number Identification. In addition, this Solution Set can be used for advanced call management (using non-EKTS voice features, such as Flexible Calling, Additional Call Offering, and Calling Number Identification). *Desktop Collaboration 2b* is equivalent to Desktop Collaboration 2a, with the change that CACH EKTS service is used for the voice service. (Note that Additional Call Offering functionality is included in the EKTS service.)

Desktop Collaboration 3a is a multimedia desktop Solution Set. It includes alternate voice/circuit-switched data on one B channel, circuit-switched data on the other B channel, and D channel packet. This Solution Set can be used for high-speed (at speeds up to 128 kbps) file transfer between individuals, video, and possibly for access to electronic mail, LAN access, and access to remote databases. The added packet functionality provides a multisession environment which allows for concurrent database queries. Circuit-switched data capabilities also include Calling Number Identification. In addition, this Solution Set can be used for advanced call management (using non-EKTS voice features, such as Flexible Calling, Additional Call Offering, and Calling Number Identification). *Desktop Collaboration 3b* is equivalent to Desktop Collaboration 3a, with the change that CACH EKTS service is used for the voice service. (Note that Additional Call Offering functionality is incorporated in the EKTS service.)

Telemetry 1 is a basic telemetry Solution Set. It includes D channel packet. This Solution Set can be used for basic telemetry functions, such as meter reading. *Telemetry 2* is an advanced telemetry Solution Set. It includes circuit-switched data on one B channel. This Solution Set can be used for low-speed (at speeds up to 64 kbps) video. Data capabilities also include Calling Number Identification. *Telemetry 3* is an advanced telemetry Solution Set. It includes circuit-switched data on two B channels. This Solution Set can be used for high-speed (at speeds up to 128 kbps) video. Data capabilities also include Calling Number Identification.

Distance Learning 1 is a basic instruction Solution Set. It includes D channel packet. This Solution Set provides the ability to share resources and information by providing students access to networked databases from remote locations, using packet service with speeds up to 9.6 kbps. Databases could include educational databases, electronic mail services, bulletin boards, and corporate databases. No voice capabilities are provided; voice services can be provided by POTS.

Distance Learning 2a is a basic instruction Solution Set that includes voice capabilities. It includes voice on one B channel and D channel packet. This Solution Set provides the ability to share resources and information by providing students access to networked databases from remote locations, using packet service with speeds up to 9.6 kbps. Databases could include educational databases, electronic mail services, bulletin boards, and corporate databases. This Solution Set also includes voice features for advanced call management, using non-EKTS voice features, including Flexible Calling, Additional Call Offering, and Calling Number Identification. *Distance Learning 2b* is equivalent to Distance Learning 2a, with the change that CACH EKTS service is used for the voice service. (Note that Additional Call Offering functionality is incorporated in the EKTS service.)

Distance Learning 3a is a basic audiographics/instruction Solution Set that includes voice capabilities. It includes voice on one B channel and circuit-switched data on the other B channel. (1) With audiographics, audio teleconferences are combined with computer-based graphics. Audiographic sessions are interactive, allowing teachers and students to speak with one another, while sharing text and pictures displayed on computer monitors. (2) With remote access to instruction, there is the ability to share resources and information by providing students access to networked databases from remote locations, using higher-speed (at speeds up to 64 kbps) data capabilities. Databases could include educational databases, electronic mail services, bulletin boards, and corporate databases. In this Solution Set, data capabilities also include Calling Number Identification. This Solution Set also includes voice features for advanced call management, using non-EKTS voice features, including Flexible Calling, Additional Call Offering, and Calling Number Identification. *Distance Learning 3b* is equivalent to Distance Learning 3a, with the change that CACH EKTS service is used for the voice service. (Note that Additional Call Offering functionality is incorporated in the EKTS service.)

Distance Learning 4 is an interactive video or computer-aided networked instruction Solution Set. It includes circuit-switched data on two B channels. (1) With interactive video, two-way interactive video can be used to extend education programs beyond traditional boundaries. (2) With computer-aided network instruction, there is the ability to share resources and large amounts of information (data and images) by providing students access to databases from remote locations. It is assumed that higher-speed data capabilities (at speeds up to 128 kbps) are required to access self-paced training located in educational and corporate databases. In this Solution Set, data capabilities also include Calling Number Identification. No voice capabilities are provided.

Fax 1 is a basic fax Solution Set. It includes alternate voice/circuit-switched data on one B channel. This allows the capability for either Group 3 or Group 4 fax. Data and voice capabilities also include Calling Number Identification. *Fax 2* is an advanced fax Solution Set. It includes alternate voice/circuit-switched data on two B channels. This allows the capability for two simultaneous Group 3 or Group 4 faxes. Data and voice capabilities also include Calling Number.

Telemedicine 1a is a low-speed telemedicine Solution Set. It includes voice on one B channel and circuit-switched data on the other B channel. The circuit-switched data can be used for low-speed (at speeds up to 64 kbps) image transfer, as well as claims processing. Data capabilities also include Calling Number Identification. This Solution Set can also be used for advanced call management (using non-EKTS voice features, including Flexible Calling, Additional Call Offering, and Calling Number Identification). *Telemedicine 1b* is equivalent to Telemedicine 1a with the change that CACH EKTS service is used for the voice service. (Note that Additional Call Offering functionality is incorporated in the EKTS service.)

Telemedicine 2a is a high-speed telemedicine Solution Set. It includes alternate voice/circuit-switched data on one B channel and circuit-switched data on the other B channel. The circuit-switched data can be used for high-speed (at speeds up to 128 kbps) image transfer, as well as claims processing. Data capabilities also include Calling Number Identification. This Solution Set can also be used for advanced call management (using non-EKTS voice features, including Flexible Calling, Additional Call Offering, and Calling Number Identification). *Telemedicine 2b* is equivalent to Telemedicine 2a with the change that CACH EKTS service is used for the voice service. (Note that Additional Call Offering functionality is incorporated in the EKTS service.)

Telemedicine 3a is a high-speed telemedicine Solution Set. It includes alternate voice/circuit-switched data on one B channel, circuit-switched data on the other B channel, and D channel packet. The circuit-switched data can be used for high-speed (at speeds up to 128 kbps) image transfer, as well as claims processing. The added packet functionality provides a multisession environment which allows for concurrent database queries or medical claims processing. Circuit-switched data capabilities also include Calling Number Identification. This Solution Set can also be used for advanced call management (using non-EKTS voice features, including Flexible Calling, Additional Call Offering, and Calling Number Identification). *Telemedicine 3b* is equivalent to Telemedicine 3a with the change that CACH EKTS service is used for the voice service. (Note that Additional Call Offering functionality is incorporated in the EKTS service.)

LAN-to-LAN 1 is a low-bandwidth LAN-to-LAN Solution Set. It includes circuit-switched data on one B channel. It is designed for LAN-to-LAN or LAN-to-PC communications with low-speed (at speeds up to 64 kbps) requirements. Data capabilities also include Calling Number Identification.

LAN-to-LAN 2 is a high-bandwidth LAN-to-LAN Solution Set. It includes circuit-switched data on two B channels. It is designed for LAN-to-LAN or LAN-to-PC communications with high-speed (at speeds up to 128 kbps) requirements. Data capabilities also include Calling Number Identification. The goal is to create two specific ISDN configuration features, namely, automatic switch (type) detection and service profile identifier capabilities.

The next set of ordering codes from the ISDN Consortium is concise. There are only four codes. I suggest you use these as a basis for any ISDN installation. I equate it to ordering a car from the manufacturer by specifying the basic car and the handful of options I desire. Since ISDN pricing is fixed by formal tariffs, getting a "good deal" by buying the closet match off the dealership lot is not a useful metaphor. You buy ISDN from the "factory" with a basic chassis and specifically selected options, such as air conditioning and seat treatments.

ISDN Consortium Codes

The problem is that the NIUF codes are not an effective simplification for most users, let alone a vendor. As a result, another consortium was created to define a fewer number of configurations which are marketing oriented. The four basic configurations from the ISDN Consortium (lead by BellSouth, AT&T, Ameritech, Nynex, PacBell, Boca Research, IBM, Intel, Motorola, and U.S. Robotics) are listed in the table in Figure 3.3.

ISDN Order Code	Service Definition
Package 1: 2B+D	The first bearer channel includes one primary directory number, one additional call offering such as ISDN call waiting, calling party number identification and visual indicators of waiting callers. The second bearer channel includes voice and data with calling party number identification. This is mainly suited to Internet access and work-at-home telecommuters who can throttle down to only one B channel.
Package 1A	Same as Package 1, but internetworks with voice messaging.
Package 2	Same as Package 1. Also includes two directory numbers on first bearer channel. Assigns numbers and routes calls to analog ports for phone, fax, modem, etc.
Package 2A	Same as Package 2, but internetworks with voice messaging.

Figure 3.3 ISDN Consortium standard configurations.

Note this is solely for ISDN BRI, but since ISDN PRI is a bundle of that many more channels, this is still a good starting point. For example, specify 1A for a typical office environment with a PBX, and add in hunt groups, EKTS, and conferencing. The PRI tariff is often more than 60 pages, but you will be able to deal with the manufacturer for the services you want and will work within your environment.

Current Information

ISDN is here to stay. The importance of information about ISDN is reflected by the scores of web and file storage at file transfer protocol (FTP) download sites full of material about ISDN ordering, provisioning, installation, hardware selection, and performance. Refer to Appendix A with its addresses to good ISDN-related web sites. You can also access the pages for Lycos, Web Crawler, the robots page, InfoSearch, Alta Vista, Magellan, or others and perform a search based on the keyword *ISDN* to uncover many additional sites. Most RBOCs and ISDN hardware vendors include copious information about ISDN. Overall, you can get a good picture of the service, the hardware, and the service providers. Most of this information is repeated again and again in different formats and with different wording, so the material gets tedious quickly. However, the information is apt to be more current than printed publications, this book included.

The best sites include Dan Kegel's ISDN pages, Intel's ISDN site, PacBell's provisioning information, and Microsoft huge sites. Microsoft is committed to supporting

ISDN with Windows 95, Windows NT, and future operating systems. Getting information that is useful and practical to you is rather time-consuming because of the abundant but dispersed information about using ISDN. See the screen shot in Figure 3.4 showing Microsoft's ISDN site.

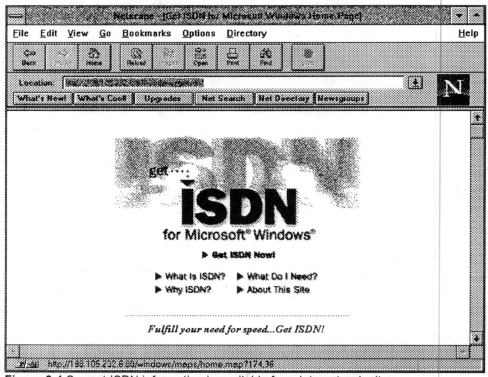

Figure 3.4 Current ISDN information is available from Internet web sites.

The main entry point for the Microsoft web site is www.microsoft.com. Additional information and other uniform resource locators (URLs) are listed in Appendix A.

Conclusion

This chapter provided all the current information on ISDN ordering codes. It lived up to the title of this book, particularly in that it presented detailed facts. This information is useful primarily for understanding the current applications and capabilities for ISDN rather than simplifying the actual ordering and provisioning process from a user's perspective. There are too many ISDN features and functions, as the first part of the chapter listed, but this is even further complicated by dueling ordering codes from the NIUF, the NIUF ISDN application codes, and the ISDN order codes from the ISDN Consortium. The proliferation of competing ordering codes is bound to be confusing, but in reality, ISDN installation is usually a simple process, as Chapter 4 details.

4

ISDN Physical Installations

Introduction

This chapter describes the process of ordering, provisioning, installing, configuring, and maintaining ISDN. Details about pricing (at the time of writing), performance, and physical wiring are also included. ISDN often is as simple as ordering ISDN, buying ISDN equipment, installing the ISDN line, connecting the equipment to the line, and configuring the connection. The wide range of ISDN equipment, the wider range of software enablement of ISDN features, and coordination with the ISDN service provider can also be a simple process. It has been for me for most sites. Very rarely do I have to plow through the service features lists or ordering simplification codes found in Chapter 3. I would equate it to the process of setting up an internal analog modem and getting all the interrupt values (IRQs), base memory input/output (I/O) addresses, and software settings to recognize the hardware. In fact, some ISDN terminal adapters conform to the Hayes modem set and work just like a modem to the communication software. Although, I did run into provisioning and configuration problems at some sites, most of the problems were my fault; ignorance, sloppiness, or just getting anxious and not working through the problem. This book details how you handle these problems more intelligently than I did.

Sometimes, ISDN configuration settings can represent a difficult and aggravating process for some organizations. Technical support people at some of the equipment vendors who individually had 10 or more years experience with ISDN stated that "it used to take 50 people to provision just one ISDN loop" whereas now it is very easy, or at worst, merely inconvenient. The problems are subtle and represent integration oversights.

My best advice is to plan for several hours over several weeks on the phone with ISDN service providers and ISDN equipment vendors. Allocate time to work out the bugs with communication software, dial-up scripts, WINSOCK.DLL versions, and service issues with Internet service providers (ISPs) if you anticipate connecting into the World Wide Web. If you are installing a basic rate interface (BRI) line, insist on coordinating the process with a senior-level installer. Here is another common-sense tip:

While that person is on-site (and hence a captive audience), test the line, the wiring to the outlet jacks, and the ISDN terminal equipment. If you are installing a primary rate interface (PRI) line or multiple ISDN lines that extend beyond a regional Bell operating company (RBOC) or independent operating company (IOC) coverage, insist that all installation teams coordinate the provisioning and installation effort.

In Europe, getting ISDN BRI may be a major effort because it is not a common service. If your organization requires ISDN connectivity and has use for most of the other 22 or 27 channels (for standard phones, analog modems, fax, and X.25), install ISDN PRI. The crossover point is likely to equal nine regular phone lines in Benelux, Denmark, Finland, France, or Germany. The crossover economics are a little different for Japan, Singapore, Australia, New Zealand, and other parts of Asia.

If you have a choice between installing a residential or a business line because it is being installed in a house or apartment for telecommuting and the actual user is not technically sophisticated, pay the *premium* difference and install a *business* ISDN line. Most service carriers take business users more seriously and provide better technical and customer support. The extra difference ranges from 10 to 18 percent; this is from $48 to $204 per year in the United States. That small difference is well worth the cost of a frantic technical support or basic service support call. This premium doesn't include inside wiring or device connectivity support (most telcos charge $71 to $158 just to start on inside wiring problems with rates about $100 per hour thereafter). However, the material in this book should help you deal with most of those simple wiring issues.

The last part of my opening statement is that you understand the difference between a communications task and a business solution. ISDN is not the solution to every connectivity situation; it is not always the cheapest, easiest, most reliable, or fastest. If you are responsible for providing remote site LAN connectivity, that is a *business solution*. The choices you make to implement it are the *communications task*. You have the options of using ISDN for dual analog modem lines, for single digital lines, for connection to switched digital lines, for X.25, for X.31, and even for frame relay. Understand what you need to accomplish. Communicate this with the service provider and potential equipment vendors. Review the practicality and suitability of chosen solution so that you obtain the correct circuit provisioning from the vendor. Most service providers charge extra for changing the provisioning later on; it is much like adding a new service, such as Caller ID or call forwarding, to an analog service after initial installation.

ISDN Physical Structure

As you learned in Chapter 2, ISDN is a *circuit*. It is a dedicated connection from the service provider's central office switch, called the Subscriber-Line Interface Card (SLIC), to a *demarcation point* (demarc) at your site. There are at least eight ways ISDN can be "delivered" from an ISDN-ready digital switch. These alternatives are:

- Through a direct BRI connection from an ISDN switch
- Through ISDN Centrex service
- Through a PRI connection through a switch or on T-1 (and E-1)
- ISDN as a second line
- ISDN as the only line
- ISDN as the only line, using a "digital modem"
- Indirectly through a satellite downlink
- Encapsulation within frame relay or asynchronous transfer mode (ATM)

One or more standard BRI (2B+D) connections can be used to link a company directly to an ISDN-ready switch in a central office. These lines can connect directly to ISDN equipment in a small office or residence or can be connected through a PBX or key system so that devices can communicate with one another without having to call through an outside connection. One or more BRIs can also be linked to ISDN Centrex service. This arrangement offers several advantages for an individual or company. Since the ISDN switch at the central office functions as an organization's switching system, it does not have to own or maintain a PBX or key system on site. It also offers a low-cost, virtually unlimited growth path because you can add ISDN lines instead of adding or replacing PBX hardware. A PRI delivers 23 to 29 (in Europe and Far East) B channels plus one D channel from the telephone company to a PBX or other control device, which then distributes the individual or bonded B channels as needed throughout an organization. Configuration varies greatly. Users with heavy data traffic, for example, might configure the connection through an ISDN router, multiplexer, or controller rather than through a PBX to reduce the chance of congestion through the switch or PBX.

While ISDN is specifically designed to deliver digital connections through existing copper twisted-pair lines, many smaller locations (and many larger ones as well) will have to decide whether or not to use ISDN as the only telephone connection or to install it as a separate, second, supplemental line. Depending on existing equipment and usage, a small or home office might configure ISDN in several ways. Consider using ISDN as the only line. This requires that all telephones and fax machines be ISDN systems or be linked through analog ports on terminal adapters or network termination devices and that all PCs or other computers have special ISDN cards installed. Depending on requirements and expectations, these alternatives let users migrate to ISDN in one sweeping move or in measured steps to replace older equipment (primarily antiquated or unserviceable PBX equipment) in a systematic, economical way.

Point-to-Point

ISDN connections can be established between two points. You can install an ISDN terminal adapter at home or in your office strictly to dial the corporate site or an Internet provider. This is a trivial application of ISDN, one that telecommuters and web surfers are using. In effect, this is a single-purpose use of ISDN to provide remote connections that are faster than 33.6 kbps, less prone to line noise, and the setup call is much faster to initiate. There is no magic here, no special configuration. The primary concern is that you match hardware at both ends, system drivers, dial-up scripts, and connection protocols and that your end of the ISDN line is configured properly for your terminal adapter. For example, if you are communicating with CompuServe, the best service you can get at the current time is 64 kbps. They do not support compression or bonding. On the other hand, some Internet providers do support bonding, but you need to know what type and whether the drivers and terminal adapter software will support that connection.

Mesh

ISDN supports mesh connectivity in two distinctly different methods. First, call-conferencing supports multiple interconnections using existing central office switching technology. The current limitation is six-way calling. Although I have mentioned this limitation repeatedly with the ISDN-H support for videoconferencing, the implementation is a little different. In this instance, separate calls are made and the end-user hardware splits the signals on different lines. Second, ISDN also can create separate connections with each B channel. You can use B1 to call one location and B2 to call another. In fact, you can also use a packet-data-enabled D channel to connect into a third location. Calls on each channel are independent of each other and can be initiated and terminated at different times. Typically, routing between sites is handled by an external router. This mesh can be extended for multiple lines and bonded lines and can be as extensive as a single PRI circuit or multiple PRI circuits. Third, X.25 is the winner in mesh connectivity because you do not need the switched virtual circuits (that is, calls) in place to connect each party you want to communicate with; the X.25 cloud handles the connectivity and mesh delivery. These options are illustrated in Figure 4.1.

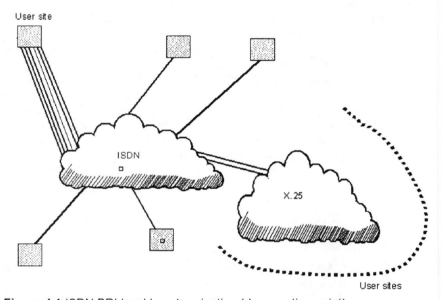

Figure 4.1 ISDN BRI local loop termination (demarcation point).

Easy Steps for ISDN

Clearly, you get the sense that ISDN is flexible and can do many things that plain old telephone service (POTS), frame relay, X.25, asynchronous mode transfer (ATM), Asymmetric Digital Subscriber Line (ADSL), and other services cannot do or that they do provide them but at higher costs. Here are four easy steps to ISDN service:

1. Order your ISDN service.
2. Configure (that is, provision) your ISDN line.
3. Get these critical descriptions:
   A. Service profile IDs (SPIDs), and there may be more than one.
   B. Your ISDN phone numbers (may look like SPIDs).
   C. Your ISDN switch type at the service provider's central office.
4. Install the CPE device at the jack or wire the jack into the line termination and connect the CPE device into the jack.

Easy Steps for ISDN Internet Access

Internet access is separate from ISDN installation. It is a separate service that POTS, ISDN, T-1, T-3, and frame relay can provide. However, when you are planning to connect to the Internet with ISDN, here are the four steps:

1. Provisioned and installed ISDN CPE equipment.
2. Service with Internet service provider (ISP) that provides ISDN connectivity.
3. ISDN service from the local provider.
4. ISDN telephone number(s) for the ISDN access lines at the ISP.

How to Order ISDN

There is no such thing as a standard ISDN configuration. The closest to that is the *Intel Blue* definition for ProShare videoconferencing. Ordering ISDN is as much about "provisioning"—the process of selecting the exact switch parameters at the central office—and feature selection as it is about pricing. However, confusion over ISDN provisioning and matching hardware with central office switches has led to several efforts to create standardized and meaningful ISDN ordering codes (IOCs) as listed in Chapter 3. There are also Capability Packages which are different from ordering codes. Intel Blue does not conform to any of these codes or packages. It is a little bit different; most carriers know what it is. However, you do need to understand the performance requirements for ISDN connections and the pricing ramifications. There are many surprises. Microsoft Corporation has tried to simplify the ordering and provision process by matching hardware, service providers, and locations in an Internet-enabled web page (http://www.microsoft.com/windows/getisdn). Basically, you fill out the forms, and the database server replies with enough information to order ISDN and often get it working on the first try. Think of this application as something that was marketed as a shrink-wrapped product and is presented more as a public service. This site is illustrated in Figure 4.2 below:

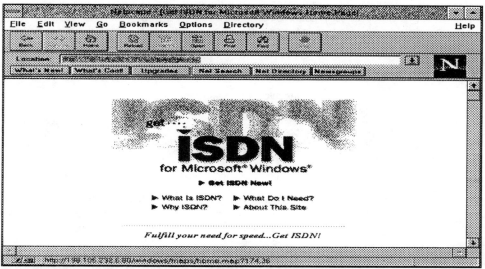

Figure 4.2 Microsoft web-based application for provisioning ISDN.

In fact, the local phone company may quote you rates for ISDN BRI as a straight installation and monthly rate or as an "add-on" to analog service. Watch out for that add-on rate because that quote might not include the $12 to $24 you normally pay for a basic analog link. In other words, ISDN costs X but the entire circuit costs $X + $Y, for the POTS connection, which is something of a hidden cost. Business service is usually more expensive by 10 percent than residential connections. Nothing is really clear until you see the first bill. Question the bill since the first one is usually wrong and different from the tariff quoted to you or faxed to you. ISDN pricing has four parts:

- Installation
- Basic service
- Supplemental services
- Per-minute costs

Installation is a one-time fee. Basic service is typically billed monthly. Supplemental services are usually part of the initial provisioning and often do not cost more than the basic service. If you want to add them at a later date, you may get charged addition installation and configuration fees. Per-minute charges apply mostly to commercial users but also to residential customers in some service regions. Some ISDN service providers include 200 hours free in the basic monthly service charges. Others include a measured business rate for minute usage. Some ISDN providers do not charge anything for minute usage. However, realize that long-distance point-to-point ISDN crosses local areas, states, RBOCs and IOCs.

You will get charged per-minute costs for the long distance for each channel used. For example, if you create a two-channel bonded call from Europe to the United States, you will be charged for two long-distance calls. If the service is within the RBOC or IOC, this will appear as part of the monthly bill. If you arrange for another carrier to provide integration or long-distance connectivity, this may be a separate bill. Long-distance ISDN rates are comparable to standard analog phone and fax calls; they may be slightly more expensive because you may require special carrier provision so that you get clear channel (64 kbps per line) and because you get 64 kbps rather than 56 kbps standard with SS7 circuits. Figure 4.3 lists common components in ISDN pricing.

Provider	Installation	Monthly	Free Time	Cost per minute
AmeriTech	$135	$28 to $34	0	$0.02
Bell Atlantic	$135	$24	0	$0.05
BellSouth	$206	$24 to $95	unlimited	$0
Nevada Bell	$227	$80	unlimited	$0
NYNEX	$270	$36	0	$0.032
PacBell	$125 + $35	$25	0	$0.012
SNET	$50 to $200	$45	o	$0.035
Southwestern Bell	$400	$50 to $60	flat rate per call	
US West	$110	$69	200	

Figure 4.3 ISDN BRI tariff rates.

Unfortunately, ISDN PRI tariffs are based on T-1 rates in the United States. Although the service requirement for PRI is not much more than two BRI lines (that is, two pairs of wire instead of one pair), the price break varies by the basic cost for BRI. In Europe, BRI is likely to be more expensive (if even available) than in the United States, but the break point is likely to occur at four or six B channels. Figure 4.4 illustrates PRI rates in the United States. Most services have a breakeven point at 4 to 6 lines. For example, 4 to 6 BRI lines will justify a PRI circuit. Similarly, 4 to 6 T-1 lines can justify the cost of a T-3 circuit. Transitions in Europe and Asia vary.

Provider	Monthly	Free Time	Cost per minute
AmeriTech	$635 to $825	0	$0.08
Bell Atlantic	$300 to $450	0	$0.02
BellSouth	$1800	Unlimited	$0
Nevada Bell	N.A.		
NYNEX	$900 for voice and data but only $400 for data only	0	$0.08/$0.01
PacBell	$545	0	$0.03
SNET	$1628	Unlimited	$0
Southwestern Bell	$1040 to $2890	Unlimited	$0
US West	$1800 for voice and data but only $1100 for data	Unlimited	$0

Figure 4.4 ISDN PRI tariff rates.

Figure 4.5 shows ISDN availability by year as a function of RBOC or IOC carrier. This is useful to gauge ISDN availability growth and also the commitment of the service carriers to ISDN and other advanced services. These numbers are plotted for the 1995 availability in Figure 4.7.

Provider	1992	1993	1994	1995
AmeriTech	25%	68%	78%	78%
Bell Atlantic	47%	59%	87%	90%
Bell South	30%	46%	53%	59%
GTE	9%	15%	16%	18%
NYNEX	15%	34%	58%	70%
Pac Bell	44%	56%	68%	80%
SW Bell	16%	22%	23%	23%
US West	36%	46%	64%	69%

Figure 4.5 ISDN availability (courtesy of Intel Corporation).

Figure 4.6 illustrates coverage areas for ISDN service carriers in the United States. Figure 4.7 shows coverage areas matched with current ISDN available as a percentage of telephone users. It is disproportionately available in urban areas.

This isn't the entire picture. I surveyed the several local service carriers in Florida and created the map in Figure 4.8 showing that service is mostly available in metropolitan areas, absolutely none in and around swampy areas, rural, and difficult to find in less-populated towns.

If you want more concise and exact information about the availability for ISDN in your exchange area, you can access the Intel web site and search for this on-line application shown in the screen capture in Figure 4.9.

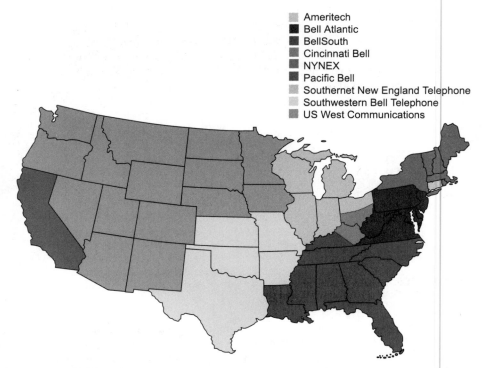

Figure 4.6 ISDN carriers by coverage areas.

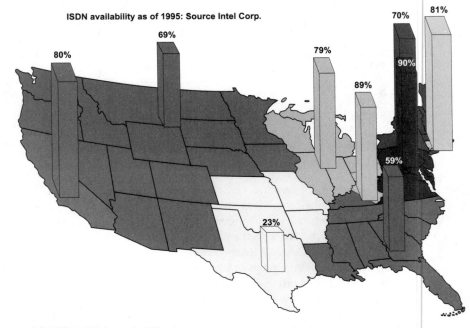

Figure 4.7 ISDN 1995 availability by coverage areas and carriers (courtesy of Intel).

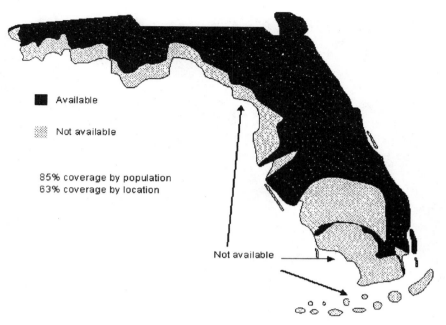

Figure 4.8 ISDN coverage in Florida by location.

Figure 4.9 ISDN availability by area code and phone exchange (courtesy of Intel).

When you fill out this simple form, the host server looks up the information in its database and provides availability information, the local carrier, and even a contact number for the local carrier, as shown in Figure 4.10. If you like web surfing, this tool at http://www.intel.com/isdn may be easier than looking up the contact numbers in the various tables in this book.

Cost of course is a different matter. Costs vary by carrier and location. You can get a copy of the official tariffs from Bellcore, your local public service commission (listings

Figure 4.10 Availability results (courtesy of Intel Corporation).

with names, addresses, and phone numbers are in the appendix), or you get some rough approximations from Figure 4.11.

Provider	Range	Average	Range	Average
AmeriTech	$14	$158.00	$39 to $97	$52.00
Bell Atlantic	$11	$181.00	$45 to $57	$47.33
Bell South	$16	$204.88	$26 to 46	$37.55
GTE	N/A	N/A	N/A	N/A
NYNEX	$14	$163.75	$32 to $82	$51.50
Pac Bell	$70	$70.00	$27	$27.00
SW Bell	$15	$154.00	$56	$56.00
US West	$85	$85.00	$67	$67.00

Figure 4.11 Average base monthly ISDN costs (courtesy of Intel Corporation).

After you decide to move forward with ISDN installation, ask the local service provider to install it. You should receive a confirmation on the order. Not only does this letter or facsimile confirm installation and a likely installation date, but the order number (Service Order Number) is useful for tracking all correspondence. This form should include the type of service, the switch type, the circuit ID, and both the directory numbers and SPID. The directory numbers are referred to as telephone numbers. I filed this work order for my home office and initially ignored the switch type and format of the SPIDs. This caused me some operational grief because I provisioned ISDN terminal adapters as 5ESS National ISDN-1 rather than 5ESS custom (multipoint) and used the format of 01999-999-9999000 instead of 01999-99990 where the extra two zeros at the end of the string correspond to the terminal identifier required for National ISDN and a daisy chain of analog devices. By the way, not all SPIDs include the area code, which explains the difference in the string of 9s. The dashes in the SPID are only for human interpretation and consumption; if you include any delimiters (dashes, parentheses, or spaces) in the terminal adapter setup strings or configurations, the ISDN circuit will not work. Figure 4.12 shows one of my ISDN work orders.

You might note several details on this work order. First, the service number should be filed away in your rolodex. You are likely to need this during the first day or two after your service is installed, although this is much less likely now than in 1994 or 1995. Second, note that this line does not include a directory number or SPID for the D channel, which would be necessary if you were using the D channel for X.25 or dial-up on-demand packet data. The D channel can have its own number, but it cannot support a standard analog phone line because that requires the full bandwidth of the B channel.

Date	*1-12-96*	# of pages:
To	*MARTIN*	
Fax#	*305 868-0530*	

From: **Nancy Kelly**
        Service Representative
        BellSouth Business Systems
        Phone: (770) 496-2706
        Fax: (770) 496-2760

BELLSOUTH BUSINESS SYSTEMS
ISDN CUSTOMER ORDERS & BILLING CENTER
(770) 496-2925 / (800) 858-9413 / FAX (770) 496-2760

ISDN ORDER CONFIRMATION

ORDER NUMBER _NQB14WM2_   DUE DATE _1-19-96_

CUSTOMER NAME _NETWORK PERFORMANCE INSTITUTE INC_

SERVICE ADDRESS _1630 DAYTONIA RD_

BILLING ACCOUNT NUMBER _305 531-9682_

MONTHLY RATE _____   INSTALLATION _____

TYPE OF SERVICE: SWITCH TYPE:
☒ BUSINESS   ☐ RESIDENCE   ☐ DMS 100   ☒ 5ESS
☐ FLAT RATE ☐ MEASURED   ☐ NATIONAL*   ☒ CUSTOM

TELEPHONE NUMBERS AND SPIDS:
  (B CHANNELS)
  (1) NUMBER _305 531-9682_   SPID* _01-531-9682-0_
  (2) NUMBER _305 531- 8193_ SPID* _01-531-8993-0_

  (D CHANNEL)
  NUMBER _____   SPID _____

CIRCUIT ID (B CHANNEL) _____
            (D CHANNEL) _____

LONG DISTANCE CARRIER
PIC _____ RPOA _____

REPAIR _800 247-2020_
TECHNICAL SUPPORT   (770) 496-2901 OR (800) 858-9413 (select 3)

*NOTICE: National ISDN service requires that an additional 2-digit
Terminal ID of 00 (numeric) be added to your SPID.

Figure 4.12 BellSouth ISDN BRI Service Order Number (SON).

Third, the circuit ID is left blank. This will be provided when the ISDN local loop is wired to your site. I discovered this information is more useful when dealing with Bell-South billing, than the first B channel SPID.

The installer wired a line from the nearest pole to my home office and installed the termination box. He handed me the toe-tag, shown in Figure 4.13, with the important circuit ID information, the SPIDs, and a place to call for help.

Figure 4.13 ISDN installation tag.

I laughed when I received my first bill on my home office. It was about double the quote on the confirmation. Although it is normal to get upset and angry, laughter seemed the better approach. I promptly called the business office for business lines and got a run-around. The third person I talked to finally explained that no one had any idea what I was talking about. Then I read a line about the billing account number that included "CKT: 70.ISBD..SB" and I was switched to another customer support person who set up an appointment with my account representative to go over each line item 2 days later. Figures 4.14, 4.15, and 4.16 illustrate my first multiple-page bill for a single simple BRI line.

About 20 percent of the bill included items that corresponded to local, state, and federal taxes. The tariff information does not include tax information. Since I have reviewed various local and long-distance charges for ordinary analog bills, I have become accustomed to the similar tax items—and have simply ignored them in the past. I suggest you ask for specifics or factor in some additional overhead cost for various taxation on services. Included in the monthly statements are such unexpected surcharges as a franchise charge, a gross receipts surcharge, and county manhole ordinance. When your accounting department gets the bill, complains, and runs it past your management, you will look foolish or at least deficient in providing effective business planning. I know that taxes were not part of my process and financial planning for my ISDN trials at my home office. The taxes did not even occur to me because I assumed I would just have to pay them as a matter of course. However, when I quote lines for clients, I also remind them about the additional taxes. You will know better now and include them as well.

⊕ BELLSOUTH

U S A
ℚℚℚ
Official Sponsor of the
1996 U.S. Olympic Team

NTWRK
Account Number: 305
Bill Period Date: Jan 29, 1996

Important Notice(s) (continued)

Nonpayment of Regulated Charges may result in discontinuance of service. Failure to pay unregulated and certain other charges, all of which are identified by ** on your bill, will not result in an interruption of local service. The amount of Regulated Charges may be obtained by calling 780-2800.

Helpful Numbers

BellSouth Telecommunications, Inc. (BST)

NOTE: Numbers for other companies are listed on their bill pages.

Billing Questions or to Place an Order:
  If calling from within the Florida BellSouth
  service area 780-2800
  If calling from outside Florida or outside the
  Florida BellSouth service area 1-800-753-8172
Repair - If calling from within the Florida BellSouth
  service area 780-2222
Text Telephone (TTY) Users 7:00 AM - 7:00 PM (CST) Monday - Friday:
  If calling from within the Florida BellSouth
  service area 780-2274
  If calling from outside Florida or outside the
  Florida BellSouth service area 1 800 251-5325

Detailed Statement of Charges

Monthly Service Charges			*Amount*
Monthly Service - Jan 29 thru Feb 28			
Basic Services		*Quantity*	
1. Emergency 911 Charge. This charge is billed on behalf of Dade County.		1 ... **	.50
2. FCC Charge for Interstate Toll Access		1 ...	3.50
3. Telecommunications Access System Act Surcharge		1 ...	.10
Total Basic Services			4.10
Optional Services		*Quantity*	
4. Channels Activated B Channel Flat Rate Circuit Switched Voice/Data		2 ...	32.50

** Unregulated Charge

CP   E028398   (continued)▶

Figure 4.14 A real ISDN bill (page 2).

@ BELLSOUTH

USA

Official Sponsor of the
1996 U.S. Olympic Team

NTWRK
Account Number: 305
Bill Period Date: Jan 29, 1996

Page 3

Detailed Statement of Charges

Monthly Service Charges (continued)

Amount

Optional Services (continued)   Quantity

  5. Individual Line ISDN Business - Low Volume
     Access/Digital Subscriber Line (DSL)   1 ...   55.00
Total Optional Services 87.50
Total Monthly Service Charges 91.60

Other Charges and Credits

Amount

Work Completed On Jan 19, 1996
SO: NQ814WM2

Your InterLATA Long Distance Company is AT&T Communications
One-time charge for deregulated wiring JAN 19,1996 TN 305
  6. Premise Visit Charge, De-Regulated .... **   35.00
  7. Service Charge .... **   70.00
One-time charge for
  8. Jack equipment 55.25
  9. ISDN User Profile Flat Rate, Includes Caller ID 10.00
 10. Charge for service connected - first line 56.00
Circuit Number:
Charge for new service (01/20/96 - 01/28/96)
 11.   2 Channels Activated B Channel Flat Rate Circuit
          Switched Voice/Data ($32.50/mo) 9.75
 12. Individual Line ISDN Business - Low Volume
     Access/Digital Subscriber Line (DSL) ($55.00/mo) 16.50
 13. FCC Charge for Network Access ($3.50/mo) 1.05
One-time charge for
 14. Individual Line ISDN Business - Low Volume
     Access/Digital Subscriber Line (DSL) 130.00

Your InterLATA Long Distance Company for   is AT&T
Communications
One-time charge for
 15. ISDN User Profile Flat Rate, Includes Caller ID 10.00

** Unregulated Charge

                                      CP   E028398   (continued)▶

Figure 4.15 A real ISDN bill (page 3).

⊕ BELLSOUTH   USA

Official Sponsor of the
1996 U.S. Olympic Team

NTWRK
Account Number: 305
Bill Period Date: Jan 29, 1996

Detailed Statement of Charges

Other Charges and Credits (continued)   *Amount*

Work Completed On Jan 29, 1996
SO: MIA MANHOLE

16. Cost of Dade County manhole ordinance #83-3 11
531-8993
17. Cost of Dade County manhole ordinance #83-3 11
Total Other Charges and Credits 393.77

Taxes   *Amount*
18. Federal Tax 3.63
19. State/Local Tax 28.65
20. Florida Gross Receipts Surcharge 1.28
21. City Tax 8.14
22. Franchise Charge 1.14
Total Taxes 42.84

Total BellSouth Current Charges 528.21

CP   E028398   (continued)▶

Figure 4.16 A real ISDN bill (page 4).

Most providers have dropped the extra fee for providing service beyond 15,000 ft (that is *15 kilofeet* in U.S. provider/installer vernacular) because ISDN signal line extensions (that is, *line extenders*) are required for each segment longer than 15,000 ft. On average, 10 to 15 percent of customers are further than that from the switching site. Rural sites may require two or more signal repeaters or a fiber interface. ISDN is available on fiber, but the service provider usually will terminate the line at your site with a two- or four-wire set. If they terminate with fiber, very possibly with PRI service or greater, know that ahead of time so that you can have appropriate client-side equipment to convert the signal into something useful.

Installation charges may also include the addition of the line extender and a monthly recurring charge for this signal booster. You might also be assessed an extra service fee because the ISDN presents "no dial tone." Make sure the installer and you know that *ISDN has no dial tone.* ISDN sounds like static or a popping noise from its carrier signal heartbeat. However, when you add a handset to a correctly provisioned ISDN line through an NT-1 with an analog port, you should get a dial tone. Service and installation fees are fixed by a tariff of the public utility commission (PUC) and by monopoly in most of Europe, Australia, Japan, and the Near East. I am surprised at what a provider will try to charge for even when they have sent a faxed quotation. Get that price quote on paper. The confirmation should include a summary of expected charges.

Many providers charge a recurring monthly rate plus minute or hourly message unit rates. PacBell charges 1.05 cents/minute from 7 a.m. to 5 p.m. and 0.73 cents/minute until 11 p.m. and 0.42 cents/minute nights and weekends. By the way, they (and most other service providers) have filed for a tariff increase for monthly service and per-minute charges. It is likely they will be denied, but ultimately analog, ISDN, and Internet simplex and duplex phone service will equalize in terms of price, reliability, and performance. As such, analog rates are likely to drop and ISDN rates are likely to increase. This is in addition to any long-distance charges.

When you perform your economic analysis comparing analog phone lines, analog lease lines, dedicated digital lines, switched digital lines, T-1, fractional T-1, SMDS, frame relay, SONET, ATM, and X.25 with ISDN, it is important that you consider realistic transmission rates, point-to-point distances, and usage (the number of hours utilized per month). There is likely to be a *crossover* point or even several ranges where other services yield better economic justification. It is important to recognize that a T-1 has a different capacity than an ISDN PRI line even though ISDN is carried as the protocol inside a T-1 line. Specifically, while both the T-1 and ISDN provide 24 channels, you can get 24 analog circuits with T-1 using the SS7 protocol but only 23 with ISDN. You can get 24 modem lines with a maximum capacity of 33.6 kbps (an aggregate of 608.4 kbps) with an analog T-1 while you get 23 lines at 64 kbps (an aggregate of 1472 kbps) with ISDN. You are more likely to see single-line analog capacities of 19.2 or 28.8 kbps in contrast to ISDN's 56 or 64 kbps. As a result, even though ISDN PRI is carried by a T-1 circuit, ISDN PRI provides more than 3 or 4 times the data bandwidth than an analog T-1 does.

Hardware

Hardware is basically the terminal adapter and NT-1, and this plugs into a service jack. It is as simple as is shown in Figure 4.17.

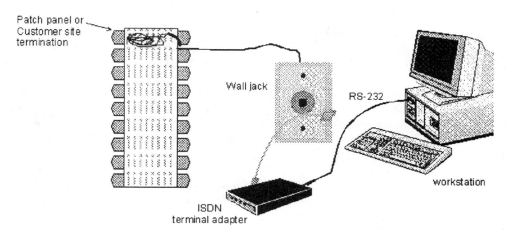

Patch panel or
Customer site
termination

Wall jack

RS-232

workstation

ISDN
terminal adapter

Figure 4.17 Minimally required ISDN hardware.

When ISDN is combined with the wiring in sites with preexisting phones, private branch exchange (PBXs), and local area networks (LANs), the hardware and the wiring can be more complicated. There are functionally two types of ISDN equipment, BRI and PRI. The difference is mainly an issue for support for more than two lines. ISDN BRI devices ostensibly support two B channels and the D channel. If you need a single B channel support for switched digital or packet data on the D channel only, the NT-1 is a sufficient interface. More complex connectivity options require a terminal adapter. The question is, which one? Perhaps, part of the answer is understanding your operational requirements. Consider the services that you might need:

- Support for more than one B channel?
- Support for more than one B channel for digital services?
- Support for B channel bonding?
- Does it support D channel services?
- Support for a POTS interface or an external S/T port?
- Support for multiple analog connections and directory numbers?
- Uninterrupted power supply (UPS) or battery backup integration?
- U or S/T interface, or both?
- Support for analog modem?
- Software support for Internet protocol (IP), or ISDN bonding (PPP, ML-PPP)?
- Support for telephony (TAPI, TSAPI, or WinISDN) interfaces?
- Does the terminal adapter include RS-232, RS-422, or V.35?
- Support for network management (SNMP or SNMP-II) agents?
- ISDN circuit and connection diagnostics?

Some of the more advanced BRI support features represent software implementations. Most ISDN terminal adapters, whether internal or external, provide roughly the same mix of functions. Some include digital signal processing (DSP), which typically means that the device will also provide supplemental analog modem features (including 28.8 or 33.6 kbps) for dialing to BBSs and Internet providers that do not have ISDN support. I recommend this for convenience, but also because it reduces the complexity of your hardware integration problems of having a modem and a terminal adapter on the same PC or Macintosh. It also unifies the dial-up scripts and command strings. And it means that operating systems that do not have built-in support for ISDN can use an ISDN device with Hayes modem commands. In addition, DSP-based devices typically can be upgraded with new features with a flash BIOS upgrade; this means, assuming the vendor is committed to the product line and its reputation, that new software will increase the feature set of the device.

The outstanding difference between ISDN terminal adapters is in how they integrate with your operations. Integration for the most part is a feature of the sophistication of the software. The software issue is more than future upgrades. Rather, the software defines the flexibility of the device in terms of how it works with dial-in and dial-out connectivity, how easily it is to configure, how ready it is to work with Windows 3.x, Windows NT, Windows 95 , a Macintosh, or Unix. It determines whether the device supports rate adaptation (V.110 and V.120) or can bond two lines together for 112 and 128 kbps. Most important, in my view, is how good the software is at diagnosing line problems, connection problems, and the integration between the PC and the terminal adapter hardware. It doesn't take long for connectivity problems to eat up a few hours here and there, sending you scrambling needlessly to vendor or service provider technical support when good software can easily adjust for and correct many of these problems. The low-cost *slave* ISDN adapters are a mixed bag. Even though they rob CPU performance from the host PC, some actually have good software to simplify the provisioning, configuration, and operational process.

Operational requirements for PRI connections are likely to be more complex in that you might well want to split out BRI services from the PRI for Internet connections, remote dial-out, dial-in communications, telephony, BBS or web site support, videoconferencing, advanced digital PBX services, and electronic key telephone systems (EKTS).

ISDN Termination

ISDN wiring is a mix of inside wiring that often overlaps analog phone lines and high-speed LAN connections. As a matter of course, I suggest that you wire sites with Category 5 wire to provide the flexibility for interchangeability and mixing of Fast Ethernet, Fiber Distributed Data Interchange (FDDI), ATM, POTS, and ISDN lines. Most telcos will not automatically provide this type of wiring; they use standard voice-grade wire (Category 1). You will likely need to do it yourself or contract out for these services. Although Category 5 wiring is sufficient for ISDN, wiring suitable for ISDN, which is Category 1, is not inherently suitable for Fast Ethernet. In fact, Category 5 wiring requirements include attention to detail on connectors, telecommunication closet panels, and the actual installation process.

While I would not suggest that Category 5 is absolutely necessary for the small office, home office, or occasional web surfer, sites wired for long-term tenancy and sites with a large number of employees will benefit from the installation of Category 5 cabling. In addition, a unified wiring infrastructure has benefits when you migrate analog phone services to ISDN and perhaps to ADSL, or to ATM and B-ISDN in 4 or 5 years. Wiring infrastructures tend to have a useful life span of at least 15 years if you do more than match current minimum requirements. The following few sections detail the wire installation process. By the way, about 80 percent of all networking problems usually get traced to wiring flaws. As the bandwidth and transmission speed increases, wiring becomes a more critical component. Wiring is not insignificant, and you should be aware of its implications to your ISDN installations. Even small office/home office (SOHO) installations can create reliability problems when wires are run under gutters that leak or along the ground line that gets flooded in heavy rains.

ISDN is wired from the central office as a two-wire pair for ISDN BRI (U interface) or a four-wire pair for ISDN BRI when wired for the S/T interface or ISDN PRI (T-1). Some Magellan T-1 lines carrying the ISDN PRI protocol are delivered on fiber to the demarcation point. The demarcation point for small businesses or residences is usually outside the building. The local telephone company may prefer to terminate the local loop at larger sites in a telecommunication service building or closet. This is inherently more secure. If the demarcation is outside, put a tough padlock on the box. The local telephone company may even run the termination to your floor or a telecommunication closet on the floor.

Fiber local loops may be converted to copper at that demarcation point. Typically, the telephone company will terminate the ISDN loop at a junction box. I suggest you ask for a new one (even if you only have a single analog phone line or even a handful of lines) so that in a year's time, someone will not confuse the ISDN circuit with the PBX lines. Even if the demarcation is *inside*, put a tough padlock on the box. There should be no extra cost for the new junction box. Such a unit is illustrated in Figure 4.18.

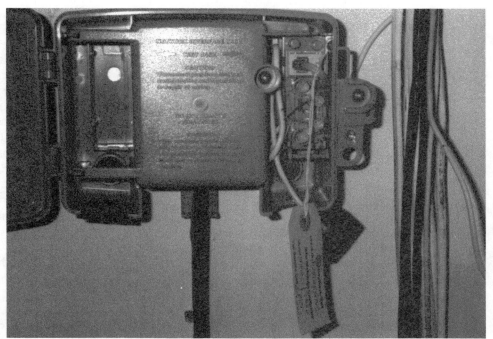

Figure 4.18 ISDN BRI local loop termination (demarcation point).

This photo is useful to show the official point after which the local ISDN service provider is no longer responsible (unless you pay them to install and service premises or inside wiring) for quality of service (QOS). That is why this is called the service *demarcation* point. Note that the ISDN "wire" is actually connected under a cover as two separate wires and is wired into the RJ-11 jack. By the way, Microsoft (on their Internet site for *Get ISDN Home Page*) and many other vendors are incorrect about ISDN jacks. The

RJ-11 jack has four wires (red, green, black, and yellow), the U.S. version of the RJ-45 jack has eight wires (blue, orange, black, red, green, yellow, brown, and gray), whereas the European 4X6 RJ-45 jack has six wires (orange, black, red, green, yellow, and brown). This color coding is standardized by Electronics Industry Association/Telecommunication Industry Association (EIA/TIA) 568 recommendations, although it is also possible to have candy-striped color jackets on the pairs as well. Try to conform to the wiring scheme because it will minimize conflicts when wires cross or create reverse polarity problems. The pinouts are defined by wire usage in this section. By the way, the less common RJ-48 jack is also interchangeable with the RJ-45. Both are also interchangeable with Acronym for the International Standards Organization (ISO) 11801 connectors.

You can always connect your ISDN terminal adapter to this jack in the demarcation box (with a female-to-female modular plug) if you suspect inside wiring problems. You wire a red and green (the colors indicate power and ground) wire with ordinary telephone wire and almost any other two-wire cable to a jack wherever you want it. Surface-mounted jacks work well; there are virtually no electrical or zoning limitations. If you plan for more involved raceways, wiring inside walls, ceilings, and flush-mounted boxes, check the local building code for compliance. You can also cut a hole into sheet rock, insert a metal outlet box, nail it to the stud, and fish the wire through the wall.

Before installing any new or repaired wire or adding twisted-pair, inspect it for visual defects. The wiring bundle should have no obvious physical defects like cuts, tears, or bulges in the outer jacket. Electrically, the individual wires in the bundle should conduct electricity without discontinuities. Each wire in a pair should be electrically isolated and not short out to each other. It is also very important to test that pairs are not separated. In other words, you do not want to mix a wire from a red-striped pair with a wire from a brown-striped pair; this defeats the noise cancellation of the twists and could mean crosstalk and excessive near-end crosstalk (NEXT). Near-end crosstalk occurs because the impedance or power of the signal in one pair overwhelms the signal on another pair. This usually results from substandard wire, connectors, excessive untwisting of pairs, excessive stripping of jackets, misuse of pairs, or use of the wrong materials. Techniques for testing with a cable tester are presented later in this chapter. To install the twisted-pair bundle you will need the tool kit, the preliminary blueprint, the cable, cable ties, and colored electrical tape for labeling.

Category 5 wiring does not guarantee 100-Mbps performance all by itself. Connectors, panels, and jumpers must conform also to Category 5 standards. Installation per recommendations is also essential. Proper installation means more than passing a continuity test. Installing unshielded twisted-pair cabling is a highly skilled art, requiring installation practices specified in industry standards. In some states, the Union of the Brotherhood of Electrical Workers (UBEW) or other trade groups have lobbied successfully for restraints on the legal right to install Category 5 wiring. In any case, not all installers are competent, and you should realize that Category 5 wiring requires significantly more installation care than voice-grade lines.

The primary standard applicable to cabling installers is the EIA/TIA-569. This is the telecommunication wiring standard for commercial buildings. A draft revision labeled SP 2840A is due to be issued soon as TIA 568A. American National Standards Institute (ANSI), General Assembly of European Computer Manufacturers Association (ECMA), and other trade groups have also ratified this standard, and there are similar international standards issued by the International Organization for Standardization and the International Electromechanical Commission as ISO 11801 and DIS-11801. These are basically the same as the EIA/TIA-568 standard for Category 5 issues; they do not cover all the material of the EIA/TIA standard. Most major suppliers of telecommunications components provide installer training based on these standards. The cabling issues that must be addressed by the installer include:

- Minimum bend radius
- Maximum pull force
- Jacket removal
- Untwisting of pairs
- Maximum cable length
- Avoiding sources of electromagnetic interference
- Cable termination methods
- Patch cord assembly

These issues are addressed in this and subsequent paragraphs. The TIA's draft SP 2840A, also referred to as Annex E to EIA/TIA 568, specifies that the minimum bend radius of Categories 3, 4 and 5 UTP cable shall not exceed 4 times the diameter of the jacketed cable. This requirement is less severe than that published in the earlier EIA/TIA-568 specifications. Avoid exceeding minimum bend radius when cable is pulled around corners or coiled for storage behind wall plates. Bend radius is important because excessive bending of a UTP cable may change the geometry of the twisted pairs within the jacketed cable and cause crimps that alter signal transmission propagation speeds and accuracy. This can adversely affect the near-end crosstalk and attenuation performance of the cable.

Pull force is exerted on a cable when it is being installed in a ceiling, raceway, cable tray, or conduit. The effects of exceeding the recommended pull force include stretching the cable, which can change the geometry of the twisted pairs within it. This, in turn, can adversely affect the transmission performance of the medium. Although no standards requirement for pull force currently exists, SP 2840A recommends that *four pairs* of UTP cables be able to withstand a pull force of 110 newtons (25 foot pounds) without damage. The same maximum pull force should suffice for Categories 3 and 5. You may need separate risers and intermediate pull boxes to stay within these limits. Pull force can be reduced with a lubricant. Water is usually sufficient, but tension reduction with silicone-based lubricant could be significant in multiple-bend pulls. This would include pulls where "bending" is caused by natural undulations of inner duct contained in conduit, plenum, or tight corners.

High-speed wiring is very sensitive to irregularities in the physical media caused by kinking, stretching, binding, and general rough handling. Wire pairs in a typical Category 5 cable comprise two insulated conductors very tightly twisted around each other. This twisting minimizes not only susceptibility to outside electrical noise, but also crosstalk from one pair to an adjacent pair. To meet Category 5 parameters, it is imperative that installation techniques be used that will maintain the integrity of this twist throughout the entire system. Problems that can be localized to the cable installation itself can usually be attributed to the disruption of the integrity of the twist. Stretching, a common issue with softer, nonplenum cables, can actually cause the wire thickness to fluctuate from normal to thinner-than-normal, thus lessening the available signal path and increasing attenuation.

Kinking of cable is sometimes difficult to avoid. However, impedance mismatching and excessive NEXT are two problems that can arise from a kinked cable. Tight radius bends cause similar problems. For example, an impedance mismatch occurs when the cable is so tightly bent that the relationship between the conductors is disturbed, resulting in fluctuating capacitive and inductive characteristics at that particular point. SP-2840A requires that bends in Category 5 cable installations be no less than 1.25 in diameter. If binding occurs when the cable is pulled tightly around a sharp object such as a support beam, hanging ceiling hardware, or ventilation equipment, damage can range from a slight flattening of the cable pairs to complete sheath destruction and removal of individual conductor insulation. Problems caused can range from mild increases in crosstalk to open and shorted conductors.

When terminating UTP cable, you should only remove as much cable jacket as needed to terminate properly to the connecting hardware. This is the TIA's recommendation in SP 2840A, which allows for installer discretion, because various types of connecting hardware require more or less jacket removal. Minimizing jacket removal maintains a barrier that protects individual conductors from physical damage and electrical interference. Also note that the telecommunication hardware is different from standard analog phone patch panel and blocks. Figure 4.19 shows the height difference between the wire retainers for a "standard" block and a Category 5-certified unit.

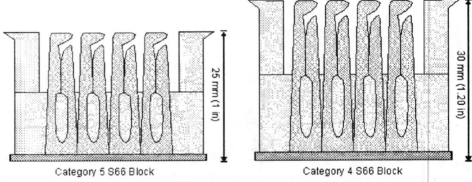

Category 5 S66 Block   Category 4 S66 Block

Figure 4.19 Wire retainer height differences for Category 5 punch-down hardware.

If you wire your ISDN lines through a wiring closet, many sites have gang connectors for the punch-down block. These connectors are sometimes called *hydras* because they have so many heads (that is, contacts) relative to the four pairs used for wiring each office. They are beneficial in sites with hundreds or thousands of connections wired into the same closet because they save space. Figure 4.20 shows this D-plug.

You also need to inspect the wiring method for these harnesses and connectors. There is at least one formal wiring scheme, and several others that are not so unusual to discover in practice. Figure 4.21 illustrates these pinouts.

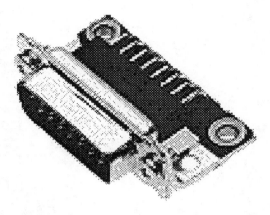

Figure 4.20 Hydra D-plug for wiring the typical wiring harness.

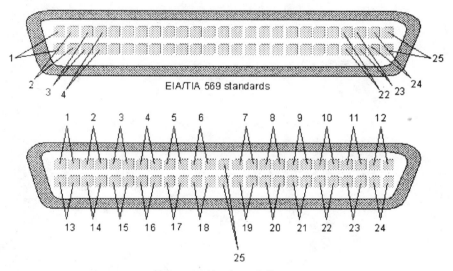

Figure 4.21 Hydra D-plug pinouts.

While this technology has long been standard with Types 66 and 110 punch-down blocks, finally this technology has migrated to the jumper panels. You no longer need to strip and individually (or even as a gang) punch down the individual pairs into the retaining clips. New modular technology includes connections for modular patching so that you can take the wiring harness with the hydra at one end and terminate into a single gang jumper connection. Figure 4.22 shows the gang RJ-45 block and Figure 4.23 illustrates a comparable Type 66 termination.

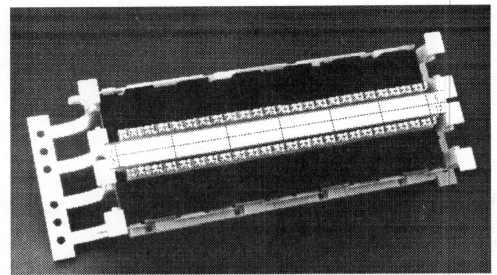

Figure 4.22 Wire closet gang jumper block (courtesy of the Siemon Company).

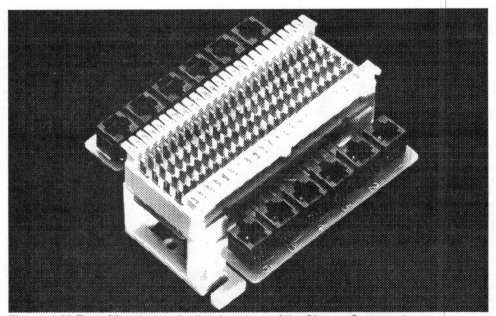

Figure 4.23 Type 66 gang termination (courtesy of the Siemon Company).

Category 5 cable pairs shall not be untwisted more than 13 mm (0.5 in) and Category 4 cable pairs shall not be untwisted more than 25 mm (1 inch) at the point of termination, according to SP 2840A. There is no requirement stated for Category 3 cables. Common sense dictates that no more than 25 mm (1 in) should be untwisted. This is illustrated in Figure 4.24.

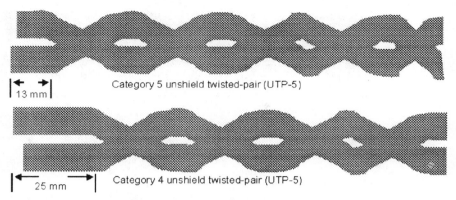

Category 5 unshield twisted-pair (UTP-5)

13 mm

Category 4 unshield twisted-pair (UTP-5)

25 mm

Figure 4.24 Do not exceed the recommended limits for untwisted UTP.

Complying with these cabling requirements maintains proper link performance to reduce near-end crosstalk. You do not want or need to strip the wire. Common punchdown methods (and tools) and modular connectors pierce the wire insulation as part of the installation process.

Maximum cable length is specified in the original EIA/TIA 568 standard as 90 m (295 ft) for horizontal cable runs, regardless of cabling medium. The maximum length for backbone cable varies by medium and is specified in SP 2840A. This recommendation states that backbone cable length may vary by application. If the spectral bandwidth of the application that will operate over the backbone is greater than 4 MHz (IBM 4-Mbps Token Ring is 8 MHz), the total backbone cable length should be limited to less than 90 m (295 ft) for UTP and shielded twisted pair (STP) A cabling. 100BaseVG supports about double this. Jumpers or cross-connects at the wiring closet can cumulatively extend to 6 m (19.44 ft), while wall connections to workstations typically cannot exceed 3 m (9.72 ft).

Transformers and other energy sources can alter the impedance of the wire. Minimize the wiring bundle lengths, and do not leave tails of wire (coils of extra wire) above the wiring closet; cut the wiring bundle to the shortest possible length but leave sufficient slack for temperature-induced expansion and contraction, as well as for the possibility that the wiring punch-down panel or office partitions may get moved a few feet.

If you anticipate that the distance limitations or external inference may cause transmission problems, select the highest quality UTP. Do not overlook vertical crossconnects and drops when estimating cable or wiring measurements, because the extra lengths increase the potential that external interference will disrupt data communications. You can offset the extra length by substituting a higher grade of cable. For example, if the normal grade is Category 3, substitute Category 4. Shielded twisted-pair is available (at a substantial premium, since it is not as widely used and not as common). Category 5 and STP Category 5 will perform better for all network applications because crosstalk, signal degradation, and insertion to capacitance are improved; however, this extra service is not required for any current networking technology, and your network will not be better for it.

Until recently the cabling itself was thought to be the primary cause of NEXT interference. However, current manufacturing techniques are producing very consistent, tightly twisted cabling that is capable of handling extremely high frequencies with minimal NEXT interference. Now the modular connectors and termination facilities are the points in the link at which the cable structural integrity is compromised. The problems occur when the cable must be untwisted and spread apart to allow connection to a plug, jack or punchdown block when interfacing to active or passive interconnecting hardware (e.g., patch panels, hub, concentrators, or switches). The more connecting and termination points that exist, the more susceptible to interference and signal degradation your system will be.

Inside wiring by its very nature connects to the outside world and typically creates more opportunities to interact with nature. Keeping this in mind, consider the following safety issues:

- Never install wiring during a lightning storm.
- Do not install wiring in wet locations unless both wire and termination are protected from water or designed specifically for wet locations.
- Never touch uninsulated ISDN lines or terminals (the wire could be cross-connected with AC power or have acquired a significant static charge).
- Exercise care when using ladders, stages, or climbing to heights.
- Do not work on your telephone wiring at all if you wear a pacemaker. Telephone lines carry electrical current at voltages to 96 VDC.
- Avoid contact with electrical current.
- Use a screwdriver and other tools without insulated handles.
- Do not work on telephone or electrical wiring while a thunderstorm (electrical storm) is in the vicinity.
- Wear safety glasses or goggles.
- Be sure that your inside wire is not connected to the access line while you are working on your telephone wiring.
- Do not place telephone wiring or connections in any conduit, outlet, or junction box containing electrical wiring.
- Installation of inside wire may bring you in proximity to electrical wire, lightning rods and associated wires, and at least 15 cm (6 in) from other wire (antenna wires, doorbell wires, wires from transformers to neon signs), steam or hot water pipes, and heating ducts.
- Before working with existing inside wiring, check all electrical outlets for a square telephone dial light transformer and unplug it from the electrical outlet. Failure to unplug all telephone transformers can cause electrical shock.
- Protectors and grounding wire placed by the telecom service provider must not be connected to, removed from, or modified by the customer beyond the demarcation point.

The photograph in Figure 4.25 shows a surface-mounted jack designed for analog telephone lines, about 1.5 cm (0.5 in) thick, bought at a dollar store.

The plastic face plate is removed to expose the connections to the modular jack. The two pairs of standard telephone wiring with a gray vinyl sheath have been threaded through the ceiling plenum and pulled through the walls into the back plate. The red and green leads have been stripped and are ready to be attached to the corresponding leads on the jack.

Figure 4.26 shows another site with the flush-mounted jack installed and the red and green wires (annotated) connected to the leads. I like this method of installation because some municipalities require a certified electrician to perform any in-the-wall wiring. A handy do-it-yourselfer can run the ISDN wire from the demarcation point, through a window jamb, and down a wall (on the surface of the wall) to the surface plate.

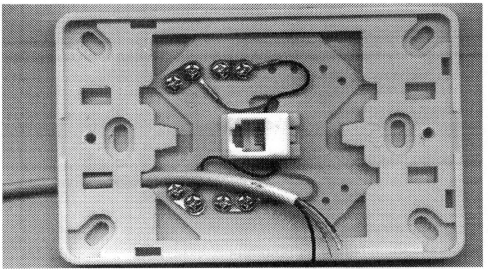

Figure 4.25 Typical small office or home user ISDN inside wiring termination using a surface mount jack to avoid any violations of building codes.

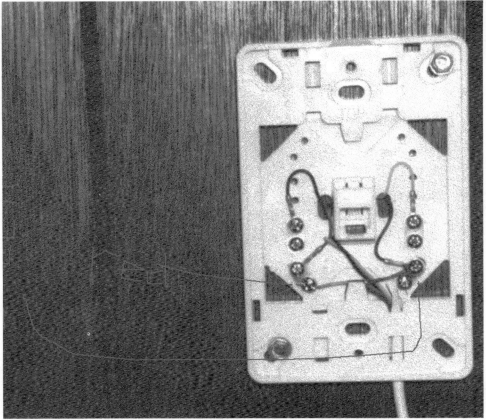

Figure 4.26 A surface-mounted jack serves as a functional ISDN outlet.

If you wire a daisy chain into the ISDN line with multiple jacks, terminate the remaining jacks with a four or six pin modular terminating resistor in each jack. This serves two purposes, ensuring that the someone will not plug a standard handset in the line provided for ISDN (which will create a trouble condition on the line and typically cause the central office to disable it) and maintaining the proper impedance for the ISDN line. You can convert the analog local loop to ISDN with the consent of your telecom service provider when they support ISDN and you have a special line installed to your location (because you are far from the central office or telephones are unique in your area). Most of the preexisting analog lines do not require any special conditioning unless the line has *load coils,* in which case these balums must be removed. However, realize that load coils are usually only found on existing lines that are 5 k (18,000 ft) or longer and local loops of this length may require Basic Rate Interface Terminal Extender (BRITE) or other signal repeaters or special extenders. The 2B1Q line transmission scheme (not to be confused with 2B+D channelization) is tolerant of a certain amount of bridge taps, and therefore only a minimal subset of existing lines [lines with bridge taps whose total length is greater than 925 m (3000 ft) for the bridge taps] would require this "deconditioning."

However, ISDN is not always a simple two-wire installation process. This is the standard for ISDN BRI U interface in the United States, but older sites and European users will also see the S/T interface, common when the service provider includes the NT-1 as part of their basic service. In Europe, the telecom provider includes the network termination, and this yields four wires of the S/T interface. If the ISDN service includes BRITE support for extended lengths in local loops, the service includes the NT-1 before the demarcation point. An NT-1 device is pictured in Figure 4.27.

Figure 4.28 illustrates how the NT-1 connects with the ISDN services and the different types of devices that it can support. At the customer premises the U-loop is terminated by a network termination 1 (NT-1) device. (By the way, there is no market and need for NT-1 devices in Europe because this service is provided by the telecom carrier.) The NT-1 drives an S/T-bus which is usually four wires, but in some cases it may be six or eight wires. When the NT-1 unit is internal to the ISDN terminal adapter, the connection is the same except you typically do not need the RJ-45 jumper between the NT-1 and the separate terminal adapter. Two such units are shown in Figure 4.29.

Figure 4.27 Typical NT-1 (courtesy of Adtran Communications).

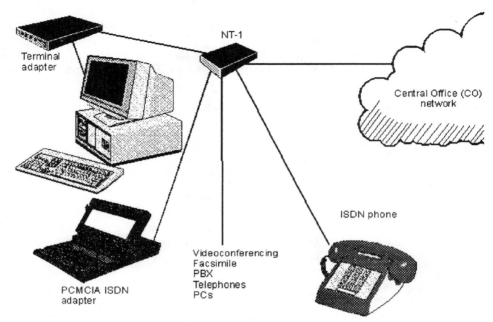

Figure 4.28 NT-1 connectivity.

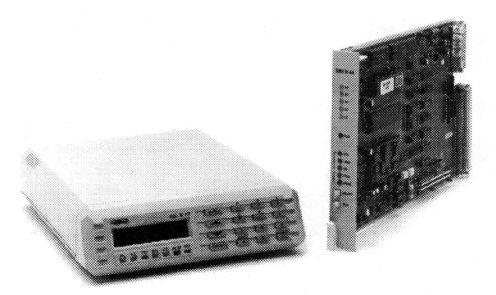

Figure 4.29 Typical ISDN adapters with integrated NT-1; external BRI unit on left, and BRI module for a multislot chassis on the right (courtesy of Adtran Communications).

In optional cases, the extra wires provide power to operate telephones when normal power fails. Alternately, "phantom" power may be derived from the standard four wires. Outside of North America, emergency-mode operation provides power for basic voice service only in the case of loss of local power. In emergency-mode operation the NT-1 receives up to 1.2 watts from the central office. In North America there is no provision for emergency-mode operation.

The following pinout tables explain the wiring schematics for the common implementations for ISDN BRI wiring. Figure 4.30 shows the common two-wire local loop connection with a four-wire RJ-11 jack and plug.

Pin	Signal
1	—
2	U connection
3	U connection
4	—

Figure 4.30 U interface cable pinouts with RJ-11 with four wires.

To add confusion to something that really should be simple, RJ-11 plugs and jacks can also have six wires. The outer size is exactly the same and the two center pins align correctly with each variation. If you have a six wire configuration, let the wiring scheme slip by one. This is shown in Figure 4.31.

Pin	Signal
1	—
2	—
3	U connection (tip)
4	U connection (ring)
5	—
6	—

Figure 4.31 U interface cable pinouts with RJ-11 with six wires.

Figure 4.32 shows how the two wires are mated for RJ-45 jacks and plugs.

Pin	Signal
1	—
2	—
3	—
4	U connection (tip)
5	U connection (ring)
6	—
7	—
8	—

Figure 4.32 U interface cable pinouts with RJ-45.

Figure 4.33 shows how the two wires mated for RJ-45 jacks and plugs in Europe, but not Germany, France, or former Eastern Bloc Soviet countries.

Pin	Signal
1	—
2	—
3	U connection (tip)
4	U connection (ring)
5	—
6	—

Figure 4.33 U interface cable pinouts with RJ-45 for most of Europe, excluding France, Germany, and former Soviet-Bloc countries.

Soviet Bloc countries use a low-current analog system called a Stroëger rotary switch that is absolutely at odds with ISDN. The rotary switch is in the central office, but you can test for the POTS lines with a multimeter (see Chapter 5). You might note that the wiring scheme is identical to the six-wire RJ-11. This six-wire RJ-45 connector is wider than the six-wire RJ-11 and identical in outer size to the eight pin RJ-45. It is also the same size as the RJ-48. Figure 4.34 shows how two wires are mated for RJ-45 jacks and plugs in Germany.

Pin	Signal
1	U connection (tip)
2	—
3	—
4	—
5	—
6	U connection (ring)

Figure 4.34 U interface cable pinouts with RJ-45 for Germany.

If you think that this is the end of the wiring complexities, consider that the European Union has the gall for yet another pinout. Figure 4.35 shows how the two wires are mated for RJ-45 jacks and plugs in France.

Pin	Signal
1	—
2	U connection (tip)
3	—
4	—
5	U connection (ring)
6	—

Figure 4.35 U interface cable pinouts with RJ-45 for France.

Figure 4.36 shows how backup power is often fed into U connections with RJ-11 connections as is most common in the United States, Canada, Japan, and Mexico.

Pin	Signal
1	-48 VDC
2	U connection (tip)
3	U connection (ring)
4	-48 VDC retain

Figure 4.36 U interface cable pinouts with RJ-11 with power cable.

Figure 4.37 shows how backup power is often fed into U connections with RJ-45 connections which might be necessary with some NT-1 and ISDN terminal adapters.

Pin	Signal
1	—
2	—
3	—
4	U connection
5	U connection
6	—
7	-48 VDC
8	-48 VDC return

Figure 4.37 U interface cable pinouts with RJ-45 with power cable.

The U signal is split into S/T channels and delivered on a four-pair wire with this RJ-45 pinout, as shown by Figure 4.38.

Pin	Signal
1	—
2	—
3	S/T receive+ (rcv+)
4	S/T transmit+ (xmt+)
5	S/T receive- (rcv-)
6	S/T transmit- (xmt+)
7	-48 VDC
8	-48 VDC return

Figure 4.38 S/T interface cable pinouts with RJ-45.

The RJ-45 pinout is different in Europe, as shown by Figure 4.39, where the RJ-45 is six instead of eight wires.

Pin	Signal
1	S/T receive+ (rcv+)
2	S/T transmit+ (xmt+)
3	S/T receive- (rcv-)
4	S/T transmit- (xmt+)
5	-48 VDC
6	-48 VDC return

Figure 4.39 S/T pinouts with the common six-wire European RJ-45.

Manufacturers do not conform to any design for the ISDN terminal devices. This includes NT-1 devices as well. I recommend wiring the incoming ISDN as only two wires. The connection between a jack and NT-1 or terminal adapter is shown in Figure 4.40. Note that internal ISDN terminal adapters are wired in the same way.

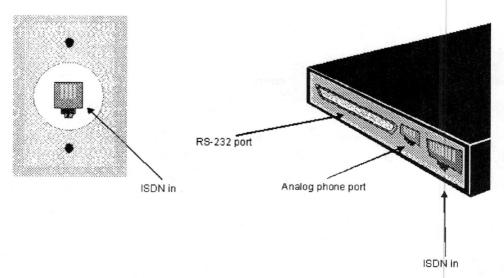

ISDN in

RS-232 port

Analog phone port

ISDN in

Figure 4.40 ISDN user jack and ISDN device connection.

However, you may run into a mating problem. The ISDN jack may be RJ-11, RJ-45, or the European RJ-45 with six pins. You will need to match the plugs. Figures 4.41 and 4.42 show the function differences between the RJ-11 and the RJ-45 male plug.

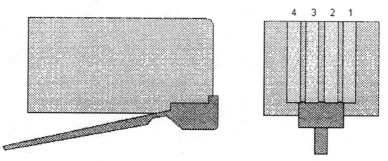

Figure 4.41 RJ-11 schematic for four wires.

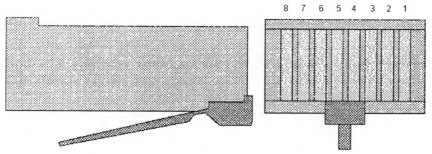

Figure 4.42 RJ-45 schematic with eight wires. The width is distinctly different.

You might note that 10Base-T eight-wire RJ-45 jumpers (Category 3) or 100Base-T jumpers (Category 5) are very suitable for ISDN. However, when you need to mate RJ-11 to RJ-45, there are three approaches you can take. First you can plug a standard RJ-11 into the center of both RJ-45 jacks. This works because the center pins (2 and 3 or 3 and 4 on RJ-11, and 3 and 4 or 4 and 5 on the various versions of the RJ-45 modular plug) align correctly.

This is really the simplest approach because analog handset jumpers (straight through) work well. You can also go to Radio Shack, buy a snap-on RJ-45 plug, cut off one end of the analog handset jumper, and work the red and green wires into the snap-on plug. Alternatively, in order to comply with the ANSI T1.601 standard, install a four-pin male to eight-pin female adapter. Place the end for the four pins on the ISDN access jack on the wall, so the eight-pin female end is exposed. These adapters are not easily located. Instead, you could install a female RJ-45 coupler to a RJ-11 jumper, except that these converters are more difficult to locate and acquire. You might try specialty telephonic parts catalogs. Figure 4.43 illustrates these mating options.

Some vendors understand the user better than others. For example, the 3Com Impact unit with the U.S. standard U interface typically comes with a six-pin RJ-11 patch cord with an eight-pin RJ-45 on the other end. Of course, 3Com also sent external units that required a boat anchor power supply with 12-VAC output that did not fit the jack on the back of the unit. A razor blade and alligator clips solved that problem once the case for the devices was opened. However, these surprises only go to show the types of problems you really are likely to confront and need to overcome.

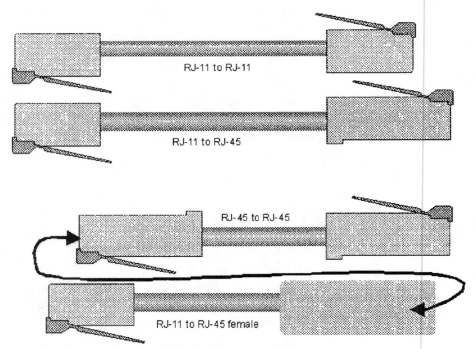

Figure 4.43 ISDN user jack and ISDN device connection.

The Last Mile

The phrase, *the last mile*, refers to the length limitation of the connection from the central office to the demarcation point. This is typically 3.7 km (2.3 mi), but it can be extended with *midspan repeaters* which must be placed at intervals of 2.8 km (1.7 mi) to regenerate and retime the signal. Midspan repeaters are also called *line extenders*. The functional limitation for using these repeaters is available power for the repeaters and the willingness of the customer to pay the monthly fees for each repeater. Another method uses two signal converters that convert the U reference point transmission signal into a new format, called the Simple Coded-Pulse Amplitude Modulation (SC-PAM), that will carry up to 6.1 km (3.8 mi). The converters, a pair of them, are positioned after the central office switch and before the demarcation point. These last mile limitations are represented in Figure 4.44. These limitations are not the last word on limitations of ISDN delivery. There are a handful of other service delivery options including wireless or satellite delivery as described later in this chapter.

ISDN Routers and Firewalls

It is inefficient and insecure to provide remote connectivity with individual modems and terminal adapters. This has become pronounced as users request connectivity to the Internet. Although users did request individual outgoing connectivity to bulletin board systems (BBS) services, such as America Online and CompuServe, in the past, and this was traditionally addressed with banks of analog modems and DOS INT-12 network redirection software, this configuration is more difficult to implement for Internet connectivity. As a result, the network architecture in practice has reverted to what is shown in Figure 4.45.

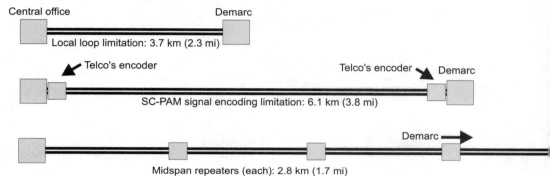

Figure 4.44 Last mile wiring connections for ISDN.

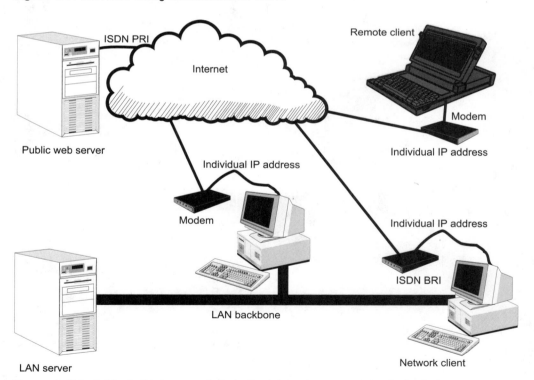

Figure 4.45 Individual client connectivity to the Internet opens a security keyhole.

The reason is that every Internet connection requires an Internet address, an IP address, which gets mapped for e-mail and FTP (file transfer) activity. You might note that the IP address does not have to be dedicated to each process or user. IP addresses are often allocated as needed and randomly (within a range) to users just as long as they need them. In some cases, a single IP address is assigned to a communication server, and user processes are routed with only that single IP address.

This architecture raises the possibility of a security breach within the network environment and also from outside the corporate Intranet. The technology has evolved sufficiently to create a firewall to control all internal links to the Internet and filter all external Internet requests. In addition, network bandwidth is conserved by processing requests in the HyperText Transport Protocol (HTTP) proxy server. The HTTP proxy server is a cache of previously (although recently) retrieved web pages through the HTTP. HTTP is the primary delivery mechanism for all Internet web page access, although other protocols are also used for file access, messages, mail, and management. In effect, bandwidth between the corporate site and the Internet can be added as needed to cut connectivity costs. The HTTP proxy server can censor access to web sites, cache frequently requested web pages, and consolidate LAN-Intranet traffic. Remote clients are protected and secured with encryption, while all publicly accessible web material is located outside the firewall. The Internet addressing requirement, corresponding loosely to INT-12 redirection, is handled by the LAN or HTTP proxy server by assigning temporary IP addresses. The dynamic host configuration protocol (DHCP) is the Internet Engineering Task Force (IETF) specification which enables clients to borrow a uniquely assigned IP address for a short period. This configuration is illustrated by Figure 4.46.

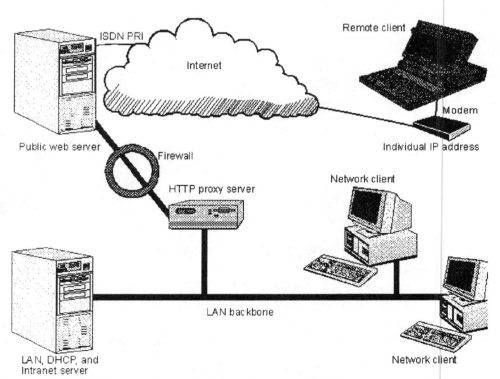

Figure 4.46 Installation of firewall and support services adds security for local networks.

When you want to ship packet data in the X.25 format, this is partially a function of provisioning. Specifically, you need to indicate to the local ISDN service provider that a D channel (or possibly even both BRI B channels or some combination of the 23 PRI B channels) connects into a public or private X.25 network. Point-of-sale (POS) terminals, cash registers, credit card readers, and automatic teller machines can connect directly into the ISDN terminal adapter. Pharmacies such as Rexall are using ISDN and X.25 to shunt prescription information between stores and to insurance companies for third-party payment authorization. This configuration presumes ISDN 0B+D and is expensive unless the X.25 traffic utilizes a significant portion of the available bandwidth. Run the financial comparison to ascertain the viability of ISDN over a leased digital or T-1 line. Do not overlook the per-mile costs that often make dedicated digital and T-1 lines expensive. This configuration is illustrated in Figure 4.47.

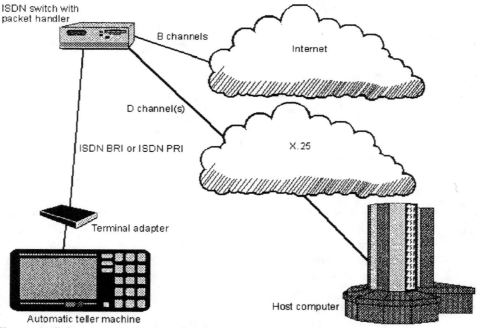

Figure 4.47 Connectivity for a packet data channel.

On the other hand, you can split both B channels or a single B channel out of ISDN specifically for data, facsimile, or phone service. If you have cost-justified ISDN on the cost savings over switched digital or a dedicated lease line for the packet data, if you need some service for packet data and do not currently have it, or if you need additional phones lines, look at the next diagram. In fact, if you need more lines or can remove lines, you can migrate to the illustration shown in Figure 4.48. The limitation to this design is that you may need an external NT-1, a special ISDN PBX, or an external adapter with two external lines, such as the Motorola BitSurfr Pro, so that all the channels can be utilized to the full potential. Also, realize that you are limited to 100 m (324 ft) between the user devices once individual lines have been extracted.

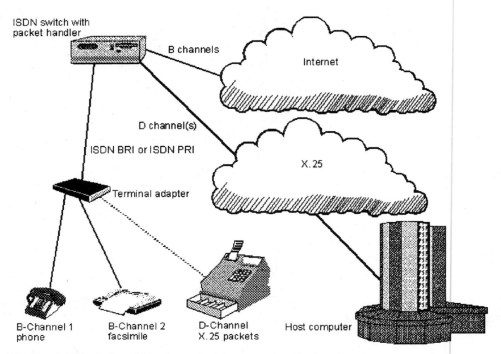

Figure 4.48 Split the ISDN channels for economic gain for other uses even when you are installing ISDN only for X.25 packet data.

Satellite Connections

Transporting ISDN over wireless or satellite digital links solves a great many connectivity and service problems. In areas where ISDN is not available, Very Small Aperture Terminal (VSAT) is a satellite-based service that provides from 128 to 384 kbps of raw transmission bandwidth, corresponding to ISDN BRI and ISDN-H services. You only need a VSAT service carrier or a point-to-point or multipoint wireless service and the gateway equipment for the connection. Refer to my book, *Implementing Wireless Networks*, for concise and practical information for wireless networking. Figure 4.49 illustrates a configuration from Data Telemark showing wireless or satellite service.

The following examples detail situations where ISDN over wireless or satellite digital links might provide connections where it is seemingly unlikely or impossible or where service is too expensive, unreliable, or difficult to provision. For example, consider the site that is outside the maximum distance for ISDN, even beyond the SC-PAM, which is used to extend the ISDN signal up to 9.3 km (33,000 ft), or repeater service. This is referred to as the *last mile* problem. Although ISDN is available in your area, your site is too far from the nearest switch. In such instances, ISDN over wireless and satellite linkages can overcome this limitation by hauling ISDN from a serving point within range of the central office.

Consider also a site that has no ISDN, only regional ISDN, or ISDN is available only in the local exchange. This means that your site can be an ISDN island lacking Interexchange Carrier or local loop access to the long-distance provider. ISDN over wireless and satellite linkages bypasses the local carrier and provide direct connectivity.

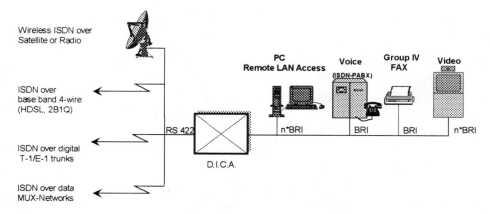

Figure 4.49 Satellite connectivity to ISDN (courtesy of Data TeleMark Gmbh).

When a loop qualification, the carrier's test to see if the lines in the area can support the ISDN local loop, does not show possible capacity; when midspan repeaters are not possible; and when an alternate network service agreement (ANSA) cannot be arranged, this wireless or satellite loop may represent the only viable alternate. This is also relevant when you need backup ISDN or just plain backup data services but cannot get a secure alternate wire path for the backup connection.

Mobility gives ISDN an important advantage when it comes to providing redundancy. Redundant telephone circuits are not that common, and creating multipoint drops when most central offices do not have alternate routes is nigh impossible. Even when businesses order separate primary and secondary circuits, the circuits often run in the same cable sheath or over the same carrier. If a cable is cut, or a tandem telephone switch inexplicably goes down, redundancy means little to maintaining that critical connectivity. Mobile ISDN is inherently redundant. Mobile ISDN can supply services during terrestrial outages because it operates independently from the telephone network. If the terrestrial ISDN fails, a mobile VSAT unit can be dispatched as emergency backup. Figure 4.50 shows this point nicely.

Wireless and satellite integration to wired ISDN makes it possible to extend ISDN connections to anywhere in the world. Similarly, ISDN over wireless and satellite linkages is valuable for rural connections. It might be used for ISDN satellite videoconferencing by companies to overseas branches, for satellite Internet access within third world countries, or for distance education in rural communities. ISDN over wireless and satellite linkages is also providing a platform for environmental research and mobile medical service in Arctic and Antarctic regions and other remote locations.

Depending on the circumstances, ISDN can be costly. I think ISDN is overpriced by BellSouth when I compare my basic costs to those of just about any other RBOC; they are about three times as much. Using ISDN over existing wireless and satellite linkages can provide a less expensive alternative by saving on usage charges. This becomes relevant when basic monthly fees plus per minute connection charges exceed the VSAT downlink fees. VSAT downlink site charges run about $200 per site per month, but air time is usually included and connection is distance insensitive. It costs the same for a 2-km (6480- ft) link as for one around the world.

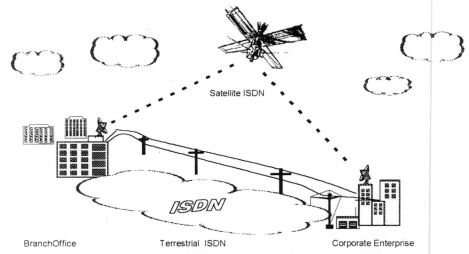

Figure 4.50 Redundant wide area network (WAN) satellite connections. Satellite connection can provide backup for terrestrial connections (courtesy of Data TeleMark Gmbh).

The correct term for the actual wireless or satellite interface is *gateway*. This is typically a box that connects between the wireless transmission equipment and the network routers. Some network routers, those based on hub chassis, may support the interconnection directly, but most often you need specialized equipment to bridge the wireless component into the wired network. Such a device is shown in Figure 4.51.

Figure 4.51 Wireless gateway device (courtesy of Data TeleMark Gmbh).

The Other End

ISDN represents at least two local loops; one for the calling party and another for the receiving party. Even when both parties are located in the same service region and are handled by the same RBOC or independent operating company (IOC) and in fact might be serviced by the same central office, recognize that variances in provisioning and equipment present subtle incompatibilities. When the calling and receiving party are in different RBOCs or IOCs, typically a long-distance carrier will provide interconnectivity. This means that there are now *three* carriers involved and the likelihood of 64-kbps clear channel is substantially reduced while provisioning differences are magnified. Although the long-distance connection is invisible for the most part, bonded service often requires special provisioning with the long-distance provider to minimize synchronization delays when channels are switched through different hubs and along routes.

In fact, Motorola BitSurfr products are configured by default out of the box for 56 kbps to lower the incidence of user technical support calls. The assumption is that most users will not notice this default value or notice the difference between 56 and 64 kbps. Rate adaptation is automatic for most central office switches. You actually have to look carefully to confirm the connection speed for each B channel and perform some complex configuration settings to increase the default connection speed to 64 kbps. Figure 4.52 illustrates the other end and any intermediate points in a long-distance ISDN link.

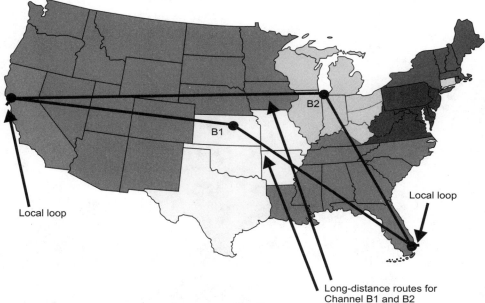

Figure 4.52 Long-distance ISDN connectivity raises issues of end-to-end clear channel.

Telecommuting

NYNEX Corporation commissioned a study from the Gartner Group to compare the benefits and costs for establishing a telecommuting policy. Although I suspect the complete authenticity of the source and its results, and the extraction of conclusions out of context, the report does conclude that telecommuting is the killer application for ISDN. Specifically, organizations save $5000 in office space per telecommuter, assuming a reduction in office space needs, gain on average 2 hours daily in increased productivity otherwise lost to commuting in traffic, and have happier and more loyal employees. However, the organization often has to initially invest $3000 to $5000 for computer equipment, terminal adapters (at both ends), and network security for each employee.

The depreciation and useful life span for this equipment is at most 3 years. For an employee earning $50,000 per year, the direct financial payback is less than 3 months. It is difficult to assess the value of increased loyalty except by factoring turnover and headhunting costs for replacements.

Although security may be an issue with remote users and telecommuters, realize that remote connectivity with ISDN includes by definition *Caller ID*. This service is really called *number calling* or *automatic number identification* (ANI). This can be used to initiate a call-back from the corporate communication server to the telecommuter. Even if a caller were to counterfeit the ISDN call header, the ANI would still be matched against an administered table of allowed phone numbers, and the communication server would return the call to one of these numbers.

Mobile ISDN

The reality is that few places (other than some hotels and businesses in Japan or Germany) have outlets for ISDN or even pay-phone service. This is gaining momentum as more users ask for ISDN and as hotels set up teleconferencing sites for shows, conferences, retreats, and special training sessions. In any event, you will need a mobile ISDN adapter that is compatible with the laptop credit card called the Personal Computer Memory Card International Association (PCMCIA) format as shown in Figure 4.53.

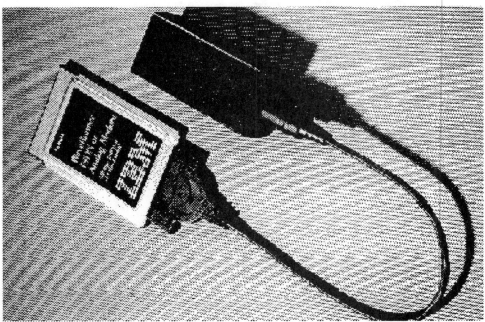

Figure 4.53 Long-distance ISDN connectivity (Courtesy of IBM).

Also, if you want to cart around the Data TeleMark ISDN-Fun gateway and a satellite dish in the briefcase, you could always have ISDN service via a satellite downlink. This is not convenient, but when you need real mobile linkage regardless of the telecom infrastructure, this is the way around the problem. You might refer to my book *Implementing Wireless Networks* for more information about costs, hardware, and realities of satellite services for remote LAN connectivity.

Internet Services

Internet access occurs in two basic formats. You can have a dial-up ISDN BRI connection analogous to a dial-up analog connection with a minimum bandwidth of 56 kbps or a dedicated ISDN BRI and even PRI connection to a local web server through the Internet service provider. Expect to pay $30 per month. The basics for a dial-up connection include WinSock, Eudora, and a browser, or a transaction control protocol (TCP) driver set such as NetManage's Chameleon (which has just about everything you need to connect to the Internet). The only disadvantage to Chameleon is the 10 floppy disks and the 21 MB of hard disk space required for installation, the desktop real estate, and the desktop rearrangement to PROGMAN.INI during installation. Like the TurboCom/2 replacement drivers delivered with the 3Com Impact terminal adapter, Chameleon improves Windows Interrupt Request Line (IRQ) performance, and it also provides a more efficient TCP/IP stack. It can be difficult to install with existing LANs. I have ran into conflicts with various NIC drivers that need to be resolved for coexistence. By the way, this tool kit, as shown in Figure 4.54, works with all versions of MS Windows.

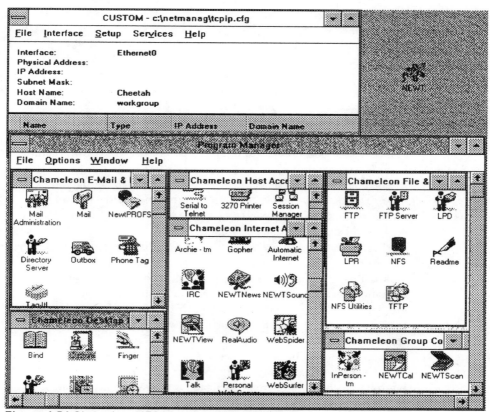

Figure 4.54 Chameleon*NFS* is a complete ISDN/Internet connection tool kit.

If you have this for an analog dial-up connection, everything should be the same for ISDN. Note that real-world performance may not be so different. Bonded connections with both B channels often require a new driver set (from the terminal adapter manufacturer) or special software (such as the MS ISDN accelerator pack). I have tried this with many providers, but the general result is that high ISP server loads or long network delays keeps FTP download times and web site browsing down to the slow speed of most analog connections. There were some bright spots with some providers at odd hours, but basically the ISDN connection was not warranted.

Internet connection for a dedicated link between the ISP and a corporate web site, Intranet, or communication server requires a dedicated network communication server, a web server software, and of course the ISDN linkage. You will actually get better utilization with this connection. This also costs more. Figure on $400 to $1000 for installation and $300 per month for service. You may get InterNIC registration and other services as part of these fees; they are not really "free." It makes the most of a single B channel or bonded B channels because the linkage is between the local server and the service provider where most Internet network bottlenecks occur. Because many browsers, Intranet users, and users internal to the organization might be using the link at the same time, the bandwidth is useful and proves its value. However, at a minimum, you need to establish a dedicated communication server with or without web server software.

Most Internet service providers (ISPs) are moving to support ISDN access for dial-up connections and dedicated connections for corporate customers. Prices range from $35 per month to more than $500 per month with sometimes hefty installation fees. There are differences in the services, how they get bundled, and how fast the connection and the host server actually is. Price seems to have nothing to do with the quality. Hunt around for a local provider or a national one with 800 number access or a local access number. Also, see if you can try the service and test it for performance. ISPs are an experience to deal with (some might say a scream). ISPs are often easier than a national service provider, such as PSInet or SpryNet, because they may actually answer the phone or respond to e-mail. The growth in the last year has simply been overwhelming for most of them, and the newest ones just do not have the infrastructure to deal with new customers, or *newbies.*

For example, my local service provider has been providing me Internet services since 1993. They are not the fastest, easiest with which to deal, or my idea of a commercial service provider in general, athough they seem to be improving. Problems have included a general lack of customer (as opposed to technical) service and several server disk crashes that wiped out my web pages and required a complete download several times. However, they have not supported ISDN until recently and have priced it at $500 per month. Realistic performance was no better than 28.8 kbps, or even 14.4 kbps. This is a limitation you should keep in mind before promising users more bandwidth. The problem for most web surfing is not so much the bandwidth but the network delays at the ISP's server or the host sites of the actual web pages. My ISP is not a bad provider, per se, but like most local providers, will have to adapt to compete with national service providers in terms of pricing, service, technical support, and new marketing services that will soon overwhelm the web. See my book *Building Cyberstores* (1996) for more information about commercial applications of the Internet and web servers.

Conclusion

This chapter described the process of ordering, provisioning, installing, configuring, and maintaining ISDN. The details about pricing, performance, and physical wiring were presented as well. ISDN does have its quirks and does provide many failure points and service problems. To this end, Chapter 5 describes troubleshooting and debugging.

ISDN Maintenance and Debugging

Introduction

It is important to understand the ISDN connection process to debug any type of ISDN problem hobbling your connectivity. The general process of ISDN maintenance and debugging is to prove what things are good and by a process of elimination determine what does not work correctly. Since the ISDN connection consists of two ends and a series of connections in the middle, you want to discover which side has the problem and focus in on that cause of the flawed connection. You want to validate (prove what things are good) from lower International Standards Organization (ISO) levels to higher ones. Figure 5.1 shows the Open Systems Interconnection (OSI) reference model protocol layer stack often used to intelligently describe protocol connectivity.

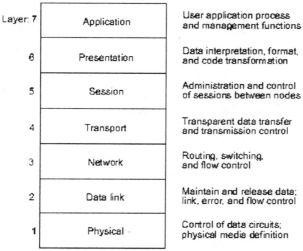

Layer: 7	Application	User application process and management functions
6	Presentation	Data interpretation, format, and code transformation
5	Session	Administration and control of sessions between nodes
4	Transport	Transparent data transfer and transmission control
3	Network	Routing, switching, and flow control
2	Data link	Maintain and release data; link, error, and flow control
1	Physical	Control of data circuits; physical media definition

Figure 5.1 ISO protocol stack definition.

Most problems relate to the physical layers. This is ISO layer 1, which corresponds to the actual wiring and hardware that you see in the ISDN setup. Most ISDN setup problems are caused by the ISDN terminal adapter or the ISDN line being provisioned to mismatched configurations at installation. This chapter describes some of the things you need to understand when setting up an ISDN line and describes the most common areas of confusion. Although much of this information is anecdotal, culled from conversations with experts, realize that situations described here are fluid and can vary with every situation. The basic process, as reinforced by every vendor of ISDN test equipment, is that you need to understand the physical and logical connections so that you can test and confirm what works and thus rule out what does not. Figure 5.2 illustrates the logical and physical connections required for an ISDN connection.

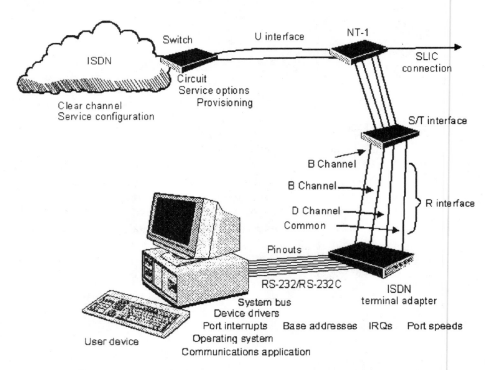

Figure 5.2 ISDN logical and physical layers and related components. A remote mirror connection must exist on the other side of the ISDN cloud and correspond to the local connection. You can have two independently-working ISDN setups that will not talk to each other.

Protocol Errors

ISDN is not just the pipe for data transmission. It is also the protocols to set up a connection, negotiate capabilities, set up the actual call, and select the channels used in that call. This process is part of the ISDN protocol. For example, the following list details the process of the ISDN call:

- Call setup
- Dial connection

- Establish connection
- Negotiate connection capabilities
    Channel selection (B1 or B30?)
    Channel speed
    Usage
        Circuit
        Stream
        Voice
        Packet
- D channel session control

Problems with the line, the switch provisioning, customer premise equipment (CPE) configuration, and synchronization errors can prevent you from getting a valid connection. Synchronization errors—when the two ISDN streams do not mesh with each other—are more often a function of bonding protocols. If you cannot get ISDN terminal adapters to connect at analog levels (if they support digital signal processing (DSP) or modem functions), try basic 56-kbps connectivity before assuming that something is seriously wrong. Work up to 64, 112, and finally 128 kbps or greater with multiple bonded channels. When ISDN is an integral part of wide area network (WAN) communications, you will want to analyze this process with a protocol analyzer outfitted with an ISDN interface. In addition, complex devices, such as routers, use the D channel to monitor and control the process, reset the connection, or make it dormant.

ISDN provides a naked digital pipe over which any digital signal can be transmitted. ISDN transmissions are generally defined by layers 1 through 4. You are likely to see Point-to-Point Protocol (PPP), MultiLink Point-to-Point Protocol (MP or MLPPP), Bandwidth Allocation Control Protocol (BACP), High-Level Data Link Control (HDLC), Synchronous Data Link Control (SDLC), Link Access Procedure, Balanced (LAPB), X.25, X.31, Internet Packet Exchange (IPX), and Internet Protocol (IP) running on various B channels at the same time. Although the complexity seems much greater with ISDN Primary Rate Interface (PRI), just realize that each channel is a separate (or bonded) conversation and view a single channel at a time. Do not get overly stressed by a problem; you are going to work it out if you read through this chapter and understand the principles of protocol analysis and traffic decoding.

Realize that the ISDN debugging technology and the tools are not as advanced as a Network General Sniffer for Ethernet or similar artificially intelligent local area network (LAN) protocol analyzers from Hewlett-Packard. However, many of these same products can also monitor and capture data from ISDN linkages by simply adding an ISDN adapter to an Ethernet protocol analyzer. Understand clearly, however, that the capability to decode and figure out protocol errors and OSI level 1 hardware problems in ISDN is not as advanced as with Ethernet. You might also note the importance of debugging an RS-232 serial port when using external ISDN terminal adapters. Similarly, you may need to decode ISDN signals before and after an NT-2 interface (within a private branch exchange [PBX]), the wiring to the NT-1 interface, and the NT-1 interface itself.

You will want to explore the proper operation of handshaking, port speed, the PC's Interrupt Request Line (IRQ) status, and actual throughput. I made the assertion that many client PCs cannot sustain even a single 56-kbps serial connection, let alone a bonded, dual B channel 128-kbps service through a serial port control under most versions of Microsoft Windows. Although many software tools exist (and some are built into the terminal adapter software) to show connection speeds and actual throughputs, your applications also need to be able to sustain those loads. You will need to test true throughput with a protocol analyzer with supplemental RS-232 decodes. For example, on-board COM1 and COM3 ports typically share the same IRQ, while COM2 and COM4 ports share a different one. If you set the mouse as COM1 and the ISDN terminal adapter as COM3, you are likely to experience a clash of IRQ requests that will not cause overt system errors but will create many busy signals on the PC bus. A typical symptom is a disabled mouse.

Typically, once a line is provisioned, look at the ready lights on the NT-1, the terminal adapter, or the ISDN PRI hub. Usually you look at the lights and everything is okay (that is, the idiot lights are green) and the line works. However, when the line doesn't work and you have tested the patience of both the ISDN service provider and the equipment vendor, you need to solve the problem independently. The ISDN connection process for voice, data streams, and packet data is outlined as follows:

- Identify itself (the connecting hardware)
- Request SPID
- Exchange Terminal ID
- Get a dial tone (if needed)
- Establish a LAPD link (polling) for a receiver ready signal
- Send a setup message for call characteristics
    Point-to-point
    Multidrop
- Check for all mandatory call setup elements
- Receive a line "OK" or "Failed" response
- Validate message (setting which B channel and the encoding standard)
- Create message for control method on the D channel

Most of the higher-level protocols are actually implemented in the ISDN terminal adapter, as device drivers, or in system software. If you discover that the ISDN connection works most of the time and you can narrow the connection problems to a particular service or ISDN number, most likely a configuration at this level is preventing the terminal adapters from talking to each other. In other words, you have established the ISDN call, but communication is not occurring. Likely causes include line speed differences, line speed adaptation differences by the terminal adapters, protocol communication differences, authentication problems, or failure for line bonding. If the connection problems are not repeatable, just sporadic, review the line provision and terminal adapter configuration. This screen shot (Figure 5.3) with information from the US Robotics Courier I modem shows that SPIDs do not match the information provided to me for my BellSouth home office circuit (shown in Chapter 4), and assuming I took the time to review my documentation and enter the correct SPIDs, the intermittent connectivity problems would disappear.

If you are not proficient in sorting through the complexities of the hardware and software from other applications, you can always use the configuration utilities bundled with the hardware which is usually easier to use and may even be convenient as well. Here is the same information displayed with the US Robotics software configuration tool in Figure 5.4.

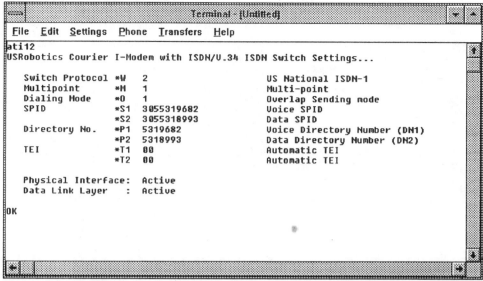

Figure 5.3 US Robotics command line attention (AT) functions show configuration information using the standard Hayes-compatible command set.

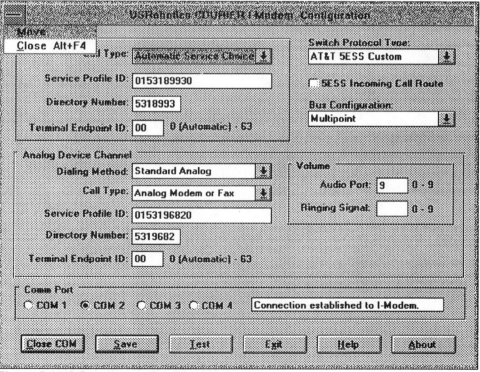

Figure 5.4 US Robotics configuration information.

Protocol Analysis

Although ISDN is described as a clear and unencumbered channel within various carrier media, it still is a protocol on top of other ones and requires yet other protocols to establish a connection and transport data. Few tools provide much insight into malfunctions on ISDN circuits or provide performance measurement. Many network failures can be traced to software that transcends the physical layers of ISDN, bridged or routed segments, circuit provisioning, or network loading characteristics. Because of this, comprehensive testing requires *protocol analysis*, which is the process of capturing packets from the ISDN channels and decoding them so that a person can read and make sense of them. Protocol analysis also includes calculating traffic load and counting packets to generate performance statistics.

Protocol analysis is based on the collection of relevant network operating statistics like throughput rates, collision rates, source and destination addresses, and overhead. Protocol analysis is a process where you view the transmission protocols, as described in Chapter 3. However, in multiprotocol environments supporting multiple NOSs, you need to decode protocols such as IPX/SPX, RIP, SAP, Ethernet, frame relay, or Token Ring or a mix of protocols tunneled inside ISDN data streams or X.25 and frame relay packets. You will want to decode packets and frames based on addresses to track traffic loading and latency patterns. Figure 5.5 illustrates the layers of protocols necessary for most ISDN transmissions.

User data
SLIP, PPP. or MLP PP (MP or MP+)
LAPB
ISDN
SS7
T-1

Figure 5.5 ISDN protocols sandwich.

Although you might think that the *protocol analyzer* is the only tool for protocol analysis, there are a number of options now available. There are built-in partial solutions to Unix, such as the system commands, *perfmon, netstat,* and *trace.* More functional software solutions include a number of software-only protocol analysis tools, such as LAN Pharaoh and LANalyzer for Windows which now include support for ISDN Basic Rate Interface (BRI) or ISDN PRI. Other protocol analysis tools consist of network communication protocol-based tools such as Simple Network Management Protocol (SNMP) and Remote Monitoring (RMON). Simple Network Management Protocol version 2 (SNMPv2) has extensions which are useful for extended and complex networks and for capturing switch information. Although most Fast Ethernet implementations are not yet compatible with this standard, some are, and you should invest in this technology. These services are typically built into the NOS software or hub-based hardware. Additionally, network management stations (NMSs), such as NetView, OpenView, Hermes (Microsoft SMS), or Spectrum, provide an integrated Unix or NT platform for protocol analysis and network management control. For the most part, however, ISDN is a telecommunication apart from data communication technology tools. RAD, Network General, and Network Communications Corporation are adding ISDN functionality to preexisting analysis products. Figure 5.6 shows a typical protocol analyzer with an ISDN capture module.

Figure 5.6 ISDN line protocol analyzer (courtesy of NCC).

Protocol analysis for PRI is more complex than BRI because there are from 24 to 30 channels instead of only three. Most of the BRI decoding skills are relevant to PRI lines, but there is considerable more analysis required for PRI. At the top level, PRI runs on the T-1 or E-1 circuit and requires line analysis testing, bit error rate testing (BERT), wide band jitter, and spectrum analysis. Jitter is a fluctuation in the bit rate or transmission frequency. The spectrum defines the wavelength and frequency of the signal used for transmitting the digital stream. Ideally, the protocol analyzer will capture the signal from the local loop (and repeat it) and contrast the demultiplexed signals with the CPE equipment. These multiple connection points are illustrated in Figure 5.7.

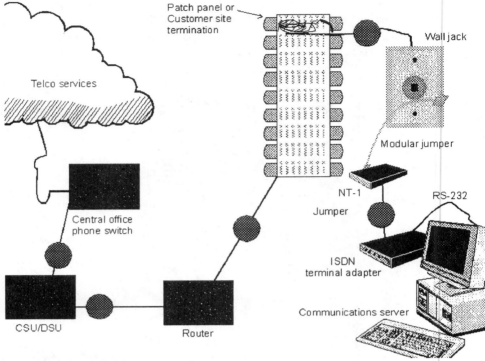

Figure 5.7 ISDN PRI protocols analysis points.

Ideally, you want to be able to capture and decode the primary line for pulse code modulated signals, signals at the digital signal cross-connect points (CSX-1), and view all the channels as described in Chapter 3. Common errors include those listed below:

- Out-of-frame failure
- Superframe remote error
- Unframed ones signal (also known as the carrier "keep alive")
- Signal frequency deviation problems when the signal frequency varies by more than 1 percent
- Excessive zeros (bipolar coding errors)
- No frame synchronization
- No signal
- Severe error second count (high data to error ratio)
- Frame slips (buffer under- or overflow)
- Framing errors
- Bipolar violations (consecutive bit errors)
- Bit received variance

Common tests include BERT and loopback tests. The primary purpose of these hardware tests is to show what is working correctly and narrow possible errors to specific components or loops. End-to-end testing requires multiple analyzers and typically two trained people at the remote sites to validate operation through the telecom service clouds. More specific line testing includes:

- Continuity testing
- Gain testing
- Gain slope testing
- Crosstalk testing
- In-service testing
- Wink measurements
- Jitter testing
- Path delays
- Spectrum analysis

ISDN BRI differs in that the primary interface is not T-1, but rather U. As such, the corresponding protocol analysis points are illustrated in Figure 5.8.

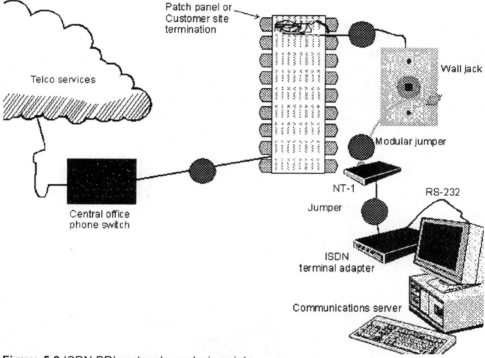

Figure 5.8 ISDN BRI protocols analysis points.

ISDN PRI testing services that differ from BRI include:

- Spectrum analysis
- NT-1 emulation
- Temporary replacement of the NT-1/TA
- Full drop and insert
- End-of-command (EOC) and M-bit command support
  Cyclic redundancy check (CRC)
  Far end block error
  Power status
  Cold start/reset
- Analysis of the NT-1 or TA devices

- Full Q and S maintenance channel support at the S/T interface
- Line activation and synchronization
- Voice communication (drop and insert)
- Tone over B1 and B2
- Basic Rate Interface Terminal Extender (BRITE) BERT
- Switch emulation
- X.25 decode
- Frame relay decode

Frame relay and X.25 decodes are important when the BRI line is used for D channel packet-data transmission. In general, most of the traffic is LAPB, and protocol analysis is based on counting frames (which are the transmission data packaged inside a defined communication format). Channel utilization is also a useful measurement. This includes good frames, bad frames, total frames received or sent, Backward Explicit Congestion Notification (BECN) frames, Forward Explicit Congestion Notification (FECN) frames, DE frames, Frame Check Sequence (FCS) errors, and abort frames. As with ISDN data streams, you can capture the data and view it directly. This becomes useful when there are many ISO layer 3 and layer 4 protocols encapsulated with the ISDN streams, the X.25 and frame relay frames, and multiple network management protocols. However, since most ISDN communication is based on SLIP, PPP, or MLPPP between a small office/home office (SOHO) user and an Internet Service Provider (ISP) or between routers at remote corporate sites, you will want to decode those protocols. The NCC protocol analyzer can decode PPP transmissions which are common with Internet connections for both single dial-up users and low-activity communication servers. This decode is shown in Figure 5.9.

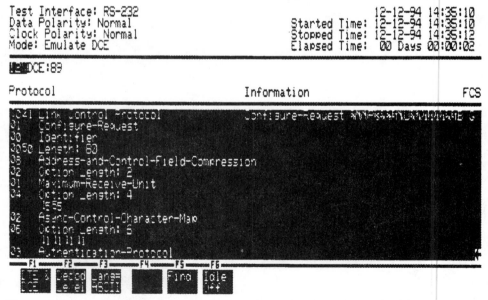

Figure 5.9 Protocol analysis of the point-to-point protocol (courtesy of NCC).

The PPP protocol consists of an encapsulation method for datagrams, the link control protocol (LCP), and the network control protocol (NCP). These are really frames of data within the ISDN clear-channel stream. You will want to validate the ISDN connection process with the protocol analyzer by decoding the connection sequencing or emulating it directly. These upper-layer processes were described in Chapter 3 but are expanded to include:

- Link dead phase (absence of protocol)
- Link establishment phase (LCP configure request packets)
- Authentication phase (peer link establishment)
- Network layer protocol phase (set appropriate NCP such as IP or IPX)
- Data transfer phase (data is sent in NCP-specified encapsulation)
- Link termination phase (call is terminated to link dead phase)

You can use the protocol analyzer for performance monitoring and analysis by tracking bit rates, transmission speeds, compression efficiencies, latencies, frame utilization, quality reports, and overhead for various protocols. In addition, you can also track D channel setup information that is typically invisible to the user or office router.

If you are using ISDN for connecting remote sites, let me give you a word of advice. Most LAN protocols are very chatty and can keep an ISDN connection active 24 hours a day, 7 days a week whether you need it or not. Since most ISDN have a per-minute pricing component, you want to check that packet filtering and spoofing actually work. Look for broadcast IP and IPX packets, AppleTalk keep alives, and even Token Ring watchdogs. If you can identify particular types of packets by purpose, you can configure the router to filter them. If traffic levels drop, do nothing; the ISDN connection can be "torn down" and reestablished only when there is real traffic to transport over the WAN.

SNMP and RMON Protocols

If you are applying ISDN BRI or PRI as remote connectivity technology, you will have bridges, routers, edge routers, or gateways providing the connecting points for the ISDN lines. These devices are best managed with a uniform and global protocol such as SNMP or RMON. If you are linking Ethernets or Fiber Distributed Data Interchange (FDDI) LANs together, ISDN is just one more part of the network and really just one more protocol in the network soup. A centralized network management station is the master repository for all managed agents and the management information (data) base (MIB). It controls network agents via commands issued and transported at the datagram level. No logical connections are maintained between the NMS and any agents; this lowers network overhead. SNMP can trigger alarms based on captured agent information in the SNMP MIB and unsolicited alert information from a network agent. However, the implementation of SNMP does not detract from the more rigorous goal of a comprehensive network management tool to provide management information on complex networks. Ordinarily, SNMP does not provide information across intermediate nodes. To meet this limitation, remote monitor-based tools collect and transport network performance data over various communications channels to bypass bridge or gateway limitations. The table in Figure 5.10 shows the functional support and overlap in the different SNMP and the RMON MIBs. The growth in internetworks and more complex (that is, "switched") Ethernet or Token Ring LANs typically requires protocol analysis across the intermediate nodes.

Service	RMON	MIB II	Hub	Bridge	Host
Interface statistics		•			
IP, TCP, UDP statistics		•			
SNMP statistics		•			
Host job counts					•
Host file system information					•
Link testing			•	•	
Network traffic statistics	•		•	•	
Host table of all addresses	•		•		
Host statistics	•		•		
Historical statistics	•				
Spanning tree performance				•	
Wide area link performance				•	
Thresholds for any variable	•				
Configurable statistics	•				
Traffic matrix for all nodes	•				
Host Top N studies	•				
Packet/protocol analysis	•				
Distributed logging	•				

Figure 5.10 The focus of various common MIBs.

The remote (network) monitoring management information (data) base (RMON MIB) defines the next stage for network monitoring with more internetwork fault diagnosis, planning, and performance tuning features than any current monitoring solution. RMON uses SNMP and its standard MIB design to provide multivendor interoperability between monitoring products and management stations, allowing users to mix and match network monitors and management stations from different vendors. The RMON MIB enhances current remote monitoring agents with:

- Additional packet error counters
- Historical trend graphing and statistical analysis
- Traffic matrixes
- Thresholds and alarms
- Filters to capture and analyze individual packets

RMON MIB agents work on a variety of intermediate devices, such as bridges, routers, switches, or hubs, and dedicated or nondedicated hosts. It is available on platforms specifically designed as intermediate network management stations. An organization may employ many devices with RMON MIB agents (one or more per network segment or wide area link) to manage its enterprise network. Each solution may offer different portions of the RMON MIB. Figure 5.11 shows a stand-alone RMON hardware solution.

Many devices, especially older or non-IP products, do not include network management agents and cannot communicate with SNMP consoles except through echo packets. RMON MIB agents monitor every network device and are often the only way to extend network management to such otherwise unmanageable devices. Although many SNMP management stations periodically send echo packets to check the status of each device, this large number of echo packets does increase base-level traffic and potentially could adversely affect performance of the network and the SNMP console. The proxy management capability of an RMON MIB remote monitor is useful when the remote network is connected over a wide area link where periodic polls of devices would consume an unacceptably large percentage of bandwidth.

Figure 5.11 A typical RMON hardware agent (courtesy of TEC).

The Internet Engineering Task Force (IETF) has planned for future growth in the RMON MIBs by designing its format for easy extendibility to other types of local area networks (such as Token Ring and FDDI) and wide area networks. Because an organization may have several management stations and there can be multiple people managing parts of a large network, the remote monitoring agent must work with several management stations concurrently. Using the RMON MIB, each management station identifies the resources it is using in the agent so that multiple tasks are completed in a timely manner.

The MIBs are not simple databases easily presented in a book. The specifications for MIBs typically require 40 or more pages of pseudo-code in an IETF request for proposal (RFC). Furthermore, a user of a network management station cannot get to a particular value in a MIB unless it has been coded into the network management software (NMS). In fact, MIBs are complex trees of objects where each single object may be represented by multiple tables. However, each row in a table of MIB values must include an index or a value that uniquely identifies the row so that it is addressable. The RMON MIB specifics that row indexes in many tables start with *1* and increase sequentially. Because the management station can usually predict the range of desired indexes, a single SNMP command can fetch multiple data requests for better bandwidth efficiency.

The RMON MIB was developed by a working group of the IETF. The working group included Carnegie-Mellon University staff and people from vendors of network management software and hardware. The RMON MIB is organized into nine optional groups. Compliance with the RMON MIB standard requires only support for every object within a selected group (RFC 1213). Because each group is optional, a number of RMON MIB agents may can be used for different purposes; one can be used to trace the network structure, another for tracking alerts, and one for maintaining a complete host table. Other uses are possible. Figure 5.12 shows the tree structure of the RMON MIB and the nine leaves that refer to the nine RMON groups.

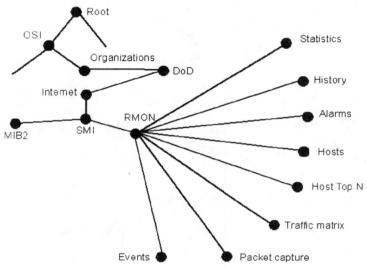

Figure 5.12 The tree structure of the RMON MIB.

The statistics group provides segment-level network statistics. These statistics show packets, bytes, broadcasts, multicasts, and Ethernet collisions or FDDI and Token Ring alerts on the local segment, as well as the number of occurrences of dropped packets. Each statistic is maintained in a 32-bit cumulative counter. The number of Ethernet collisions that are detected by the agent depends on whether the multistation access unit (MAU) is receiver-based, as it should be for accurate data gathering. Although this book is about ISDN, when you are using ISDN to interconnect remote LANs based on Ethernet, Fast Ethernet, Token Ring, or FDDI, problems with these segments tend to obscure the functionality or problems with the ISDN linkages. You do not want to be the doctor that just treats a cold-sore on the lip and misses the fact that the patient has pneumonia, too. Similarly, ISDN problems may feed into more serious LAN performance problems, or LAN problems may be causing the recognition of the ISDN flaws.

This indirection becomes even more acute when you mix frame relay, X.25, Asynchronous Transfer Mode (ATM), and LAN traffic into the same internetwork. Like a general practitioner, you have to look holistically at the patient, which is the entire internetwork and not just the ISDN problems. Problems with one component can feed back into other more obvious symptoms that are not the prime causes of ISDN performance or provisioning problems. For this reason, the next few paragraphs explore the RMON functions which are frequently important when debugging "ISDN problems" between remote LANs.

The RMON MIB has error counters for five different types of packets: undersized packets, fragments, CRC/alignment errors, jabbers, and oversized packets. These counters provide useful network management information not provided by typical interface cards. For example, industry-standard carts will provide only two separate counts of CRC and alignment and will not count well-formed packets that are either too small or too large. These under- and oversized packets are counted by the RMON MIB agent because they usually indicate configuration problems in the transmitting station. Such packets will usually not be passed from the receiving NIC software driver; this results in failed transmissions that are otherwise not indicated.

The history group provides views of the statistics maintained in the statistics group with the exception of packet size distribution, which is provided only on a real-time basis. The history group features user-defined sample intervals and bucket counters for customization of trend analysis. The RMON MIB specifications recommend two trend analyses. The first recommendation is for 50 buckets (or samples) of 30-s sample intervals, for a total time interval of 25 minutes. The second recommendation is for 50 buckets of 30-minute intervals, for a total time interval of 25 hours. Users can modify either of these or add additional historical time interval studies; the total number of such statistics is limited by the resources available to a specific agent. The sample interval can range from 1 s to 1 hour, creating the opportunity for long historical studies.

A host table is a standard feature of most monitoring devices. The RMON MIB specifies a host table that includes node traffic statistics: packets sent and received, bytes sent and received, as well as broadcasts, multicasts, and error packets sent. In the host table, the classification "errors sent" is the combination of under- or oversized packets, fragments, CRC/alignment errors, and jabbers by each node. In addition, the RMON MIB includes a host timetable that shows the relative order in which each host was discovered by the agent. This index improves performance and reduces network traffic.

The Top N group extents the host table by providing sorted host statistics, such as the top 20 nodes sending packets or an ordered list of all nodes according to the errors they sent over the last 24 hours. This extensive processing is performed remotely at the agent based on user-defined parameters. This minimizes traffic load on the network.

The RMON MIB includes a traffic matrix at the 802.2 media access control (MAC) layer (common with Token Ring, ARCNET, FDDI, and Ethernet even if you are using ISDN as the WAN connect). A traffic matrix shows the amount of traffic and number of errors between pairs of nodes, with one source and one destination address per pair. For each pair, the RMON MIB maintains counters for the number of packets, bytes, and error packets between the nodes. Users can sort data by source or destination address.

The alarms group provides a mechanism for setting thresholds and sampling intervals to generate events on any counter or integer maintained by the agent, such as segment statistics, node traffic statistics defined in the host table, or any user-defined packet match counter defined in the filters group. Both rising and falling thresholds may be set, and in fact both can indicate network faults. For example, crossing a high threshold can indicate network performance problems; crossing a low threshold may point out the failure of a network backup scheduled to occur at midnight.

Thresholds can be established on both the absolute value of a statistic or a "delta" value so that the network team is notified of rapid spikes or drops in a monitored value. This is true for SNMP as well as for RMON. The filters group features a generic filter engine that activates all packet capture functions and events. All the other groups depend on the filters group, and it is critical to many of the advanced functions of an

RMON MIB agent. The filter engine fills the packet capture buffer with packets that match filters activated by the user. Any individual packet match filter can activate a start or stop trace trigger. The contents of the trace buffer are controlled by any combination of user-selected filters. Sophisticated filtering provides distributed protocol analysis to supplement the use of protocol analyzers. The monitor also maintains counters for each packet match for statistical analysis. Alarms can trigger an event to the log using an SNMP trap. Although these counters are not available to the history group for trend analysis, a station might identify trends from these counters through regular polling.

The packet capture group depends on the filter group. The packet capture group allows users to create a number of capture buffers and to control whether the trace buffers will wrap when full or will stop capturing. Buffers can be allocated in various ways by user parameters. The RMON MIB includes configurable capture slice sizes to store either the first few bytes of a packet where the various protocol headers are located or, at the limit, to store the entire packet. The default slice setting is the first 100 bytes.

The events group provides users with the ability to create entries in the monitor log and/or SNMP traps from the agent on any event of the user's choice. Events can originate from a crossed threshold on any integer or counter or from any packet match count. Vendors often add features through the network management station software or through proprietary extensions to the MIBs. The log includes the time of day for each event and a description of the event. The log wraps when memory is full, so events may be lost if not uploaded to the management station periodically. The rate at which the log fills depends on the resources the monitor dedicates to the log and the number of notifications the user has sent to the log. It is also a function of how much memory is allocated to network management overhead and the number of user-configured thresholds, traps, and alarms.

Traps can be delivered by the agent to multiple management stations that each match the single community name destination specified for the trap. An RMON MIB agent will support each of the five traps required by SNMP (RFC 1157): link up, link down, warm start, cold start, and authentication failure. Three additional traps are specified in the RMON MIB: rising threshold, falling threshold, and packet match. By the way, it is a waste of resources to set thresholds, traps, and alarms already hard-coded into the basic management features. Make sure you know what features are already coded so that you do not create double (or more) work for not only the remote agents but also the network management station.

However, most LAN monitors do not have the capability to provide WAN cell and stream decodes, although some vendors sell multiprotocol portable decodes for LAN and WAN traffic. SNMP is useful because it supports RMON, alerts, alarms, and thresholds. The inherent limitation of any in-band management is that bottlenecks can prevent timely delivery of information. RMON is useful because it uses remote probes beyond the view of most protocol analyzers. The agents often retain traffic and performance data on the device until specifically polled, rather than broadcasting across the network for performance reasons. RMON devices can be polled to extract historical information. There are nine groups of MIB objects:

- Statistics—measures probe-collected statistics such as the number and sizes of packets, broadcasts, and collisions.
- History—records periodic statistical samples over time that can be used for trend analysis.

- Alarms—compare statistical samples with preset thresholds, generating alarms when a particular threshold is crossed.
- Host—maintains statistics of the hosts on the network, including the media access control addresses of the active hosts.
- Host Top N—provides reports that are sorted by host table statistics, indicating which hosts are at the top of the list for a particular statistic. Note that host does not mean mainframe but any RMON server.
- Matrix—stores statistics in a matrix showing conversations between host pairs.
- Filter—allows packets to be matched according to a filter equation.
- Packet capture—allows packets to be captured when they match a particular filter or threshold value.
- Event—controls the generation and notification of events, which may also include the use of SNMP trap messages.

RMON comes with a hefty price in terms of CPU requirements, RAM, disk space for storing statistics, and substantial buffer space to actually capture packets and packet information in real time. Note that RMON (as with SNMP) does skew the performance of what you are trying to measure because the measurement process requires substantial resources. This means that if you are trying to improve the performance and efficiency of an ISDN BRI or PRI link, the process of tracking performance adds to the overall traffic load over the remote links. In fact, network management could add enough traffic to require another B channel on a temporary or permanent basis.

Physical Failure

Plain old telephone service (POTS), teletype, T-1, E-1, and ISDN lines basically support the same current parameters. It is hard to distinguish between circuits and channels. In general, all circuits from a service provider should conform to the AT&T demarcation standard of 60 milliamps or less. The first message is that you will want to label your circuits very carefully. You cannot attach a handset, call the operator, and ask for a number so that you can match up the circuit to its type. You will need an NT-1 interface to do that with ISDN circuits. In addition, the complexities double once you route the telecom circuits through a PBX. Lines from a PBX, and this includes ISDN circuits, provide from 20 to 140 milliamps. Anything above 90 milliamps is enough to damage most analog modems, and teletypes expect 30 milliamps or less. For example, you clip your leads to the phone wires to test for the line parameters as shown in Figure 5.13.

There are a number of tools, multimeters included, that will help you decipher line types. If you have a lot of lines and manage services for analog modems, analog handsets (with analog telephone interface cards in the PBX), PBX handsets, teletype services, and ISDN, you should have an accurate multimeter. It costs about $20 to $60 for this tool. Telecom lines can also have three functional conditions which you can change by mistakes in the inside wiring. Although a parity condition is normal, you could also have reversed polarity and overcurrent conditions. Radio Shack is a convenient location to find tools, parts, and wire, although most other telecommunication product catalogs include special test devices for line polarity. Inexpensive tools cost $6; the most expensive I have seen runs about $27 (which is the IBM Modem Saver). Pair scanners, wire scanners, and other more sophisticated tools provide more detailed views of the wiring plant. They tell you if the lines are reversed, the impedance and capacitance of the wires, and whether there are crosstalk and signaling problems. Such a tool is not strictly necessary for small sites or telecom-only sites.

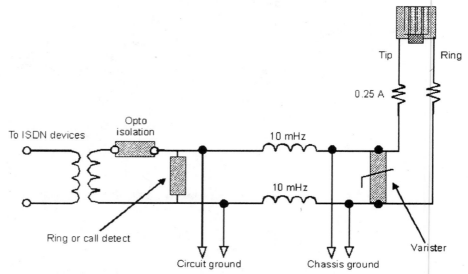

Figure 5.13 Testing ISDN lines with a multimeter.

However, if you have Ethernet, Token Ring, serial links, Fast Ethernet, and phone lines jumbled together in unified wiring infrastructure, the tool is very handy—at the price of $1500 to $8000 per unit. One less costly tool is shown in Figure 5.14.

Recall that the ISDN U and S/T interfaces are different, and problems can occur between the transition. The composite ISDN signal is split into 2 B channels and the D channel at the reference point. The ISDN PRI signal, which is the superframe, is split into frames and then into separate channels at the reference point. It is usually more interesting to perform protocol analysis at the S/T reference point because you actually have different recognizable data streams (and thus data that can be decoded and is meaningful). However, when you monitor at the U reference point, realize that the process is *intrusive* and you alter the character of the signal. Analysis of an ISDN BRI one pair local loop is not like attaching alligator clips to a line and listening in. The protocol analyzer may not have the ability to repeat the incoming local loop signal as the NT-1 receives it. This is not a negative; there are many good ISDN protocol analyzers limited in this way. You just need to understand the limitations of each tool and not get sidetracked because the ISDN connection is inactive while the analyzer is working.

You get the entire signal with the U channel and if you need to, you can watch the transmission process (and the *real* connection speed) through other equipment, such as a Motorola BitSurfr terminal adapter. When you need a protocol analyzer that can regenerate an identical signal, look to units such as the RADCom RC-1000.

A more sophisticated tool is available from Trend Communications: a handheld talk set and ISDN tester that is available for both ISDN BRI and PRI. In fact, the PRI unit can select from any one of the 24 channels and talk over that channel to someone on an ISDN phone or analog handset. This can be useful when debugging more complex channel problems. Basically, these tools can test physical line functionality, the switch setup and SPIDs, and terminal identification information and can simulate various ISDN features (both voice, data, packet, and EKTS) on a per-channel basis. The PRI tester is shown in Figure 5.15.

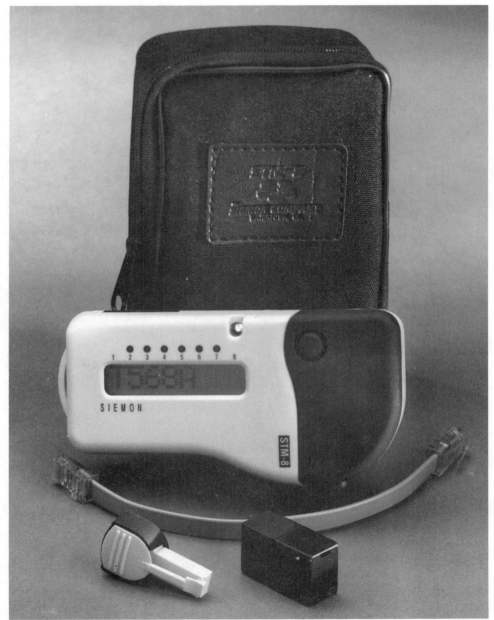

Figure 5.14 Telephone and wireline tester (courtesy of Siemon Cabling System).

ISDN supports both asynchronous and synchronous PPP. The speed of the connection (56, 57.6, or 64 kbps) is a function of the terminal adapters adjusting to local loops and any long-distance carrier's ability to transport clear channel. Automatic rate adaptation is defined by the International Telecommunications Union-Telecommunications Standards Sector (ITU/TSS) specification V.120; it is typically implemented with asynchronous PPP over ISDN.

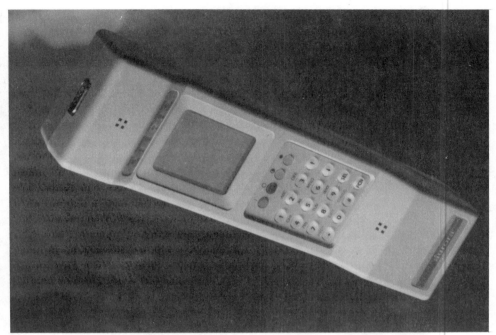

Figure 5.15 Aurora ISDN PRI line tester (courtesy of Trend Communications).

Some protocol analyzers will allow you to capture and play back streams to simulate loads, streams, or past events. Some ISDN tools also allow you to simulate certain types of equipment, provisioning expectations, and setup configurations to play back against the central office switch. Until the next ISDN national standard is in place, allowing users to query the switch for information, you will have to get the provisioning information from the service provider and play a script (from a protocol analyzer) back to the switch to verify proper operation.

There are two ways to "simulate" ISDN. The previously mentioned simulation of central office switch settings is only one method, mainly geared to the process of testing provisioning, configuration, and hunting for causes of problems. The second type of simulation is to actually simulate the performance of an ISDN linkage on the desktop. This is useful for determining the functional performance of client/server applications over ISDN links or remote LAN linkages. The box shown in Figure 5.16 simulates the performance of an ISDN BRI channel with various switch settings.

Simulation is useful during the design process of remote and mobile applications and for confirming that particular types of hardware configurations can provide the type of performance desired. However, more in line with the topic of this chapter, it is very useful for simulating failed or failing ISDN lines or when replacing them to determine true causes of ISDN connection problems. For example, when a connection will not work through a local carrier, you can repeat the event while bypassing the carrier. This way, you can narrow the problem to a carrier or adapter hardware and configuration.

A line simulator provides end-to-end real-time ISDN facilities with complete call-handling procedures without a switch. It bypasses variables and problems inherent in testing with a live switch. It can demonstrate ISDN services at trade shows and sites without working ISDN services.

Figure 5.16 Simulates ISDN line performance (courtesy of TelTone Industries).

ISDN can monitor the physical layer status so you might test go/no-go indications of physical layer problems, such as synchronization flaws, loop-back failures, and power fluctuations. A line simulator is a useful tool in a client/server or application development environment where ISDN is a primary component in the network.

Wire Scanning

Any pair scanner, telephone line tester, or similar equipment will test conformance of internal site wiring. Realize that ISDN BRI and PRI do not require the frequency range of Fast Ethernet, Token Ring, or Ethernet. You just want to test connectivity, continuity, and basic link-level support. ISDN does not need to conform even to EIA/TIA Category 3 wiring standards since it can run perfectly well on voice-grade wire conforming to Category 1. However, if you have a time domain reflectometer or wire scanner, you will want to measure the length of the internal wiring from the site termination. The internal wiring must be less than 70 m (250 ft). Wiring that breaks outs from an NT-1 or similar hub device must be less than 30 m (100 ft).

Loopback

The local loop is the responsibility of the ISDN service provider. They are responsible for maintaining the service up to the demarcation point. The ISDN switching hardware at the central office is usually robust enough to monitor the status of ISDN service and report internal errors to the telco support people. However, much like any other phone line, things go wrong. The best method to test ISDN operability is to check the status lights of the ISDN terminal adapter. You should have lights indicating service and the usage for the B channels and the D channel. Internal devices may provide external LEDs on the metal bracket that sticks through the rear of the PC. The new, inexpensive *slave* ISDN adapters (which also steal the primary PC CPU for processing) do not provide diagnostic lights. However, the internal units may include software diagnostics, or you might try to verify performance with a tool that shows functional status lights on the screen of a workstation or server. Figure 5.17 tracks the internal serial transmission lines whether they are routed through a physical serial port or a supplemental serial port on an adapter card.

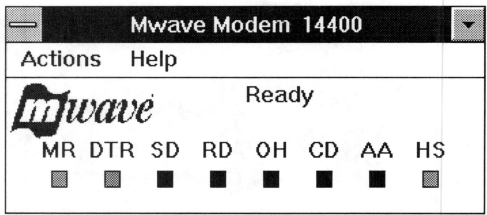

Figure 5.17 Serial line activity tracking.

In an organization committed to ISDN migration, the ISDN protocol analyzer is the more sophisticated tool of choice for verifying loopback performance. It can actually decode signals, send call setup packets, and compare switch operating with prior experience. In general, for all practical purposes, once it is clear that ISDN has been correctly provisioned and proven to be functional with the local hardware, the status lights and simple dial-up test are proof enough of local loop operation for the small office, home office, or casual user.

Logical Failure

Service Profile Identifiers (SPIDs) belong to the system configuration file and hardware initialization. They may be part of any configuration parameters loaded into nonvolatile RAM in the terminal adapter or combined ISDN adapter/NT-1. Some PBXs with NT-2 equipment store SPID information, exchange information, and preferences for call forwarding, hunt groups, and switching by call type (that is, voice, data, fax, or ISDN). The SPIDs uniquely identify the system and setup to the local switch at the local service provider. When the SPIDs are entered incorrectly, the switch will not recognize the end-user equipment, and an error message will show in the application either when the software is loaded or when a call is attempted.

What is worse, as I found out, is that you can establish a mostly correct configuration, like I did for my own home office site. The real obstacle was that the circuit worked some of the time but not all of time. It was unclear what changed between the time when it worked and when it didn't. This intermittent operation baffled technical support providers at various equipment vendors (as I tried different CPE hardware) and BellSouth as the ISDN service provider. You need to find the right technical support person who is knowledgeable about ISDN setup and provisioning and not just someone who is reading from a sequential list of likely debugging techniques. This book includes that sequential information; if this information cannot help you, then you need someone intimately familiar with the terminal adapter or the local loop.

When something works some of the time but not all of the time, the presumption is that the hardware is defective or that the ISDN provisioning is incorrect, neither of which is always true. These problems can be solved, it just takes a couple of hours to sort through the worst of them.

In my case, I had selected National ISDN-1 and the SPIDs formatted as 305999999901 where the 305 is the local telephone area code. In fact BellSouth was providing AT&T 5ESS Custom multipoint and the BellSouth SPID formats are 0199999990 without an area code. Take this error on my part as a good lesson from someone who should know better, and who has provisional experience with ISDN, and realize that simple configuration errors can create subtle performance problems. These SPID problems are supposed to disappear since the National ISDN Council, a group of regional telephone operating companies in the United States, has agreed to a uniform 14-digit format and will automate the SPID capabilities in switches by 1998 with National ISDN 3 (NI-3).

Find out if there is any documentation from the service provider and check that for SPID format information. See later in this chapter (starting on page 151) for the tedious and copious lists of SPID formats. Although many providers, vendors, and ISDN groups (under the governance of the National ISDN Council) are trying to define and implement a uniform SPID format, reality is that it is not available today and may not be implemented for years. These lists may look confusing, but the SPID formats are nothing more than lists organized by provider and switch type. You should be able to get the correct SPID format at the first or second try.

When there is limited information on the correct SPID format for the installation, check the list of known formats to find other possible SPID formats. The SPID format for a specific protocol tends to be similar for most providers, so sometimes trial testing known formats is useful. When you get ISDN service and supplemental features working and a dial tone for analog devices, note, as with my experience, that success is no guarantee of proper and finalized configuration. Follow up later and verify that the configuration is still valid and functioning. This is particularly important with remote sites and when you implement more complex protocols on top of ISDN such as the BAPC or MultiLink Point-to-Point Protocol (MP). Realize, too, that templates from one working test site are not necessarily portable to another site.

There are also some cases where the local provider hasn't given the complete format. For those cases, adding 0, 1, 2, 00, 01, or 02 to the end of the partial SPID(s) or directory numbers as provided can often solve the problem. If none of these solutions eliminate the error message, you need to call your local provider to get the correct SPID associated with the circuit. Effect changes through an ISDN terminal adapter configuration utility, or a control panel applet or run the terminal applet and set the configurations directly through the port with the Hayes or ATI command set typical with most PC-based ISDN hardware. The utilities are often installed as MS Windows applications or Control Panel applets. In Windows NT or Windows 95, the utilities are added to the system registration database and the task menu bar. If not, search the likely folders for an executable file with an expected filename.

The switch type and protocol are required for product configuration. An incorrect entry can cause a number of obscure error messages and result in failure to complete connections. If the local provider has not sent information about the SPID format, it can be helpful to check the list of known SPID formats later in this chapter to see what type of switch and what protocol are typically associated with that format. Sometimes, it is worthwhile to simply try selecting the other options to see if one of the combinations will work; you can do this while on hold waiting for technical support. If this information was never sent and other options don't solve the problem, contact the local provider to get the switch type and protocol information associated with the ISDN line. By the way, if you have installed at least one working site, do not blindly use that site as a template for additional sites. Providers, switch types, line provisioning, and hardware may differ.

ISDN lines can take on many forms, since there are approximately 400 different specific configurations involved in the way that a line is provisioned. Most CPE products have some pretty specific needs with regard to how the circuit should be provisioned. Check to find out if the ISDN circuit was ordered for the product that is being installed or if it was a previously existing line. If it is an old ISDN line, it is very likely configured inappropriately for the new usage. All it takes is 1 of 400 provisioning values to be incorrect; hence the probability of the random circuit working correctly is less than 0.000056 percent. It is important to determine if the system is able to complete a connection locally versus long-distance. If a local connection can be made, but a long-distance one cannot, you may have problems with the other "local" service or the connected long-distance provider. Likely flaws include improper end-point provisioning, inability to bond channels, lack of clear channel, different digital speeds, or synchronization errors. ISDN is more common than it once was, and you can usually resolve these problems quickly with competent technical support people—and there are more of them now at both service providers and vendors. With multiple site connections, you have many possible conflicts, as shown in Figure 5.18.

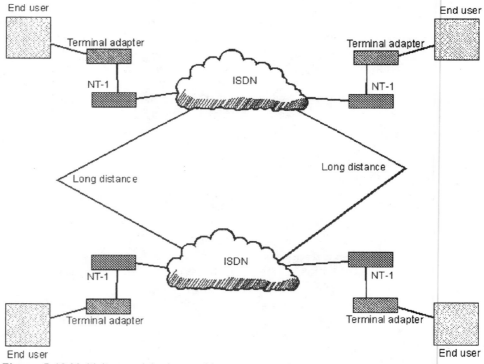

Figure 5.18 Multisite provisioning problems.

Not only can you have problems with the terminal adapter and NT-1 setup, you can also have problems with software configuration, hardware configuration discrepancies, and problems with long-distance carriers. Unfortunately, every site may be different, so what works with your trial site may not provide any insight into what works at other local or remote sites.

You can also experience ISDN speed failures. Specifically, if you expect to connect through a dial-up connection into almost any commercial Bulletin Board System (BBS) or Internet service provider, you may not achieve 128 kbps because the other end may not manage the two B channel connections and provide connection into the very same server. Large ISP providers, such as PSInet or UUnet caution against expecting dedicated service with a dial-up line when each has several hundred or a thousand different servers. The most these services guarantee is 64 kbps, and random phone calls, even placed at the same time with BAPC or MP+ protocol, easily connect into different servers. Unless you pay for and get the provider to commit to a dedicated connection (you effectively pay for the ISDN adapter and two or more lines at the ISP), you are very unlikely to see bonded connections. On the other hand, if you are calling a corporate site and expect this type of service, check the other side for lines which are actually available and comparable for protocol support.

NT-1 Problems

In many places (excluding Europe), direct BRI lines from their local provider require an NT-1. Likewise, if your ISDN BRI or PRI lines are connected through a PBX, the PBX or hub takes care of that function. When an NT-1 is involved, there is the potential for wiring problems. Ensure that all wires are correctly hooked up first, then check to see whether the lights on the NT-1 are functioning normally. A faulty NT-1 will not initialize even if the system is configured correctly, or connections will complete normally but will consistently disconnect for no apparent reason. Find out if the system is attached to a PBX or separate NT-2. If not, then ascertain if there is an NT-1 and if it is working. Many NT-1s perform a power-on self-test (POST) which can help with this process. If a problem persists, then check all wiring. Make sure that there is only one cable from the NT-1 to the ISDN adapter. Two connections from the NT-1 to the same ISDN adapter will cause failures. Lastly, try another NT-1 to eliminate the possibility of faulty equipment.

PBX Configuration Issues

An ISDN circuit attached through a PBX that functions improperly, erratically, or not at all suggests that the cause of the problem may be in the configuration of the PBX. A key symptom is not being able to place and/or receive calls. Contact the PBX administrator. If the problem cannot be located and corrected by the administrator, contact the PBX Customer Support for information on configuring that particular PBX. Some older units may not support any ISDN services or may not be fully compatible or compliant with the terminal adapter or user software. If the operation is erratic, try increasing time-out parameters to account for delays through the PBX. For example, even though ISDN connects to another ISDN line within 1 s or analog line within 3 s (as opposed to 7 to 10 s for analog-to-analog), the dial-up connection process may not be completed until serial line interface (SLIP), PPP, or LAPD is initiated too. This protocol synchronization can require a minute to reliably complete. You may need to configure redial or connection time-out parameters in the minute range.

Recall that you are expected to choose a long-distance service carrier. Problems making long-distance calls can occur even when the lines cannot support digital signals, clear channel, or sequential bonded lines. Typically the problem is either due to a long-distance carrier not having been chosen at the time the line order was placed or to the choice of a long-distance carrier that doesn't handle ISDN data services.

Cable Problems

Wiring issues occur in two different categories: Either there is a problem with the physical cabling in the ISDN network, or there is a wiring issue at the customer premises. Make sure that all of the system wiring is correct. The most common place for error is the wiring that involves the NT-1. A wiring problem will show up in the ISDN cable test. This information helps when you contact the local telco and ask them to test the line to the NT-1.

Water inside the conduit carrying the local loop or a wet cable that carries the ISDN BRI or ISDN PRI signal will cause an impedance balance mismatch between the B channels multiplexed inside the ISDN circuit. This creates crosstalk between the channels. Although it is conceivable that you can diagnose this with either a multimeter or an ISDN protocol analyzer, it really is the responsibility of the carrier if the problem is on the carrier's side of the demarcation. If you can time intermittent problems to the immediate weather or even some days after torrential rains, leakage into the cable could be a likely scenario too.

Network Termination

Install your ISDN terminal adapter or NT-1 permanently. If you unplug the ISDN line (the U interface twisted pair), it shows up as a sign of line trouble in the central office. Some telephone companies respond to this so-called trouble by disabling your ISDN line at the central office and require you to place a service call on your analog telephone to get your ISDN service restored.

Dialing String Problems

This is a broad category that covers a couple of situations. The first concerns a system attached to a Northern Telecom DMS-brand switch. Callers to these systems have to use both numbers associated with this line. This is also the case for a system that is trying to dial someone attached to a DMS switch. When there are two numbers involved in the dialing string, it is important to dial using the correct directory number format. The other possibility is that the system is attached to a PBX or a Centrex and needs an access code prefix to connect to an outside line. Customers often fail to utilize their regular telephone dialing patterns with their new ISDN line, forgetting to use access codes (like 8 or 9) or long-distance dialing prefixes. Remember to use all access code prefixes and 1 when necessary.

If the receiving end is attached to a DMS switch, the originating system needs to dial both of the numbers associated with the receiving end's ISDN line. Make sure that the format of the number is 555-555-4321: 555-555-4321. All special dialing codes must be included in the numbers on both sides of the colon. Find out if there is an access code that must be dialed before the rest of the number on outgoing calls. A preceding 9 is common in the case of a PBX or if it is a Centrex line. In some regions, other codes are necessary to place external calls.

ISDN Phone Number Problems

Periodically, there are problems with the ISDN number associated with a new line. Some problems include the customer who didn't receive the ISDN phone number information associated with the ISDN line during the ordering process or may have mistyped the number(s) during the software installation. Or, the customer entered only one ISDN number when there should be two, such as in the case of a DMS switch installation noted above. Or perhaps the customer entered two numbers when only one should be entered into the ProShare software configuration, such as in the case of an AT&T 5ESS installation with multipoint capability. The primary symptom in these cases is an inability to place or receive calls.

Contact the local provider to get the correct ISDN number information associated with the ISDN line and then correct the software configuration. If you have entered both numbers for an AT&T 5ESS multipoint configuration, simply remove one of the numbers and SPIDs from the configuration.

Routing Issues

Periodically, calls to a specific destination won't complete normally when calls to other places work correctly. Sometimes this is due to a specific area code not being accounted for in the routing tables for that local exchange carrier (LEC) or interexchange carrier (IXC) or being excluded by a PBX system. A PBX can block calls to some exchanges, or some units that predate the North American Numbering Plan (NANP) area code reorganization may not be programmed or capable of handling the new codes. The only real solution involves letting the local provider know that there is a problem. They can typically take care of it relatively quickly if it is a problem in their routing tables. They can also contact the long-distance carrier if that is necessary.

ProShare adds tools to aid in troubleshooting ISDN-related issues into the Windows Control Panel. Control Panel contains a helpful tool that can be used to help troubleshoot ISDN issues. This diagnostic, called ProShare Diags or PS Diags, divides its tests into two parts, the ISDN board tests and the Video Capture board tests. There are three possible result codes for each test:

- Passed
- Time-out
- Failed

It is important that ProShare Video and PSNOTIFY are not running when PS Diags is running. It will skew the results and possibly cause the system to lock up. Typically, consistent time-outs and failures which occur in a specific diagnostic can indicate a hardware conflict or failure for the board that is being tested. There is a thorough section in the ProShare manual which covers the possible results of each test and gives some potential causes for each result. One special feature in ProShare Diags involves the cable test in the ISDN board section. When you run the ISDN diagnostics, you can find out exactly why the cable test failed if you PRINT the results. The printout gives more detail on failed tests. Unfortunately, you can't view the results on the screen.

Voice or Data

Each B channel has to be designated at the switch by the local service provider as one of these choices:

- Circuit-switched voice
- Circuit-switched data
- Circuit-switched voice/data
- Packet-switched data

You can't *generally* make a data call over a voice-only channel or vice versa. If you have your ISDN line configured as voice/data, it can be used for either voice or data if you have the correct terminal adapter with the correct settings.

Videoconferencing and Bonded Services

You can achieve dynamic allocation of B channels in both BRI and PRI. For practical purposes, combining multiple channels in a PRI, for large videoconferences, data transfers, and remote LAN connectivity, is most often programmed into the digital switch serving the location. However, new bandwidth-on-demand controllers have begun to enable a network manager to combine larger bandwidths in real time to meet specific needs. They can also monitor quality and traffic on both corporate leased-line and ISDN networks and perform dynamic allocation of B channels to relieve bottlenecks or backup error-prone or damaged lines. When you run into problems, you need to assess the actual implementation of bonded and on-demand services. If bandwidth is allocated through the central office switch, problems are provisioning or service issues from the carrier. End-user implementations require uniform equipment, and certainly uniform implementations of the bonded channels. Check the configurations at all ends and homogenize them if you need too.

Digital Speed

Every terminal adapter has an option for speeds up to at least 64 kbps and maybe 384 or 1534 Mbps with bonding. Most often the terminal adapter supports 56 or 64 kbps for data calls automatically. However, some devices will not self-adjust for speed changes. As previously noted, the Motorola BitSurfr line is a good example. These devices are fixed to 56 kbps by manufacturer default (so that they have fewer technical support calls), and changing this parameter is very difficult. I think the configuration utility is quirky at best. If you are communicating with many Internet providers, CompuServe, and remote corporate sites with different local service capabilities, long-distance capabilities, and different types of CPE, then service problems can be a function of the type of CPE hardware you select. You want something that is more robust than the Motorola equipment which will self-adjust among analog, digital, and the supported speeds.

If your application is asynchronous (e.g., the kind of application you might have done with an analog modem), it may be designed to run at more or less than 56 kbps, but set the terminal adapter at 56,700 bps anyway because this refers to the ISDN line itself, not the amount used by your application. If you have one terminal adapter configured for 56 kbps and one for 64 kbps, you may have a situation where one side will not accept an incoming call from the other.

You can solve mixed-speed problems on both terminal adapters if both ends of the data call are served by the same switch (i.e., the phone numbers have the same area code and prefix). In this case the network limitations do not apply. If you try to make a 64-kbps call over a 56-kbps link, it will not work.

With the right equipment and software X.25 packet transmission over the D channel is possible at speeds up to 9.6 kbps on BRI and 64 kbps on a PRI B or D channel. Note that you should set your communications software at 19,200 or 56,700 bps or else Windows-centric software will establish port speeds at 9600 bps. It is important to remember that 9600 bps is the default port speed for all Windows versions. This default setting might cause D channel connectivity problems and could limit the traffic flow between the workstation and the terminal adapter. Set the port speed baud rate to 57,600 bps if you have that value; otherwise set it to 19,200 bps. Windows 3.x (and later versions) doesn't handle throughput greater than 19,200 bps very well in general unless you replace the communication driver software (TurboCom/2 is one example) or use applications that bypass it entirely. WinCim is a good example; it bypasses the Windows communications drivers with its own.

Rate Adaptation

Most ISDN adapters automatically adjust to the B channel signaling speed. However, some Motorola BitSurfr modems do not. If they were set at the factory by default to 56 kbps as mentioned in other sections, I am sure that other connections and connectivity devices will not support the V.110 and V.120 standards completely or that the IXC could route one-half of bonded call through a switch without SS7 capability. It becomes hard to figure out a connection that persists at 121.7 kbps until you recognize that the call is split at different rates. Obviously, an ISDN protocol analyzer can decode D channel setup information and provide the bit rate directly.

However, when you do not have those tools, the solution to debugging rate adaptation problems and split rates is to initiate the ISDN call from the MS terminal applet or NetManage Chameleon so that you can watch all the setup messages for the line connection and all subsequent higher-level protocols. Most advanced terminal programs, WinSock and WinCim included, mask the protocols. These might include IP, PPP, SLIP, or Compressed Serial Line Interface Protocol (CSLIP). Watch the messages. Most customer service unit/data service units (CSU/DSUs) and ISDN adapters will send a message to the screen indicating the connection speed for each channel.

Your terminal adapter or the one at the called destination is set for 64 kbps instead of 56 kbps. (This refers to the ISDN line speed, not your application's speed, which might be different.) Check for rate adaptation support, change setup strings, or reconfigure the terminal adapter. Consider problems with 64 kbps when the line supports 56 kbps only.

If you do not have this facility, set the rate for your terminal adapter with its configuration utility or modem AT setup strings. Try each speed to narrow down the size of the problem. You do not want the problem to be that ISDN just doesn't work, but rather that ISDN doesn't work at 64 kbps. For example, when you can make a data call from a 5ESS line to a DMS-100 line, but a call you originate on the DMS-100 line to the computer on the 5ESS line drops the call, set the terminal adapter configuration to 56 kbps on the DMS-100 side. It may be set for 64 kbps and doesn't adjust automatically.

You may need to create connections with both local and long-distance sites at a preset speed to ascertain if the problem is with the hardware or what side of the ISDN cloud it is on. However, by logically making calls, you can narrow the problem to a particular device or service. This is illustrated in Figure 5.19.

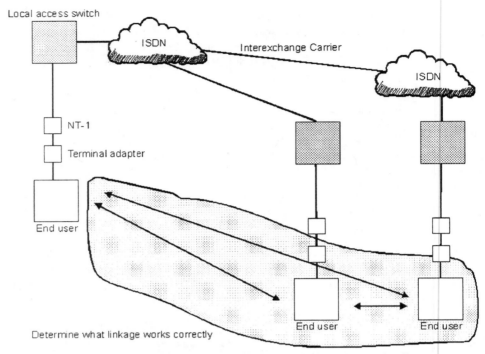

Figure 5.19 Narrow rate adaptation problems with fixed (preset) speed calls.

When the terminal adapter is performing below its specified speed for an asynchronous data call and you are running a communications application under Windows (e.g., Procomm for Windows), you are limited by the speed that Windows can handle, which is usually 40 kbps. Try running a communications program from DOS instead or test it under Windows NT. Acquire TurboCom/2, from Pacific Commware, to speed up your com port under Windows with its replacement communication drivers. The 3Com Impact unit comes bundled with this software in anticipation of the Windows driver bottleneck. By the way, I have not had good luck with TurboCom/2 for several projects, books, and articles I have done in the past because I end up with software conflicts or strange modem errors; most users find it is very functional though. Your terminal adapter may need an Erasable Programmable Read-Only Memory (EPROM) upgrade to realize the higher speeds. The Universal Asynchronous Receiver/Transmitter (UART) should be version 16650 or greater in order to have more than a single-byte buffer for serial communications I/O. Macintosh, Unix, Amiga, and other workstations do have this problem. Most PCs based on 80486 or Pentium systems have this chip, whereas older systems do not. The smaller buffer in older systems does not reliably handle speeds higher than 40 kbps.

Also realize that the terminal adapter on the far side of the call may be the limiting factor. You can only get as much throughput as the slowest link. You may have your communications software set for a lower speed than the terminal adapter can handle. Your terminal adapter may have a speed configuration setting. Terminal adapters with external buttons and LED displays can be programmed for the required speed. This setting may be misleading because the terminal adapter probably sends at its highest speed regardless of this setting. And there is also a setting for 56 or 64 kbps on your terminal adapter, although these refer to the maximum speed of your ISDN line, not the maximum speed of your terminal adapter.

When the communications software won't dial a V.120-compatible terminal adapter using the Hayes modem commands to initiate the connection, try using *ATD* instead of *ATDT*, from the Windows terminal applet. Type ATD5551234 when you can get an "ok" from the terminal adapter. Your dialing directory of the communications software may also automatically use ATDT instead of ATD. There may be a setting for changing this in your software configuration options. Try entering the phone number without any dashes or separators. The dashes may work for a standard analog modem call but may confuse the ISDN terminal adapter or the central office switch.

If speed problems persist and it is clearly neither the channel nor the hardware, verify that the PC or server can sustain higher bandwidths. Recall that Windows and older PC UARTs will not sustain high bandwidths because of interrupt overhead and lack of data buffer space.

NCOS and PIC

The ISDN line has to be configured at the switch to have access to a Switched 56 line for data calls outside of the phone number prefix area. Switched 56 line calls outside your local calling area are provided by a long-distance phone company. The symptoms of incorrect line settings are that you can't make outgoing data calls to anyplace outside of your phone number prefix. However, you will be able to receive incoming data calls from anyplace. (By the way, you can also duplicate these same symptoms with an incorrect switch type setting, service type selection, and SPID numbers on the CPE configuration.) The switch setting to interconnect ISDN with ATM, X.25, X.31, switched digital, and frame is called the Preferred Interexchange Carrier (PIC) for AT&T 5ESS switches and related Network Class of Service (NCOS) for Northern Telecom DM-100 switches.

Multipoint or Point-to-Point

Anytime your terminal adapter requires more than one identifying SPID, it is multipoint by definition, with multiple directory numbers. X.25 packet applications using multipoint delivery may provide that several devices can be active at the same time on the same channel using packet technology on either the D or B channels. If the local office switch is 5ESS, many applications do not require SPIDs; this is called point-to-point connectivity. For example, a basic ISDN application using voice and data on the two B channels really requires no SPIDs. The problems arise because some ISDN terminal adapters will not work on an AT&T 5ESS switch that is configured multipoint (multiple SPIDs). Conversely, the application may require multipoint. You need to confirm operation with other hardware or simplify the overall ISDN network infrastructure.

Directory Numbers

Even when you correctly set the SPIDs with the appropriate area codes (or lack thereof), prefixes, and suffixes, the switch and your terminal adapter need to agree on the directory number (DN). This is usually the seven-digit phone number for the first B channel (individual ones for multiple B channels or even a hunt group or specialty configuration for unique ring identifier, analog devices, or fax routing), but the terminal adapter may require you to enter a ten-digit number in some cases. Realize that the DN is only used for incoming calls. The symptom of an incorrect DN setting is that you can make data calls but you cannot receive them. The DN may be requested by your terminal adapter during configuration for use with an AT&T 5ESS switch, but it will not actually be used and can generally be omitted. However, if you supply a DN, it cannot be wrong.

Terminal End-Point Identifier

The terminal end-point identifier (TEI) is a suffix to the directory number related to supplemental ISDN services. If it is set differently in the terminal adapter than in the switch, it will prevent a dial tone for voice applications. The options are "Dynamic" or a number from 0 to 63 (also known as *Static*). Default for an AT&T 5ESS, NT DM-100 or Seimens switch is *Dynamic* unless you have provisioned packet-data application.

Terminal Type

ISDN lines from the 5ESS switch need to be configured for the terminal type (A, B, C, D, or E) of the terminal adapter you're using. Some common terminal adapter types are:

- AT&T is D except for the model 7500, which is E.
- Combinet is C, D, E (all will work).
- Fujitsu 410 and 400 and 300s are A.
- Fujitsu 1050 and 2000 are D.
- Hayes is A.
- Hitachi is C.
- Teleos is D.

Type E is generally used for data-only terminal adapters (ones that have no provision for adding voice capability). This is misleading because some terminal adapters designed for data-only use may have some voice options.

Voice or Fax Calls

Voice and facsimile require an analog connection by definition. You cannot send these calls over a digital-only line except with special CPE. If you cannot make a voice or fax call on one channel, try the other one. The channels may be backward. If you still cannot make a call, it is possible that the provisioning is incomplete or incorrect. Ask the ISDN provider for a current copy of the switch settings.

Electronic Key Telephone System

The EKTS setting is what activates the various voice features on the switch. This should correspond to the functions available on the handsets, serviced by the PBX, and provisioned for service on the ISDN line. For the DMS-100 switch, EKTS must be set to "yes" on the switch for any B channel that is to be used for voice with an ISDN phone. If you are simply plugging a non-ISDN analog phone into an ISDN terminal adapter that has an analog phone jack, set EKTS to "no" because there are no supplemental services available on an analog handset. By the way, an analog handset can be either a rotary dial phone or a touch-tone phone. Do not think that the DTMF tones on most modern handsets mean the phone is digital. In fact, the DTMF codes are analog conversions of the buttons. For data applications, set EKTS to "no" for any B channel carrying data. Some data terminal adapters will not work when EKTS is set to "yes."

Release Key

On an ISDN EKTS telephone, the number you assign at the switch for the Release Key is what maps the physical key on the ISDN phone set to be the release function (as in releasing a call). When ordering the ISDN line, have the service provider set the release key to the number specified in your ISDN phone documentation. It is different for every ISDN vendor and even for different products from the same vendor. It is highly unlikely you can mix or match products from different lines or from vendors. If you run into service problems, realize that ISDN for voice service is a new concept, that you are an early adopter (at least in the United States but not in Europe), and that ISDN voice services require very careful network integration. Data-only terminals have no release key, so order the line with the release key set to "no."

Functional or Stimulus Signaling

ISDN lines from a DMS-100 may be configured for either functional or stimulus signaling. Terminal adapters are built to work with one type or the other. Functional is the option for newer terminal adapters. Stimulus is an older signaling method.

PVER or PVC Version

Each terminal adapter that is designed to work with a DMS-100 switch has a version number labeled as a *PVER number*, which identifies which version of the DMS switch generic software it works with. Older terminal adapters will not work with newer (higher-numbered) switch generics. ISDN line can be configured to provide a signal that looks like an older generic software version if your terminal adapter requires it.

This information may not be in the terminal adapter's documentation and often is a source of problems. Contact your terminal adapter vendor if you cannot find the PVER or PVC or version information in your equipment documentation. The possible settings include:

- Version 0   BCS Level 30
- Version 1   BCS Levels 31 to 34
- Version 2   BCS Level 35

ISDN Setup Problems and Solutions

The following are common ISDN installation problems and their solutions. When you have both analog and ISDN equipment installed on the same user workstation or communication servers, you may run into call setup problems because of the robustness in support for directory numbers. Try entering phone numbers, directory numbers, and SPIDs without dashes or separators. The dashes, parentheses, and spaces might work for modem calls but may confuse the ISDN terminal adapter or the line switch if passed through directly.

As Chapter 4 detailed, SPIDs are used to identify what sort of services and features the switch provides to the ISDN device. Currently they are used only for circuit-switched service (as opposed to packet-switched data transmission). Annex A to ITU recommendation Q.932 specifies the (optional) procedures for SPIDs. They are most commonly implemented by ISDN equipment used in North America. When a new "subscriber" is added, the telco personnel allocate a SPID just as they allocate a directory number. In many cases, the SPID number is identical to the (full 10-digit with the number area code) directory number. In other cases it may be the local directory number with various other strings of prefixes and suffixes, such as digits 0100 or 0010, 1 or 2 (indicating the first or second B channel on a non-Centrex line), or 100 or 200 (same idea but on a Centrex line) or some other seemingly arbitrary string. Some SPIDs are in the form 0199999990 for AT&T custom and 019999999011 for NI-1, where the 9's represent the seven-digit directory number.

ISDN is switch- and implementation-dependent. Subscribers need to configure the SPID into their terminal (i.e., computer or telephone, etc., not the NT-1 or NT-2) before they will be able to connect to the central office switch. When the subscriber plugs in a properly configured device to the line, layer 2 initialization takes place, establishing the basic transport mechanism. However, if the subscriber has not configured the given SPID into the ISDN device, the device will not perform OSI layer 3 initialization and the subscriber will not be able to make calls or will find, as I did, that service is flaky and works intermittently.

After the SPID is configured, the terminals go through an initialization and identification state which has the terminal send the SPID to the network in a Layer 3 information message. The ISDN network responds with a message with the Equipment Identifier (EID) information element (IE). The SPID is not sent again to the switch. The switch may send the EID or the Called Party Number (CPN) in the SETUP message to the terminal for the purpose of terminal selection.

SPIDs should not be confused with TEIs. TEIs identify the terminal at layer 2 for a particular interface (line). TEIs will be unique on an interface, whereas SPIDs will be unique on the whole switch and tend to be derived from the primary directory number of the subscriber. You need the TEIs particularly for complex BRI services, multiple hunt-group BRI services, or ISDN PRI with its many lines and complex channel interactions. Although TEIs are used at different layers, they have a one-to-one correspondence, and matching them correctly with services is not a difficult task. TEIs are dynamic (different each time the terminal is plugged into the switch), but SPIDs are not. Following the initialization sequence mentioned above, the one-to-one correspondence between SPIDs and TEIs is established. TEIs are usually not visible to the ISDN user, so they are not as well known as SPIDs.

The address of the layer 3 message is usually considered to be the Call Reference Value (also dynamic but this time on a per-call basis) as opposed to the SPID, so the management entity in the ISDN device's software must associate EID/CPN on a particular TEI and Call Reference Number to a SPID. There are some standards that call for a default Service Profile, where a terminal doesn't need to provide a SPID to become active. Without the SPID, however, the switch has no way of knowing which terminal is which on the interface, so for multiple terminals an incoming call would be offered to the first terminal that responded, rather than to a specific terminal. This matters if you have multiple analog or digital devices on the NT-1 or NT-2.

As stated in Chapter 3 and previously in this chapter, SPID problems are supposed to disappear. The National ISDN Council, a group of regional telephone operating companies in the United States, have agreed to a uniform 14-digit format and to automate the SPID capabilities in network switches by 1998. However, if you are getting SPID error messages, try the examples listed under your local ISDN service provider for your switch type and the likely protocol. See the tables in Figures 5.20 through 5.29 for switch settings based on your provider. These often confirm to National ISDN 1, but *not always* as some of these tables show.

Switch type	Format	Directory number	SPID
5ESS-Custom	01 "7 digit" 0	555-555-1234	SPID: 01-555-1234-0
5ESS-NI1 (5E8 version)	01 "7 digit" 011	555-555-1234	01-555-1234-011
5ESS-NI1 (5E9 version)	"10 digit" 0111	555-555-1234	SPID: 555-555-1234-0111
DMS100-NI	"10 digit" 0111	555-555-1234	555-555-1234-0111
DMS100-Custom	"10 digit" ?? (??= 0, 1, 01, or 11)	555-555-1234	555-555-1234-0
Siemens EWSD-NI1	"10 digit" 0111	555-555-1234	555-555-1234-0111

Figure 5.20 AmeriTech SPID formats.

Switch type	Format	Directory number	SPID
5ESS-NI1	01 "7 digit" 000	555-555-1234	01-555-1234-000
DMS100-NI1	01 "7 digit" 100	555-555-1234	555-555-1234-100
Custom ISDN	01 "7 digit" 0	555-555-1234	01-555-1234-0

Figure 5.21 Bell Atlantic SPID formats.

Switch type	Format	Directory number	SPID
DMS100-NI1	7 digit number followed by 2 zeros	555-555-1234	SPID: 555-1234-00

Figure 5.22 Bell Canada SPID formats.

Switch type	Format	Directory number	SPID
5ESS NI-1	"10 digit" with 0100	555-555-1234	555-555-1234-0100
DMS100	"10 digit" with last two digits repeated	555-555-1234	555-555-123434
DMS100	"10 digit" with last digit repeated	555-555-1234	555-555-12344
DMS100 NI1	"10-digit" 0100	555-555-1234	555-555-1234-0100
NI1	"10-digit"-0	555-555-1234	555-555-1234-0
NI1	"10-digit"-00	555-555-1234	555-555-1234-00
NI1	"10-digit"-000	555-555-1234	555-555-1234-000

Figure 5.23 BellSouth SPID formats.

Switch type	Format	Directory number	SPID
DMS100 NI1	"10-digit"-0100	555-555-1234	555-555-1234-0100
DMS100 NI1	"10-digit"-0000	555-555-1234	555-555-1234-0000
NI-1 AT&T	01 "10-digit"-000	555-555-1234	01-555-1234-000
AT&T Custom	01 "10-digit"-0	555-555-1234	01-555-1234-0

Figure 5.24 GTE SPID formats.

Switch type	Format	Directory number	SPID
5ESS-NI1	"10-digit" 0000	555-555-1234	555-555-1234-0000
DMS100 NI-1	"10-digit" 0001	555-555-1234	555-555-1234-0001

Figure 5.25 Nevada Bell SPID formats.

Switch type	Format	Directory number	SPID
DMS100	"10 digit" 1 or 2	555-555-1234	555-555-1234-2 (or 20, or 200, or 2000)
DMS100	555-1234-1 (or 10, or 100, or 1000)	555-555-1234	555-555-1234-2 (or 20, or 200, or 2000)
DMS100	"10-digit" 1(both)	555-555-1234	555-555-1234-1
DMS100	"10-digit" 01	555-555-1234	555-555-1234-1 (or 01, 0100)
DMS100	"10-digit" 02	555-555-1234	555-555-1234-2 (or 02, 0200)
5ESS-Custom	01 "7 digit" 0	555-555-1234	01-555-1234-0
5ESS-NI1	01 "7 digit" 000	555-555-1234	01-555-1234-000

Figure 5.26 NYNEX SPID formats.

Switch type	Format	Directory number	SPID
5ESS-NI1	01 "7 digit" 000	555-555-1234	01-555-1234-000

Figure 5.27 Southern New England Telephone SPID formats.

Switch type	Format	Directory number	SPID
DMS100	"10 digit" - 01	555-555-1234	555-555-1234-01
Siemens-NI1	"10 digit" - 0100	555-555-1234	555-555-1234-0100
5ESS-NI1	01 "7 digit" 000	555-555-1234	01-555-1234-000

Figure 5.28 SouthWestern Bell SPID formats. (Note: You may need to add two zeros to SPID given by SouthWestern Bell.)

Switch type	Format	Directory number	SPID
5ESS-NI1	01 "7 digit" 000	555-555-1234	01-555-1234-000
5ESS-Custom	01 "7 digit" 0	555-555-1234	01-555-1234-0

Figure 5.29 US West SPID formats.

Some Troubleshooting Hints

When the NT-1 lights are red, you are not physically connected to the ISDN terminal adapter or the terminal adapter is not functioning. Check the cables. Reboot the terminal adapter by powering off and then on. When the NT-1 U line light is red, you are not connected to a working ISDN line. Check your wire connections.

When you can make ISDN data calls to other locations within your telephone prefix and can receive incoming data calls but can't make a data call outside your local calling area, the switch is not properly configured to let your line make data calls across the Switched 56 network. On the DMS-100 this is the NCOS setting; on the 5ESS it is the PIC setting. The ISDN service provider must make the change at the switch.

When everything is configured correctly for a data application but still doesn't work or it gives garbage, reboot the terminal adapter by power cycling the PC, NT-1, and terminal adapter. If that doesn't work, try replacing your data cables or phone cables. Defective cables are one of the most frequent sources of problems. Even when the cables are new, they might not be correctly wired. Also, as I discovered the hard way, just because an RS-232 cable works for an external modem does not mean that it will work for the external ISDN terminal adapter. The return-to-send (RTS) or clear-to-send (CTS) lines could be broken, miswired, or creating crosstalk at the higher-performance requirements of ISDN. Older ribbon cables that work with a modem do not typically work with ISDN. You really need a shielded foil cable. Try physically removing the phone line from the jack on the terminal adapter, then powering off and on. Reconnect the phone line. Try again.

If you've tried your line several times after powering your terminal adapter on and off several times or plugged a standard analog handset into the ISDN line, your ISDN line can get hung up at the central office switch. It might self-correct in a few minutes (maybe up to half an hour) or it might be hung until it's cleared the next morning as part of normal switch surveillance and maintenance. This generally only happens when you're playing with the line, not in normal use. Your terminal adapter may be an older type with T-Link or V.110 rate adaptation instead of V.120, the current standard.

Even if you're sure you've ordered the ISDN line with the correct configuration, realize that each ISDN line is configured by clerks. The information is not very meaningful, readable, or familiar. As such, errors are not uncommon. Request a printed confirmation of how the line is actually configured as confirmation of the planned provisioning process, and you may want to call technical support at the ISDN service provider to verify these settings. They can read most of the information remotely from the switch.

You may have ordered the line exactly as the terminal adapter vendor's documentation specified, but sometimes their documentation is incomplete. In particular, on a DMS-100 switch, make sure you have ordered the correct version for your terminal adapter. The version determines which software version of the switch is activated for your line, such as BCS 29, 30, 31, 32, 34, or 35. Also realize that NI-1 is not the same as AT&T 5ESS Custom, and that the switch may not support NI-1 but will work with a similar customer provisioning.

If your Switched 56 service is AT&T's Accunet, and you're calling a relatively new area code, the number may not be entered in AT&T's routing tables yet. Give customer support a call so that they can update their tables. Or send e-mail to their web site. Similarly, ISDN service provider's switches might require a table update for the location you're calling to because it is a new exchange.

If you're getting garbage or the system is locking up after the call is established, make sure your communications software on each end is set for hardware flow control (sometimes called RTS/CTS) instead of software flow control (sometimes called XON/XOFF). This setting only affects the buffering and flow of information between the local terminal adapter and the PC. Old UARTs, competing IRQs, and too many interrupts can create problems. Also realize that COM1 and COM3 share the same IRQ, typically IRQ3, and COM2 and COM4 share IRQ4. If you are using COM1 for a mouse and COM3 for a terminal adapter, you may create a conflict over the IRQ2 that disables the mouse or the terminal adapter or simply slows data flow. Software flow control may be too slow for your application. Try turning off the power on the terminal adapter, physically unplugging the phone wire from the back, then plug it again, and power up. This solves some problems.

You may be operating at a speed higher than the serial port on your PC can handle. Set your communications speed for 9.6 kbps and see if the errors disappear. Then try 19.2, 28.8, 38.4, or 57.6 kbps. Also realize that many Windows workstations or servers will not handle streams greater than 28,800 bps without supplemental or substitute communications drivers.

When the NT-1 light indicates a line or terminal problem, but the application works okay, the light itself or circuitry inside the NT-1 or integrated terminal adapter might be defective. You may not have a problem. Substitute a good NT-1 to see if the problem persists.

When there's no dial tone on your voice application, even though the NT-1 lights are green and the cables check out correctly, EKTS may be set to "no" on the switch. Special ISDN phone sets need to be configured (or reconfigured) using the phone's documentation to conform to the switch settings, too.

When the videoconferencing application requires you to dial two phone numbers to get both of the B channels active on a device served by a DMS-100 line, ask the telco to place the phone number for the second B channel in a hunt group with the phone number for the first B channel. When the first number is busy, the call automatically "hunts" to the second B channel so the caller need only dial the first number. This is a nice trick for both outgoing and incoming calls. It also solves the problem of not getting both lines connected properly with regularity.

When your videoconferencing unit won't dial or receive calls and it is served by a 5ESS switch, your B channels may be configured for voice/data (both voice and data on each B channel), but you also need to have the line configured by specifying "Data channels = 2" in order for both channels to work. The video-conferencing unit may not work at all unless both B channels are working. Also, try entering the phone number without dashes or separators. This information should be entered into a desktop phonebook or rolodex; make sure that fixes in one place are communicated to other users, too. The dashes may confuse the ISDN terminal adapter.

When the terminal adapter won't answer an incoming asynchronous data call, even though it "sees" it (you can see a status display or status light that tells you a call has been detected), configure the terminal adapter to turn the "autoanswer" function on. Refer to the terminal adapter documentation. Your applications software on the receiving end should also be configured for autoanswer, or the user needs to enter a command to answer the incoming call. Your application software may be expecting modemlike connect messages from the terminal adapter (e.g., CONNECT 9600). You may need to adjust your application software to recognize whatever connect messages are coming from the ISDN terminal adapter (e.g., CONNECT 57600).

If a face-to-face groupware application doesn't connect over ISDN, even though the lines and applications are correctly configured, you may require a special Hardware Handshake Cable, not a standard RS-232 cable. Since they look alike, it's easy to have a problem. You can also go out to Radio Shack (it is often close by) and get two crimp-on ribbon connectors and a length of ribbon cable. A Hardware Handshake Cable can be obtained from Hayes (part number 07-00681), telephone (404) 840-9200. By the way, I ran into a problem with 17-year-old serial cables that worked well with external modems but did not enable the RTS light on a Courier-I modem to function. The RTS wire was marginal at ISDN speeds but worked well with analog modems. It is a function of signal crosstalk with a cable designed for 300-baud modems notwithstanding the impedance at 14.4 kbps.

In the case of a V.120 data call, the connection sometimes locks up in the middle of the call, even using different terminal adapters and different lines from different ISDN switches. The callers were using slow 286-based PCs and also dumb terminals. When flow control settings and application speed were changed, it didn't help. Try faster workstations or communication servers.

Software

Although ISDN is an interface for a circuit, the connection is mostly a function of terminal hardware. However, a good part of what makes one terminal adapter better than another one, more flexible, and easier to debug is the software. ISDN devices differ markedly on the software. However, more importantly, you will want to select a device supported by your operating system. For example, because the US Robotics Courier-I product is controlled by AT commands, it will work with DOS, Windows, Windows NT, Unix, Solaris, and Windows 95 without software additions. However, many devices will require special software to configure and manage the hardware. Microsoft has released the Windows 96 ISDN Accelerator service upgrade to resolve problems like these and to provide bonded channel connectivity, which is a software interpretation of how data streams are split over two or more channels.

Conclusion

This chapter showed how to debug and resolve most ISDN problems. As you read, the general process of ISDN maintenance and debugging is to prove that things within the connection are good and by a process of elimination determine what does not work correctly. Since the ISDN connection consists of many parts, you want to discover which side has the problem and focus in on that cause of the flawed connection. You want to validate (prove what things are good) from lower ISO levels to higher ones. The next chapter describes dial-up connectivity options for ISDN.

6

ISDN Network Interconnection

Introduction

This chapter shows how to use ISDN to dial into the Internet for one user or many users, how to use ISDN as a inbound communication protocol, and also how to employ ISDN to wire together far-flung local area networks (LANs). ISDN is not the answer to every need for network interconnection. This service is priced well in many locations and the bandwidth with compression and filtering is sufficient for those needs. It certainly has benefits over dial-up analog connections, its major competitor at the low end. Figure 6.1 compares low-bandwidth connectivity options and several of the possible connections included in bonded 56- and 64-kbps channels.

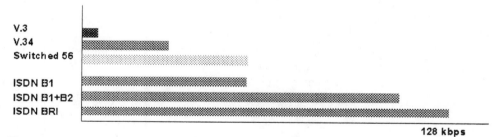

Figure 6.1 Analog and ISDN BRI connectivity bandwidth options.

Although ISDN is at least twice the bandwidth of any analog solution, you need to contrast it to LAN bandwidths. You should not expect remote users to experience the same performance as local users and you cannot expect to copy large databases or image files from site to site over ISDN at LAN speeds. Narrowband ISDN (N-ISDN) provides bandwidths to 1.534 Mbps, whereas a slow Attached Resource Computer Network (ARCNET) LAN begins to provide service at 2.5 Mbps. Fast Ethernet and Fiber Distributed Data Interchange (FDDI) yield bandwidths to 100 Mbps. High-speed LAN-to-LAN interconnectivity is the province for ATM (B-ISDN) when it becomes available and affordable. The chart in Figure 6.2 shows ISDN basic rate interface (BRI) and ISDN primary rate interface (PRI) bandwidths in contrast to common LAN protocols.

ISDN is an efficient and viable solution for many types of remote dial-in and dial-out connectivity and linkage for remote LANs. Just keep its bandwidth limitations in perspective and plan for its implementation and usage accordingly.

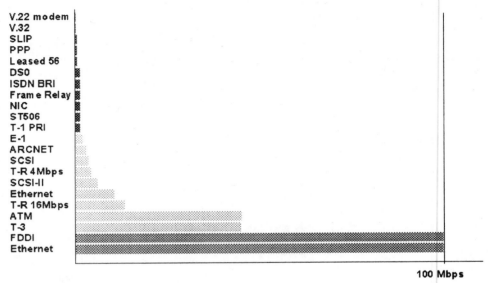

Figure 6.2 ISDN bandwidths in contrast to LAN protocols.

Remote Dial-Out

Dialing out with ISDN is simply a matter of installing a local loop, wiring to the user from the demarcation point, adding an ISDN terminal adapter, and providing some terminal software. Windows 3.x and later versions includes a terminal program which is merely sufficient to get started (when you can send Hayes commands to the ISDN adapter). Most other terminal programs, from ProComm to WinComm to specialty programs like WinCim for Windows access to CompuServe or Internet communications tools such as Chameleon's NetManage or WinSock with Navigator provide more robust and specialized functions. This ISDN substitution for a modem is trivial and even when the user workstation is a node on a network, the architecture looks just like the typical setup for a modem in a home office/small office (SOHO). This is shown in Figure 6.3.

Many ISDN modems, the US Robotics Courier-I modem and the IBM WaveRunner as examples, support the normal Hayes analog modem command set so that WinCim, Winsock, NetManage, and most other telecommunication managers will work as easily with digital servers as they do with analog connections. The Hayes command set begins with *AT* meaning *attention* and uses strings such as ATDT and AT&F1&F2*S1. By the way, the command set is not really case-sensitive unless you *mix* both upper- and lower-cases in the setup strings.

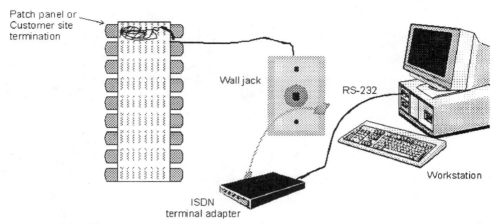

Figure 6.3 ISDN setup for remote dial-out access.

Then the results may not be what you anticipated. In fact, with devices that support both ISDN and analog you can mix and match digital and analog connection methods through the same ISDN circuit. This 1:1 ratio of users to processors and a modem or terminal adapter is expensive in quantity. It is also difficult to manage when you have many circuits and stacks of modems or terminal adapters in the telecommunications closet. When you have more than two or three users with dedicated lines for dial-out, the infrastructure cost becomes significant. You need to wire that many lines into each user's office and you need multiple ISDN devices. Since most ISDN connections are not bonded, ISDN BRI can provide at least two unique connections at the same time.

In fact, if one of those channels is connected to an independent Internet service provider, that single link can provide LAN connectivity for every network user to the Internet for mail and web surfing. Instead, consider establishing a *communications server*. This is a device that may be the file server, or it may be a separate network node dedicated solely to remote dial-out and dial-in services. Some stand-alone devices are called *hubs, concentrators, hub concentrators,* or *communication gateways,* and they may integrate directly in Unix, NetWare, or a Windows-based server. These devices connect into the network much like printer servers. Most communication servers do not include the actual modems or terminal adapters, which is just as well, since the technology changes before the server hardware itself becomes obsolete. Typical configurations are shown in Figure 6.4.

Modem (and terminal adapter) sharing, called *pooling,* multilevel security, and Simple Network Management Protocol (SNMP) management mitigate some of the management problems common with many individual lines and modems. Software, such as pcAnywhere, handles the network access and DOS redirection (under most versions of MS Windows and NetWare) for all serial I/O to the network communication server.

If you are using Windows 95, the ISDN Accelerator Pack provides a driver for Windows 95 to utilize most ISDN terminal adapters directly regardless of whether they support the Hayes AT command set.

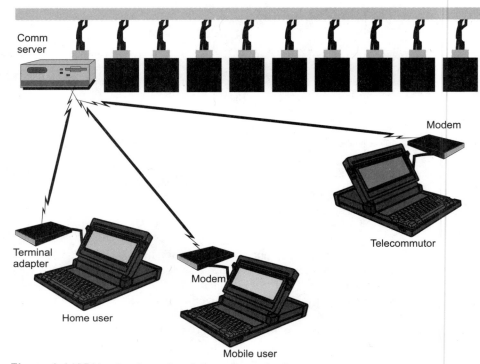

Figure 6.4 ISDN setup for network-based remote dial-out access.

Also, these drivers work with the Internet Explorer (under Windows 95). In general, this Windows 95 service pack (ISDN support was originally listed as an incentive to migrate from Windows 3.x but was not included with the initial release) includes support for Point-to-Point Protocol (PPP) and bonding. If you have an ISDN terminal adapter with V.34 or other analog capabilities, dial-out through ISDN works the same as analog calls. However, if you want PPP and do not have the software support for that or want MultiLink Point-to-Point Protocol (MP) or Bandwidth Allocation Control Protocol (BACP) bonding for 112- or 129-kbps connections, you will need special vendor drivers or this service pack.

If you are dialing into CompuServe, not only do you need the local ISDN number or a national 800 ISDN number (with its $9 per hour surcharge), you need to configure WinCIM with custom modem strings so that it can connect at 64 kbps using V.120 on a B channel. You cannot get bonded service at this time. However, CompuServe performance with one B channel is much better than analog access and typically better than Internet access through an ISP at ISDN on one or two B channels. Access through the Spry web software and CompuServe is no better than the typical ISP. The setup string is:

AT&F&C1&D2\Q3

Recall that setup strings are *case*-sensitive when you mix the cases, so use either all uppercase or all lowercase—but do not mix the cases. In addition, you need to set the modem speed to 56,600 bps; there is no 64,000 bps setting in the pick lists and the 112,200 bps setting will not work correctly. As you can see in the MS Terminal session in Figure 6.5, you can play around with setup strings and obtain different results.

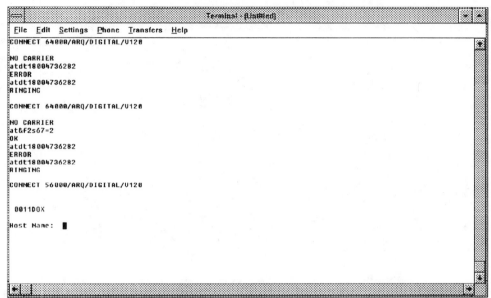

Figure 6.5 CompuServe connectivity with different setup strings. Notice the connections at both 56 and 64 kbps using different strings.

Remote Dial-In

While remote access to a LAN is as simple as attaching a modem or ISDN terminal adapter to a network node, and access for multiple remote users is as brainless as adding a modem or terminal adapter to each network node as needed for multiple users. However, most network users do not like to share desktop resources with remote users, and even if remote users connect into their own desktop machine back at the office, this configuration represents a waste of significant hardware, extra dedicated phone or digital lines, and a potential for security problems since each access point might not conform to the corporate standard. In addition, any access failure represents a dispersed management nightmare.

Instead, the preferred connectivity method is to establish a trusted communications server with a bank of modems or terminal adapters. A series of analog or digital lines is provisioned with sequential (most often, but not necessarily) numbers that form a hunt group with a single main directory number. A *hunt group* is a set of telephone numbers that allows the calling party to connect with the first available line. The set of telephone numbers can be sequential, an exchange group, or just a random collection of numbers available when the ISDN group was ordered. Figure 6.6 illustrates the structure of the hunt group into the communications server.

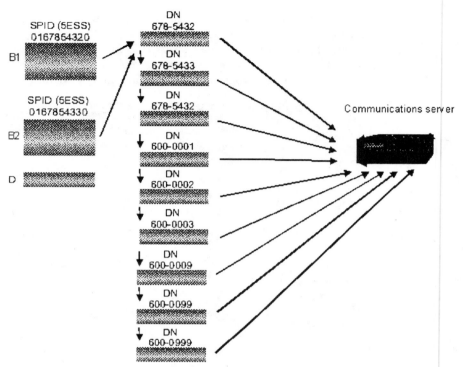

Figure 6.6 Remote access hunt group and the communications server.

While this is as useful for ISDN as with analog lines, ISDN adds a few wrinkles that are worth noting. Provision a series of 2B+D lines to meet small office dial-in requirements. The hunt group is established so that each B channel is part of the entire hunt group sequence and there is a main directory number. As Chapter 2 described, you apply the directory number and calling number mapping to distinguish the types of calls, who is calling, and the types of services you want to offer. This logical directory number and Automatic Number Identification (ANI or Caller ID) information is mapped to the physical ISDN line but can be extracted at the end-user site, as shown in Figure 6.7.

Given current ISDN BRI and PRI pricing, the financial crossover rarely occurs, so ISDN BRI is often the cheaper alternative. In Europe, PRI (E-1) is the common configuration for ISDN, and BRI is not widely available. In PacBell territory, the crossover occurs with 6 BRI lines, so PRI is cheaper after 12 B channels. The concept with PRI hunt groups is analogous. You set up the hunt group so that access is through a single main directory number. You can set up multiple main directory numbers for digital, analog mode, fax, or voice; you just need to wire the physical devices into the appropriate ports, as Figure 6.8 illustrates.

This works and may be appropriate for small offices and home offices. However, you can use the power of ISDN to distinguish the type of call and automatically switch the call into the appropriate service. US Robotics is leading the way but by no means is the only vendor of multifunction terminal adapters. Their MP/8 I and MP/16-I modems support both analog dial-up and ISDN at the same time and will switch for the communications server.

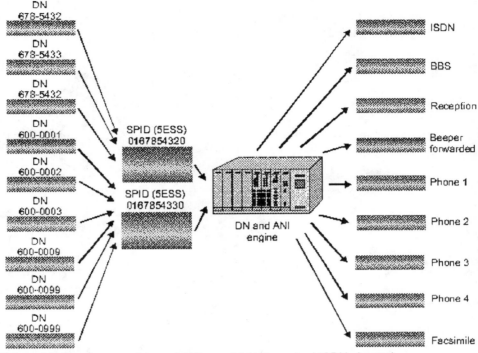

Figure 6.7 Directory number and ANI maps into the actual ISDN channel.

You can configure devices like these to support 4 or 8 ISDN BRI lines or split the 23 (on T-1 or 29 on E-1) channels with a racked hub unit. Since any ISDN terminal adapter with at least one analog port can support a connection for a standard modem, handset, or fax, these devices can direct calls based on the directory number and/or the Caller ID.

One problem with remote or dial-in connections is the cost of supplying every user with a copy of the necessary remote LAN or communication software. If you potentially have 100 remote users, paying $79 for each copy of Carbon Copy may represent an unnecessary expense. In contrast, some licensing agreements with some router and modem-pooling hardware (such as the DIAL software delivered with the IBM 8235 LAN access router) provide unlimited distribution of the remote component of the software for all users.

Security

Almost everyone knows from the press coverage about security lapses on corporate networks, bulletin boards, Internet sites, and remote access ports. Security is a basic service you should not overlook. Network access, except anonymous File Transfer Protocol (FTP) or World Wide Web (WWW) access, should be limited by an authentication process. IDs and passwords are a good first pass. Since ISDN includes the Caller ID (the calling-party information), remote communication software can match this number against the list of acceptable numbers. So far, there have been no reports of falsifying the Caller ID field or overriding the information at a central office switch. This can be as secure as establishing a single remote number for each user (for example, the home ISDN line or the phone number at home) or a list of likely locations each user may call from. Figure 6.9 shows this security.

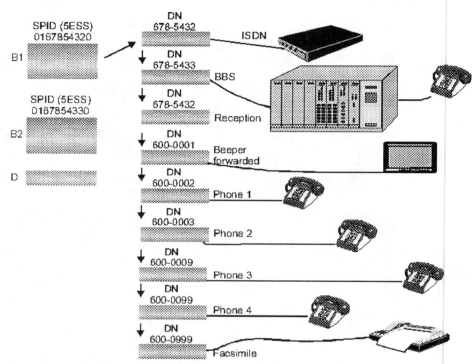

Figure 6.8 Remote access hunt groups physically wired to different devices.

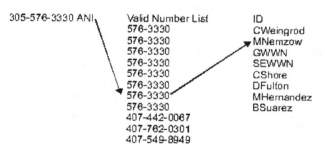

Figure 6.9 Caller ID call-in management.

As you add levels of security, it becomes more unlikely and difficult for a hacker or disgruntled former employee to breach the system. The addition of a Caller ID with a system ID and a matching password is one more level. If you want to further this, you add call-back to the remote connection. The Caller ID is then used as a dial-out phone number or a table lookup into the proper dial number to connect an analog or digital device. The advantage of the table lookup is that you are adding one level of complication that is not burdensome to a remote access user. In fact, all these steps are invisible in use; it just requires some network configuration. Call-back works this way, as shown in Figure 6.10.

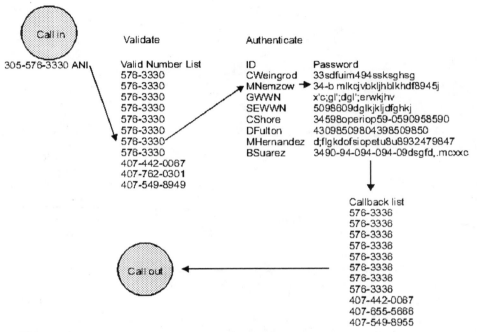

Figure 6.10 The process of call back remote access.

Backup Services

A number of products on the market are designed to provide backup service in case of primary line failure. The primary line might be a satellite link, T-1, T-3, ATM, frame relay, X.25, or dedicated lines and a hard-wired LAN backbone. In addition to fail-safe backup communication services, ISDN also provides a supplemental bandwidth in cases of peak loads, partial failures, or overloads. The economics are such that 24-hour service is cheaper with dedicated lines than ISDN. Under that scenario, you still want the basic services but need a different connection in case of a failure with that primary connection. ISDN is effective for that. Products from Haddax Electronics and others provide switchover within 1 minute by bonding both ISDN BRI B channels. Realize that 128 kbps is not the same as the 1.534 Mbps available through the T-1 line, but it is something when the frame relay, the ATM switch, or dedicated service fails. In general, you do not want the backup circuit to be wired in the same conduit or parallel to the primary lines. Ideally, the ISDN circuit would be wired to a different central office, as shown in Figure 6.11.

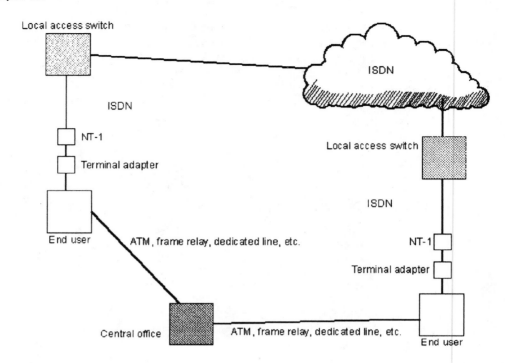

Figure 6.11 The process of call-back remote access.

One of the main advantages of ISDN as a backup method is that you typically only pay for service when you use it—by the minute. The backup lines can be as inexpensive as the cost of the ISDN installation and the fixed monthly service costs. Also, since ISDN can be configured for other uses, such as supplemental analog phone service, dial-out modems, and dial-in communications, you can use the existing ISDN lines for other productive uses. In emergencies, the backup routers switch over to the available ISDN lines. You can terminate users abruptly and just grab and terminate ongoing calls or inform them the lines are required for backup and they should terminate their calls as soon as possible. The routers will bond as many channels as are free and are needed to handle the emergency or overload situation.

Voice and Data

ISDN supports Extended Call Management (ECM). This is provided through the telco's ISDN Centrex service, or those in a larger organization or campus are served by their own comparably equipped digital switch. The range of call management features can be greatly extended. Some of these include:

- Call Forwarding. Forwards calls to a preselected number when the called number is busy or after a preset number of rings.
- Call Pickup. Lets an incoming call be picked up at another station.
- Directed Call Pickup. Lets calls from a specific extension be automatically forwarded to a second number.

- Message-waiting Indicator. Shows that an incoming voice message has been received.
- Directed Dial. Incoming calls can be automatically forwarded at the central office to a car, a cellphone, or a phone at another location.

Virtual key features make Centrex a formidable competitor to in-house PBX or key systems. The central office switch becomes, in effect, the "PBX" serving a location. It offers extended functionality and the potential for unlimited growth and enhancement with virtually no capital investment or risk of obsolescence.

Centrex systems also offer full-channel (that is, 64 kbps per B channel) data speeds within an organization, and extended control of how intercompany calls are handled. An organization with several smaller offices in an area, a real estate firm for example, might link its locations with a customized yet easy-to-use array of call-forwarding, paging and call-pick-up options.

In another twist, ISDN can multiplex voice signals into a mixed voice and data stream. The Congo Voice Router, Cisco's 735 branch office routers, and Ascend's Pipeline support voice. The voice component is routed through secondary B channels or uses the B channel if the ISDN BRI is available. This is a useful technology when combined with PBX systems but adds a layer of complexity to ISDN BRI services for home or telecommuter use that probably just isn't warranted.

LAN-to-LAN Connections

It is becoming increasingly important to interconnect LANs together and provide remote site connectivity. Analog dial-up connections are the principle low-speed connection medium, providing on average 9600 to 14,000 bps of unidirectional throughput. Connections supporting greater bandwidth include dedicated lines, Switched 56, X.25, frame relay, ATM, SMDS, and satellite links. Dedicated lines cost by the mile and average $300 per month for speeds up to 56 kbps. Dedicated lines are lines from one location specifically to another. Switched 56 service provides multipoint connectivity through a dial-up connection and costs about the same, but there is also a per-minute charge. T-1 connections are about double the cost and per-minute charges are 24 times greater. The benefit of X.25 and frame relay is that there are no per-minute charges but rather only per-packet charges. However, the round-trip latency for these protocols is very high. VSAT is actually inexpensive when you need to connect a hundred sites and can work with bandwidths within 128 to 384 kbps and 15-s round-trip latencies. ISDN provides digital services at the same cost as analog with at least twice the throughput.

ISDN provides service analogous to analog dial-up, so users can dial in as remote nodes or enable a transient connection. ISDN can provide a dedicated 7X24 connection, although the economics are usually better in those circumstances for a dedicated line. When you need a dedicated connection part of the day during workdays or for remote site backup, ISDN is the winner. When ISDN is combined with router spoofing and data compression, a bonded BRI line or BACP PRI can provide substantial bandwidth and perhaps even terminate the connection during periods of inactivity.

The ISDN LAN connection is used, and changes are incurred, only when data is sent. One of the most immediately accepted and widely used applications of ISDN is in linking LANs to each other and the outside world. Information in most local area networks travels through dedicated fiber optic or coaxial cables at speeds of from 10 to 1000 Mbps. This means that, contrary to a common misconception, ISDN was not designed to replace these LANs or bridge them into larger local or wide area networks. Rather, it is ideal for the cost-effective, temporary linking of LANs to each other, to remote hosts, or to individual non-LAN users or locations for the timely transfer of specific information or files. In fact, the growing popularity of these applications has spurred manufacturers to offer comprehensive lines of ISDN LAN-bridging and file-transfer equipment.

There are several design issues for ISDN WAN interconnections. A WAN network consists of both LAN components and the links among these. The first issue is that LAN performance is best when the interconnection is provided with routers. This is further improved when the routers are located at the edge of the network and provide *edge-routing* services. Routers lessen WAN traffic by filtering traffic not bound for the remote networks. In addition, as previously mentioned, routers should also filter a considerable amount of network management overhead and spoof the network "uh-huh" traffic, which is in the form of watchdogs, keep-alives, and broadcasts. This is described in more detail in a subsequent section. In addition, you need to plan you LAN interconnection loading. You may anticipate that a single BRI line is sufficient in terms of bandwidth; however, you need to match the configuration with the physical connections. For example, Figure 6.12 illustrates three sites interconnected by routers and ISDN lines.

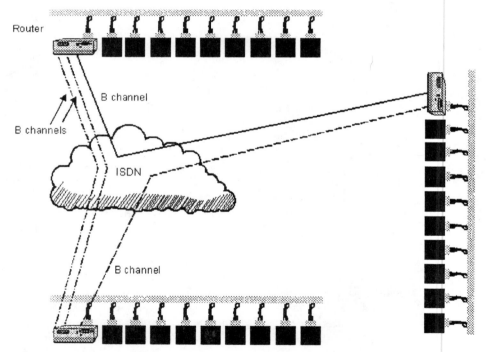

Figure 6.12 Three sites connection by more than three ISDN BRI lines.

One site fully utilizes the BRI channel, but the other two sites require more B channels than are available with one BRI circuit. The implication by this diagram is that the traffic between two sites exceeds the bandwidth of a single B channel (even when compressed). The second channel is needed either on a dedicated basis or for bandwidth on demand. The illustration also makes it very clear that you must install ISDN at all three sites and configure compatible equipment (that is, the router and external ISDN terminal adapters if the routers do not contain an internal ISDN TA).

The routers handle forwarding traffic from each site to other sites in the proper way. They measure this with a protocol analyzer on any LAN segment; the hop count on all frames should be no higher than 1. A hop count of 2 indicates that a frame was sent to another site before it was passed on to the final destination. This indicates a doubling of WAN traffic, an unnecessary delay in the delivery, and additional load on one router.

However, when traffic levels are low and you need to support multiple sites, you may not need to create a mesh between each and every site. You will not save on hardware costs, and you still need at least a BRI line for each site, but you may not need more than that unless the traffic load warrants it. Figure 6.13 shows two sites connected together through the same site. You might note that traffic from one satellite site to another must be routed through the central site. This increases traffic load on that router and potentially doubles the per-minute communication costs.

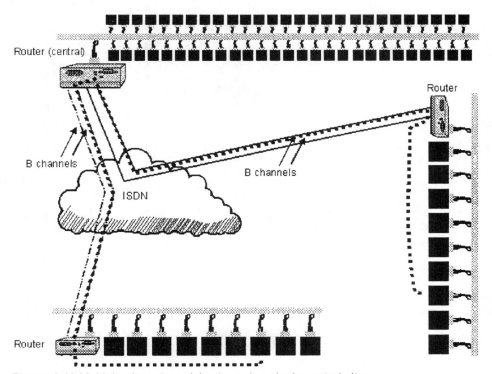

Figure 6.13 Multiple-site connectivity through a single central site.

Small sites that do not want to commit to another piece of hardware for routing can use the network server as a router. Since it can become fairly complex to configure memory settings for all the boards you would like to install in the server and still have sufficient slots, consider a higher-density solution. Companies, such as Telebit, Racal, and US Robotics, manufacture multiport serial boards and even multiport ISDN BRI adapters. This also simplifies setting the IRQs necessary for most internal adapters. A unit as shown in Figure 6.14 has all the ports necessary for the three-site network as shown in Figure 6.13 and even a port with space for other on-demand dialing services.

Figure 6.14 Quad port board simplifies in server routing.

For example, if you have two sites, a mesh connection requires two B channels or one BRI line for each site. If you have three sites, you require six B channels or three BRI lines. When you have four sites in a mesh, you need 3 B channels at each site for a total of 12 B channels and 8 BRI lines. The equation for B channels required (for just a minimal connection) is:

$$n \times (n-1) \quad \text{where n is the number of sites}$$

However, you can minimize ISDN channel requirements by creating a central hub site and routing all traffic through that centralized site. Remote sites can have remote routers, and the central site would typically include a larger organizational router or backplane hub. The typical illustration metaphor for this design is the hub and spoke of a wagon wheel. This is also referred to as a corporate headquarters design with satellite offices. Since several vendors provided presales documentation showing remote site connectivity that looked more like a ladder, I have included their metaphor in Figure 6.15. This misrepresents the line requirements, so be aware of the real needs. Realize that the ISDN line requirements for the remote sites are not really daisy-chained but rather are separate BRI circuits ordered and provisioned from the local telephone company (telco).

Just to make sure the point is clear, Figure 6.16 illustrates the hub and spoke configuration often implemented for corporate headquarters and remote offices.

Do not create an ISDN linkage between LANs as a replacement for a local network or an internetwork. In other words, do not use ISDN as an extension of the LAN but rather as a fast WAN linkage. Remote clients that need to load Windows and client portions of a client/server application will not perform well when these components are loaded over a WAN link. Although ISDN is likely to be some 4 to 8 times faster than an analog modem link, it remains 99 times slower than Ethernet. ISDN is not a replacement for LAN speeds, but a functional and speedier replacement for dial-up connections. Even ATM at 155-Mbps WAN speeds is no substitute for local networks and local resources. In fact, early tests with ATM have shown me that Transaction Control Protocol/Internet Protocol (TCP/IP) stacks and basic overhead reduce the raw 155 Mbps to a mere trickle of 2 to 6 Mbps available for data streams. Understand the network and application design before you promise unlikely answers to WAN problems you should not even create. Use ISDN to transport data or synchronize remote data. Use ISDN to access the corporate database but not to replicate your networked desktop environment.

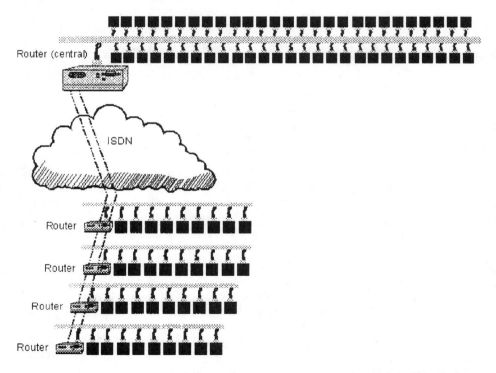

Figure 6.15 A centralized site with WAN (hub and spoke configuration) connections to remote sites.

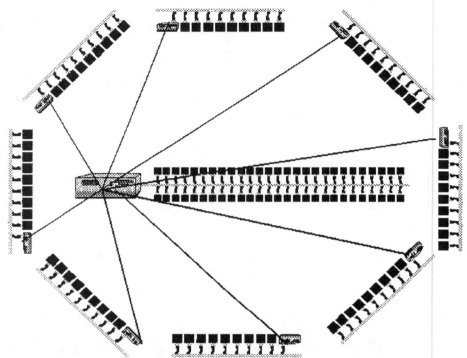

Figure 6.16 ISDN hub and spoke configuration for connectivity between corporate headquarters and satellite offices.

Bonding

One of the significant advantages of ISDN is that you can configure bandwidth on demand, when needed, for only as long as needed. Bandwidth on demand leads to the acronym *BonD*, or *BonDing*. In other words, if a connection between sites requires up to 64 kbps (or 56.7 kbps for an asynchronous connection) bidirectionally, only a single B channel is allocated for the WAN connection. If these sites are beyond a metropolitan service area, this saves on the long-distance carrier charges. When a burst or sustained loads increase beyond that, a second line is added within 1 s. Subsequent additional lines can be added when you have multiple BRI circuits or at least a PRI circuit. In addition, the routing hardware must support bonding on more than two B channels. Many routers and service backup devices do not support anything other than a single BRI line.

The protocols and algorithms used to determine when a line should be added and dropped are proprietary. In general, you will need to pair devices from the same manufacturer so that bonding works as intended. The Bandwidth Allocation Control Protocol (BACP) is a new Internet Engineering Task Force (IETF) standard for ISDN for dynamically adding and dropping channels on a MultiLink Point-to-Point Protocol (MLPPP or MP) connection. Primary authors of this standard include Ascend Communications, Microsoft Corporation, Bay Networks, Cisco Systems, and Shiva Corporation. This is an ongoing cooperation between software and hardware vendors to create interoperability between ISDN devices. In addition, this protocol is specific to ISDN but may also be applied to other technologies in the future.

You should note that high-end routers which support bonding on multiple channels up to and including full PRI support may not be able to sustain the bandwidth. As a result packets are dropped, with a significant drop in performance, to as low as 30 kbps of the 1.534 Mbps intended. This is a function of packet size, burst rates, arrival distributions, and the qualities of the routing equipment. Do not make the assumption that you can provide full 1.534 Mbps with ISDN unless you get the latest and best equipment.

You might also note that it takes about 10 s for a bridge or router to recognize that there is an overload or overrun on a single B channel and to activate a second B channel. If you are transferring files on an occasional basis, you may create this yo-yo problem of connecting in a second B channel that is no longer needed by the time it is bonded into the link. Furthermore, since most telco providers bill in minute increments (AT&T bills for an initial 30 s and then 6-s intervals thereafter if you sign up for some of the special long-distance promotions), 7 s of overrun for the bonded line can get disproportionately expensive and not really provide any added performance benefit to ISDN line throughput.

Routing

ISDN routing is important for performance and capacity reasons and also for flexibility. Many sites may support multiple LAN protocols because NetWare, NFS, Vines, W4WG, Windows 95, or LANTastic are the network operating systems for those sites. Common protocols include:

- Internet Protocol Exchange/System Protocol Exchange (IPX/SPX)
- Service Address Protocol (SAP)
- Router Information Protocol (RIP)
- Internet Protocol (IP)
- Address Resolution Protocol (ARP)
- Internet Control Message Protocol (ICMP)
- System Network Architecture (SNA)

The routers will need to support some superset of the underlying network protocols in order to connect the sites together. In addition, many remote site routers have small routing tables to cut costs. Some small office products support only four or eight addresses (for the routers at the other side of the ISDN cloud), which may cause network design problems later on. You can always create hops to handle multiple sites as illustrated in a previous section, but that increases transmission delays. Some routers can be upgraded with more memory to buffer the packet flow and increase the memory available for routing address tables. Memory is also a limitation on the number and type of routing protocols supported. Each protocol requires its own address and translation table for filtering. Routing protocols specify the shortest distance or shortest time between network addresses, node availability, and what to do when connections fail. Although there are about 30 different important routing protocols, the ones most important for ISDN include:

- 802.3
- 802.1D spanning tree
- Closed loop bridging
- Tail circuiting (LAN bypass)

The 802.1D is the protocol often implemented when the network has router loops (that is, multiple possible paths between nodes) and is important as a backup in case the primary route fails. However, recognize that 802.1D is slow to enable the cutover after a route failure because the routing tables must be laboriously and slowly rebuilt over a period of minutes. The hardware devices that recognize primary circuit failure on a dedicated or switched line cut over within a minute. There is a complete section later in this chapter that provides more information about ISDN routing.

Compression

Routers often support compression so that a bonded BRI line can provide as much as 500 kbps of data throughput and a T-1 PRI line can yield 6 to 10 Mbps of service. However, compression is typically 2:1 but may be as high as 7:1. It is a factor of the data type. Data which is already compressed increases the compression and delivery latency because it takes a long time to discover that the compressed data cannot be compressed further. Even when files have been compressed by ZIP or MPEG, they are usually as efficient as possible, so they can be compressed no further. You probably want to avoid proprietary compression schemes, since they will not interoperate between equipment from different vendors, or plan for a uniform distribution of compatible routing equipment.

One aspect you need to know about compression and its throughput and efficiency is that actual throughput drops by about half with compression enabled and latency is increased by 100 ms. You may actually utilize the ISDN channel more efficiently or use fewer B channels, but performance as far as the LANs and remote sites is concerned is actually less.

Multihoming

IP Multihoming is the ability to assign multiple IP addresses to the same physical port. This is typically applied when a communication server provides a centralized connection for many users to an Internet service provider. The provider has a limited number of IP addresses available and will assign one subaddress for the duration of the call. However, when the other side of the ISDN connection cannot support multihoming, it is possible that your users could get the wrong packets, and worse, it is also possible that other organizations could get your traffic because it was misrouted. Keep in mind the security and integrity of your data, and if you use IP and a public network, you may want to add encryption and authentication to all remote traffic.

Spoofing

ISDN spoofing is important for performance and capacity reasons because much of LAN traffic is overhead. LAN traffic does not need to be transported over the WAN. Although routing and filtering by destination will relive the WAN bandwidth requirements, some WAN traffic is specifically directed from the LAN to the remote nodes by node address. The typical router filtering does not remove this traffic. Typical protocols that are more difficult to filter include:

- TCP session keepalives
- IP RIP broadcasts

- IPX watchdogs
- IPX RIP broadcasts
- IPX serializations
- Sequencing

Even if these frames were tagged by type, format, or other characteristics and removed from the stream, this would create a secondary problem. These frames are sent because the sender anticipates a specific response, as shown by Figure 6.17.

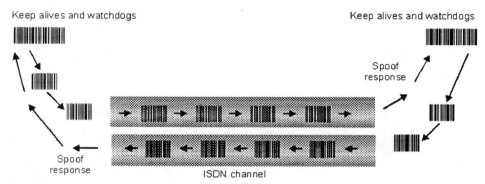

Figure 6.17 Watchdogs and keep alives may be directed over the WAN.

Lack of response to these watchdogs and keep alives initiates a flood of more requests or a cessation of the connection. That is not what you want in general over an WAN link where fees are charged by the minute whether the active circuit is idle, has some sporadic traffic, or is fully utilized. However, a more complex filtering process can remove these frames from the WAN stream and create a dummied-up response. This filtering processing and corresponding reply by the router on the near side of the WAN link is the spoofing process. It is shown in Figure 6.18.

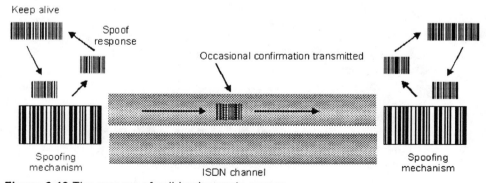

Figure 6.18 The process of call-back remote access.

Spoofing can cut line traffic by 50 percent. More important, an idle ISDN line can be "active" without actually maintaining a point-to-point connection, thereby saving measured business minute rates and long-distance charges.

Caching

AirSoft is the first of a number of software vendors providing specialized network caching. This product is specifically designed for DOS-based client/server processes and is designed to improve file system performance by interacting with the DOS INT21 and INT2F hardware calls. These correspond to delete, rename, open, close, read, lock, unlock, move next, move last, and some other common I/O calls through fetch and prefetch caching hits. Caching these calls represents from 200 to 400 percent improvement with database operations. This service is also pertinent for CC:mail sessions, logon, starting a mail application, and reading messages. The primary savings is that the latency for a round-trip message which no longer must be sent over a WAN is reduced from 600 to 800 ms to 200 ms. Analogous services are also appearing on web servers for caching of page hits and other commonly accessed information. This can reduce Intranet paging times and improve access to commonly access pages from distant sites.

Connections on Demand

One of the benefits of ISDN is its support for a connection on demand and also for bandwidth on demand. Since it takes less than 1 s to establish an ISDN link versus up to 10 s for analog modems to dial and synchronize on speeds, ISDN is practical for establishing connections for directory synchrony, credit card processing, and e-mail delivery.

In addition, you can add bandwidth with ISDN as needed. Although most vendors communicate the concept of bonding as the process of using two B channels simultaneously, you should also think of it as the process of adding and dropping channels as needed. H.384 supports six ISDN B channels, and many videoconferencing products support this level of bandwidth. Bonding is not just for BRI, and the bonding protocols are being designed for other eventualities and applications.

There are several methods of BRI bonding two B channels. Obviously, there is the primitive PPP over IP bonding and there is the MLPPP or MP technology. The goal with this protocol is to resolve some of the proprietary implementations in many ISDN terminal adapters. For example, you can send frames down each B channel and alternate the channels as the buffers are filled. In addition, you can share bytes between the channels and synchronize the transmission at the other end. This is illustrated in Figure 6.19.

The difference between bonding by filling the channels with frames rather than alternated bytes is that the overhead in the B channels is less and the process is less restrictive so that the bonded channel can provide bandwidth closer to 128 kbps. Although, in theory, both methods conglomerate two B channels for a total available bandwidth of 128 kbps, some of that bonded channel must be used for overhead in managing the data streams. NetWare 4.x is primarily used for multisite and multiserver networking. Network Directory Services (NDS) is a unified database registry for the WAN designed to eliminate the complexities of manually performance server-by-server directory synchronization. Changes in NDS now include intelligent distribution of changes only and support for on-demand connections through ISDN.

Changes to a global directory or changes in a single site's directory, which means that every other site must be updated to reflect these changes; that is, synchronized, the traffic over the standard WAN and LAN interconnects can bring the network to a halt. Instead, out-of-band, on-demand ISDN connections enable management replication without the traditional hit of network overhead.

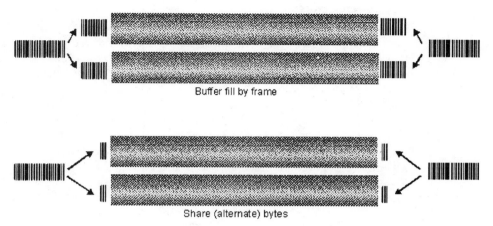

Buffer fill by frame

Share (alternate) bytes

Figure 6.19 BRI hardware bonding techniques.

Residential Connections

Residential connections based on ISDN BRI can replace existing phone lines. You can get your employer to pay for the line or split its cost because, in general, a 2B+D represents all the phone service a family needs. This is true of course unless you have a half-dozen teenagers or even just one. A teenager can tie up a modem line and a phone line at the same time. However, the 2B+D can provide analog and digital services for both voice on an analog phone line, fax on the phone line, and digital dial-up services. Basically, you establish both SPIDs on the B channels for voice and data. You get at least three directory numbers (DNs) and map them in this way:

- DN 1 for voice (mapping to B1)
- DN 2 for fax (mapping to B2)
- DN 3 for Bulletin Board Service (BBS) or ISDN (mapping to B2 and then B1)

You can use the voice and fax on B1 and B2 concurrently when the NT-1 or terminal adapter supports that technology. However, you cannot use both B channels for data and also use the phone. The line will be busy, and callers trying to reach you will also get a busy signal. You want to map DN 3 into both B1 and B2 so that callers with ISDN can get full 112- or 128-kbps service. This also precludes voice or fax during the data call. However, if you create a data connection on just B channel, you can receive faxes or phone calls on the other line; whether you receive phone or fax is a function of the configuration for the data call. If you are using the B1 channel you can still receive faxes.

If you are using the B2 channel, you can receive phone calls. If you represent a busy office and also want the ability to receive at least one phone call, one fax, and have bonded ISDN service concurrently, you will need two BRI circuits and the hardware that can route voice calls to an analog answering machine or ISDN phone with call hold, call waiting, or a voice mail facility when you are otherwise occupied. However, remember one important aspect of ISDN service. The ISDN line does not provide its own power. Thus, an independent power supply is always required.

Mixed Architectures

ISDN in all forms, BRI, PRI, and broadband, is a protocol which runs on top of the delivery pipe. ISDN might be available as a sub- or full-rate 192-kbps dedicated line known as ISDN BRI, or it can be delivered *on top of* T-1, E-1, or T-3. ISDN is delivered on top of Synchronous Optical Network (SONET) or ATM, which itself is a protocol which runs *on top of* SONET. Likewise, as has been described elsewhere in this book, X.25 and frame relay can be transmitted over ISDN lines. X.25 packet data is routed into the great X.25 packet mail delivery system represented by a cloud.

While some host-based organizations use X.25 for remote site and site-to-site connectivity, delivery is always through the cloud, and the connection to the cloud is usually a dedicated connection, although it can also be provided with a switched connection. The connection rate is fixed at 1200 baud or 9600 bps. While 56-kbps dedicated lines are available almost anywhere, they are expensive, more expensive by the mile, and often serve just the single purpose of providing a point of entry for X.25 packets. When the traffic level is bursty or sporadic, ISDN can provide 16 kbps (on the D channel) or even 64 kbps (on a B channel of BRI or PRI) as needed on demand for connection to the X.25 network. X.25 connectivity with a 64-kbps B channel is very common in Europe because of the advanced state of ISDN and digital services.

Likewise, frame relay data packet transmission is very similar to X.25 (although the packets are much larger). Connections are supported as either dedicated or switched (virtual), but frame relay often is provisioned with a base connection rate and supplemental bandwidth on demand to match load characteristics. Both X.25 and frame relay are supported over digital pipes, and ISDN is perfectly suitable for either packet-data protocol. In general, you encapsulate these packet protocols inside ISDN. This is illustrated in Figure 6.20.

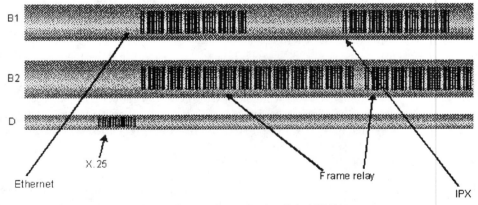

Figure 6.20 Packet-data encapsulation (tunneling) within ISDN.

Encapsulation of IP within X.25 is often referred to as X.31 data packet services because it is a common delivery scheme. Request for comment (RFC) 1357 defines IP encapsulation with frame relay. I mention this definition and the technology for it because often the uniform availability of ISDN, X.25, or frame relay may prevent simple construction of a single protocol wide area network. However, X.25 or frame relay may be already in use, widely available, and less costly, or the crossover price points may favor it, so multiple site linkage is best handled by a mix of X.25, frame relay, dedicated lines, and ISDN. Router and central office switching technology make it very possible to mix and match these protocols and delivery methods. Since IP is the standard for most Internet access, e-mail, and web site access, it is important to realize that the delivery of IP information need not occur just through dedicated lines, ISDN connections, or dial-up analog lines. If a corporate site uses IP for Intranet or Internet connectivity, and you can show the need for a constant connection, tunneling with various protocols is very effective for maintaining that constant link. Figure 6.21 shows several layers of protocol tunneling. Notice that separate channels are required for different virtual connections.

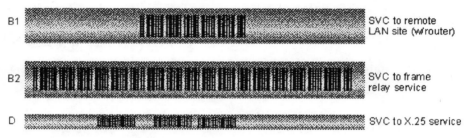

Figure 6.21 Protocol tunneling with IP, X.25, and frame relay.

The remote site Intranet and Internet connectivity with these mix and match protocols is illustrated by Figure 6.22. Service is handled by a multiprotocol router at the end side of the linkage.

Why ISDN for Packet Data?

Defined in the early 1970s, X.25 packet-switching protocols are accepted as a worldwide communications standard. Information is divided into small packets, each of which contains a complete address as well as sequencing codes to maintain the packet order. X.25 packets are not always delivered in sequential order. Packetizing is accomplished by a Packet Assembler/Disassembler (PAD) at the sending end. At the receiving end, another PAD accepts the transmission, puts it in correct order (packets can travel many routes through the network and may not arrive in the order they were sent), and forwards it intact to its destination.

It's possible to link directly to services or locations on packet networks around the world. Since the X.25 protocol was originally for often noisy and interference-prone analog lines, PADs also perform a broad range of error-checking and error-correction functions. If any packet is not received correctly, the receiving PAD signals for retransmission until the packet comes through correctly.

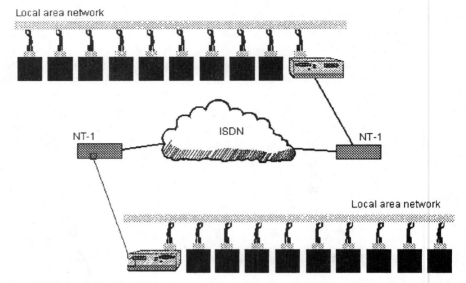

Figure 6.22 Multiprotocol Internet and Intranet tunneling with IP, X.25, and frame relay is transmitted through a multiprotocol router connecting remote sites.

The result is exceptional accuracy through lines that were, and in many areas of the world still are, less than perfect for data transmission. Today, many of the functions for which packet switching was designed, the transfer of large data files, LAN-to-LAN interconnection, and the like, have been taken over by faster, more modern communications technologies, including dedicated lines, frame relay, and the circuit-switched data connections of ISDN itself. The coupling of X.25 to the power of ISDN virtually eliminates the traditional limitations of packet switching. The combination offers:

- Simplified host connections
- Universal user connections

A single B channel of ISDN can support hundreds of user X.25 connections to a bulletin board, information service, database, or transaction host system. And while user transmissions can originate through either dialed modem links or permanent ISDN D channel connections, a host system's permanent, fully digital B channel link to the packet network totally eliminates the need for modems at the host site. Because most ISDN telephones and many ISDN terminal adapters contain PAD capabilities, these ISDN telephones, PCs, and data terminals become receiving locations on the worldwide packet network. The connection does not need to be dedicated because ISDN supports the Switched Virtual Circuit (SVC) connections. For example, it is no longer necessary to send messages to a mailbox on a host or information server. Rather, the message can be sent directly to the ISDN telephone or PC at the receiving location.

Each location is permanently connected. Every D channel terminal has its own host ID or data network address. It means that X.25 e-mail and messages can be sent directly to another location rather than to a mailbox for later retrieval.

If you need confirmation of the importance of this transition from analog to digital that prefaced Chapter 2, you need to look no further than Microsoft's ambitious computer telephony integration strategy. The vendor is adding voice-modem, digital simultaneous voice-and-data (DSVD), and ISDN integration to its development products and operating systems. You might note the WWW reference for downloading the Windows 95 drivers for ISDN. Although part of this transition is simply the shift from analog data communication to digital formats (with the Unimodem specification), the transition is also represented by the integration of PBX services with PCs and servers. Since ISDN provides integration of voice, data, and packet as part of the core services, it represents a first line in this strategy. The hierarchy of the Unimodem V standard is illustrated in Figure 6.23.

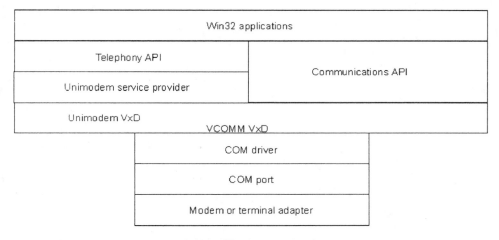

Windows CTI drivers and interfaces

Figure 6.23 Microsoft CTI stack.

ISDN is not the only integration effort to support voice and data. Radish Communications applies DSVD services with its VoiceView specification for 14.4-kbps data with a voice overlay. Creative Labs is also supporting this VoiceView standard. The forthcoming ITU DVSD standard will support 28.8-kbps data with a voice overlay. More vendors are implementing VoiceView because hardware is available. Just for reference, Figure 6.24 illustrates the current Windows support for VoiceView.

However, VoiceView is not DSVD and not a public standard. Because there are competing implementations of DSVD, the lack of universal standards puts this technology behind ISDN. If you review the services in the last figure, you can clearly see that all these services are available with ISDN, at greater bandwidths, with autoselection, and they scale to multiple lines and even greater bandwidths than just ISDN BRI. In addition, DSVD is not fully enabled by development tools. On the other hand, Telephone Services Application Programming Interface (TSAPI) maps into ISDN with some patches, and it maps into ISDN with the Windows 95 ISDN driver set. In fact, some ISDN terminal adapters (such as the I-Courier or the 3Com single-user and bridging units) are configured with standard Hayes modem setup strings. All European and most American ISDN hardware can be supported with the CAPI command set.

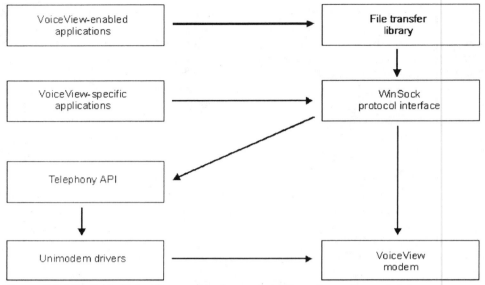

Figure 6.24 Radish's VoiceView protocols support alternating voice and data.

ISDN Bridging

If your organization is using ISDN as an on-demand or pervasive link from an Internet Service Provider (ISP) to the corporate site, you will need a domain name system (DNS) to assign an IP address for each user (or user-spawned process) to establish a link over the local area network, through the communication server, to the ISP web or mail server.

Operating System	TCP TTL	UDP TTL
AIX	60	30
Pathworks	30	30
BSD 2.1r	64	64
HP/UX	64	64
Linux	64	64
MacOS	60	60
OSF/1	60	30
Solaris	255	255
SunOS	60	60
Ultrix	60	30
VMX/Multicast	64	64
VMS/TCP	60	64
Windows 3.1	32	32
Windows for Workgroups 3.11	32	32
Windows 95	32	32
Windows NT	32	32

Figure 6.25 Default TCP and UDP stack settings for effective TCP/IP communications.

A fundamental enterprise network limitation within any bridged, routed, or switched environment is the functional efficiency of TCP/IP. One of the nasty connectivity problems that arises with LAN-to-LAN connectivity and this requires hosts, servers, or bridging, routing, or gateway devices, is the limitation in intermediate device hop counts. For example, many operating systems assign the TCP/IP parameters for the Time To Live (TTL) at default sizes of 12 or 16. Since a packet can easily hop between 40 addresses between source and destination, the default IP Time To Live parameter should be set to a minimum of 64. Realize that these settings may vary for TCP and the User Datagram Protocol (UDP) stacks. Both need to be reset or at least verified. Figure 6.25 specifies *default* settings for typical operating systems:

You set the TTL parameter in Windows 3.x by adding this line to the SYSTEM.INI file:

```
[MSTCP]
defaultTTL=64
```

The default in Windows 95 is 32 hops for TTL and UDP. Since Windows 95 migrated all the .INI settings into a registration database, you will need to access the information through that database with the registration utility. Run RegEdit, locate \Hkey_Local_Machine\System\CurrentControlSet\Services\VxD\MSTCP. Select Edit and alter the setting to look like one of these lines:

```
defaultTTL 01 00 00 00
defaultTTL 64
```

The first entry is in binary or hexadecimal format. The second is decimal. Make changes based on the current default format. Windows NT is a little bit different. The utility is called REGEDIT32.EXE and it is usually found in:

```
\root\system32.
```

Locate:

```
\Hkey_Local_Machine\System\CurrentControlSet\Services\TCPIP\parameters
```

Select Edit and alter the setting to look like one of these lines:

```
defaultTTL REG_DWORD    64
```

ISDN Routing

Some routing features become cost-critical and necessary for improved performance when you switch from dedicated lines to ISDN. These features are not always available in existing routing equipment because it was selected and installed prior to ISDN availability. You may want to rethink the edge routing equipment to make the most of the permanent, on-demand, or emergency ISDN circuits. Figure 6.26 illustrates the features of dedicated or ISDN routers.

Protocol support, primarily for spoofing and multiple protocols, becomes important when you are linking multiple LANs together or connecting remote sites to headquarters. You may need to contend with IPX support for NetWare LANs, NetBIOS with Windows clients, and IP for Internet or Intranet web services and e-mail. That represents at least seven significant protocols that you will want to filter and spoof: NetBIOS, NetBEUI, IP, TCP, User Datagram Protocol (UDP), IPX, and SPX. If you support Ethernet and FDDI, or maybe even Token Ring, you will need to contend with these protocols as well. These add overhead. NetWare, in particular, is the watchdog poll, which you really do not want running over WAN connections.

Feature	Dedicated line	ISDN
Data compression	Desirable	Desirable
Spoofing	Not applicable	Critical
Filtering	Desirable	Critical
Inactivity timer	Not applicable	Critical
Bandwidth on demand	Not applicable	Desirable
Call accounting	Not applicable	Desirable
SNMP management	Desirable	Desirable
Gateway filtering	Desirable	Desirable
Firewall	Not applicable	Desirable

Figure 6.26 Router features for ISDN.

It is easy to overwhelm any WAN bandwidth when connecting LANs with a base speed of 10 Mbps, so it becomes critical to filter traffic that does not belong on the WAN connection. In addition, since ISDN can effect a connection within 1 s, it may not be necessary to maintain an open and active link all the time. Just establish on-demand service, when needed, with as many channels as needed to handle peak loads and bursts, and enable compression for 2 to 7 times the effective throughput. Realize that older routers designed to attach to the Customer Service Units/Data/Service Units (CSU/DSUs) on dedicated lines provided a different atmosphere. The connections to the ISDN terminal adapters may still be provided through serial ports, or the interface may be different and more efficient. At least, there is the potential of being more cost-efficient with ISDN.

Internet Services

Internet access is offered at rates from 2400 bps to T-3 speeds. Bandwidth is a function of cost. Since the Internet is a vast, open mesh of servers, almost any access point provides the necessary inroad to this information superhighway. However, before you become too thrilled with ISDN access to the Internet, realize that performance is not only a function of your local access speed to the Internet but also of how that local access point performs in relation to the rest of the Internet. The local access point, and it may not even be local, is provided by an Internet service provider (ISP). This logical network design is illustrated in Figure 6.27.

If you anticipate using ISDN as the link for web hosting to an ISP such as NetCom, UUnet, or PSInet, consider a dedicated connection rather than a single ISDN B channel. On the other hand, if you will be using multiple ISDN B channels, BRI circuits, or ISDN PRI to provide sufficient bandwidth, BACP is sufficiently robust to handle occasional disconnects. Frankly, a redial time of under a second is less than most web server delays. When the financial crossover is based on utilization levels, go with ISDN only because it is cheaper and more flexible in terms of incremental bandwidths. Realize, however, that unless you have dedicated ports for all your ISDN B channels, it is unlikely that most ISPs can actually achieve 128 kbps because the other end may not manage the two B channel connections and provide connection into the very same server.

Large providers, such as PSInet or UUnet caution against expecting dedicated service with a dial-up line when each has several hundred or a thousand different servers. The most these services guarantee is 64 kbps, and random phone calls, even placed at the same time with BACP or MP+ protocol easily connect into different servers.

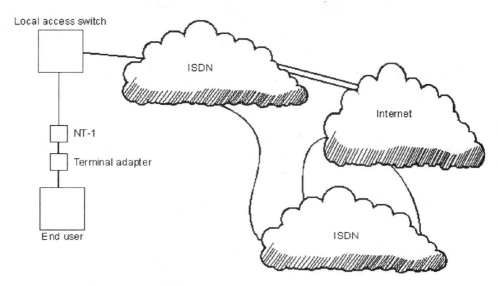

Figure 6.27 The logical Internet design with service provider access points.

There is another important issue for ISDN (as the plumbing) for Internet service. Although the term *committed information rate* (CIR) is more related to frame relay services and bandwidth-on-demand connectivity in terms of providing a minimum operational bandwidth, it has significant bearing on your comfort level with ISP services. For example, you need to know how many ISDN ports the ISP supports and the expected load level at those ports during your hours of need. It is possible that ISP will oversubscribe ISDN access, with the result that you cannot get through to the Internet server. It is also likely that ISPs will not commit service resources to the ISDN lines, with the result that when you do get a connection on one or more B channels, you still do not get a response time in aggregate faster than an analog 9.6-, 14.4-, or 28.8-kbps connection. Even when an ISP dedicates an ISDN port or multiple ports to your organization (by installing a unique phone number and a paired terminal adapter), these parts can still be overcommitted and oversubscribed. Get used to busy signals on ISDN lines.

The uptime for a dedicated line approaches 98.4 percent on the various U.S. public networks, whereas ISDN uptime approaches 97.4 percent. This difference of only 1 percent, while not disturbing, does not include connection-dropped information. The national and local carriers cannot provide call-dropped information as a percentage of calls made because they do not track or disseminate abnormal termination yet. However, a sales person at UUnet suggested that a Switched 56 or dedicated connection would be more reliable. In digging deeper into this assertion by the UUnet representative, the real issue seems to be that a dedicated web hosting costs 4 times more than dial-up links and there is a higher initial "installation" fee. Look at the financial crossover points and choose the WAN service in conjunction with the ISP service that shows a lower cost.

Conclusion

This chapter showed the use of ISDN for dialing into the Internet for one user or many users, the use of ISDN as an inbound communication protocol, and also how ISDN can connect remote LANs together. Chapter 7 shows the advances in Computer Telephone Integration (CTI) and how this relates to ISDN and how ISDN supports CTI services.

Digital Phones, PBXs, and ISDN

Introduction

This chapter describes the integration of telephony with ISDN digital services and how this can be applied to migrate Centrex and older Private Branch Exchange (PBX) services into the new digital wave. While Chapter 2 explored some of the applications for ISDN telephony and Chapter 3 defined the Common-ISDN-API (CAPI) programming interface, and Chapter 4 described the physical process of connecting data and voice services, this chapter presents the details about digital phones, phone service, and PBX facilities. The information in this chapter shows the transition from mixed PBX and analog phone services to a centralized focus for phone service through ISDN.

The fundamental difference between ISDN and analog plain old telephone service (POTS) is that many of the services added to POTS are bundled into ISDN at the ISDN base service cost. Since analog can call into an ISDN line and be handled appropriately, the selection of ISDN represents a financial decision. Figure 7.1 shows ISDN in use for corporate telephony.

The next illustration is more explicit about the possibilities of integration of ISDN with standard POTS services. ISDN does not limit service in any way; in fact, you are typically adding capabilities (such as Caller ID, hunt groups, and conferencing). Specifically, Figure 7.2 shows how the analog lines and other digital services can be routed into ISDN loops and devices. You should also note that ISDN can carry other signals internally after the NT-1, NT-2, or hub as standard analog calls. The conversion is simple.

The impetus for ISDN integration should be based on financial reasons. Just because ISDN is digital and somehow "digital" is better is an insufficient justification. If you are using 800 numbers, inbound and outbound telemarketing with Caller ID and Called ID tracking, these services typically cost more on T-1 and analog phone lines because these features are added at the telco switch. ISDN service may provide a cheaper alternative. If your organization is considering adding extra lines specifically for Internet connections and plans to add devices to the PBX to convert the key system or other in-house digital

PBX so that it flashes signaling into something functional with analog modems, you will discover that the conversion box costs about $500 to $1200 per line.

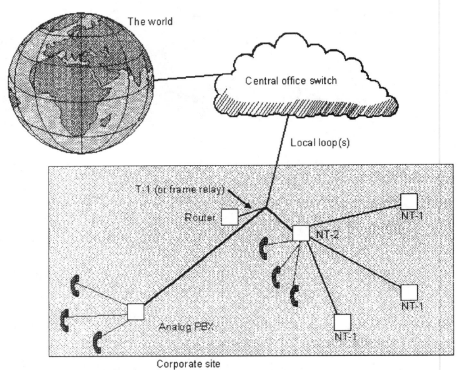

Figure 7.1 Integration of ISDN with standard POTS.

The real alternative to that is wiring around existing PBXs with direct single-purpose phone lines. This circumvents other uses for these lines and bypasses central PBX management. If traffic and line use are managed and augmented as needed through the PBX, there is no load-leveling on the new single-purpose lines. In other words, if you add 120 lines for modems and do not create a modem pool on a local area network, those 120 lines provide no utility when they are not in use. You cannot recycle them during the day because they are wired into specific end-user offices. While a modem pool could at least concentrate them and reduce the number of lines potentially needed by 60 percent or more (due to usage distributions), the lines cannot be allocated for voice from the PBX or bonded after hours for high-speed Local Area Network- (LAN-) to-LAN data replication or backup. Although this may not seem like an issue for a small office with twelve analog phone lines for voice, two analog lines for fax, and ten analog lines for modems, consider that ISDN consolidation could mean that only seven analog channels are required for voice, and three digital data channels on a hunt group (with a different number) could handle all inbound and outbound fax and data service. Some ISDN PBXs/hubs can differentiate between data, fax, and voice calls with nearly 100 percent accuracy, so you could consolidate all phone lines to nine lines [five ISDN Basic Rate Interface (BRI) lines or one Primary Rate Interface (PRI) line with lots of spare capacity] into a single access number. The voice, data, and fax number would be the same.

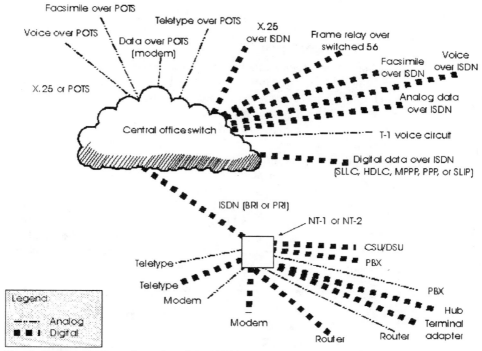

Figure 7.2 Integration of services into ISDN.

In addition, you may discover that the cost of reprogramming older PBXs to support the North America Numbering Plan or adding new features is not effective. The newer PBX equipment, especially when its based on the Systems Control Architecture (SCA) bus or is PC-compatible, is inexpensive relative to that reprogramming cost. You may be able to upgrade the entire phone system, add the conferencing features, call forwarding, and connectivity with the PC for less cost than struggling with an older key, digital (connecting to analog T-1 service), or Centrex system. If you can demonstrate that integration with client/server sales and marketing or customer accounting systems speeds up lookups, reduces billing errors, and makes your organization look very professional, these benefits can justify hardware and local loop upgrades.

For example, consider the script for an inbound customer service call:

```
<Caller ID: Unknown Number>
<Called Number: 888-456-7321>
```

An explicit Caller ID was not provided because Indiana is serviced through MCI and the Indiana legislature has not subscribed to the mandatory ID pass through; therefore, a specific call-in number was assigned to Indiana. The service representative can ask, "What is your zip code in Indiana?" to narrow the possible list of customers to one. In fact, you can use an automated attendant to query the caller entering a work or home phone number with touch-tone codes. This can be passed from the ISDN hub to the representative's desktop application through the integrated ISDN and computer network. Of course, if you get a real Caller ID or a valid number from the caller, you can look up the customer by company, name, prior orders, hotel arrival time, and other billing events such as the claim number of products.

New Wave Telephony

The major concept of computer telephony integration is summed up by increased efficiency in workflow. An inbound telephone call provides the *calling* and the *called from* number. Both pieces can be relevant when an organization establishes multiple phone numbers, 800 or the new 888 service numbers, for inbound calls as described in the prior section. These numbers may represent a region because the same 800 number was not available nationwide or worldwide, or it may indicate classes of customers, such as those who are calling for presale information or those requiring technical support. For example, a tour operator may want to provide a special phone number that is only available in the destination location for the tourists who are traveling. They may get sick on holiday and need to leave by plane immediately or may be unable to travel for several days and need to reschedule a later flight. In any event, calls can be routed more quickly without human interaction to the appropriate person. As an example, Figure 7.3 shows technical support calls routed to me because they were dialed in on the special 800 phone number established only for technical support.

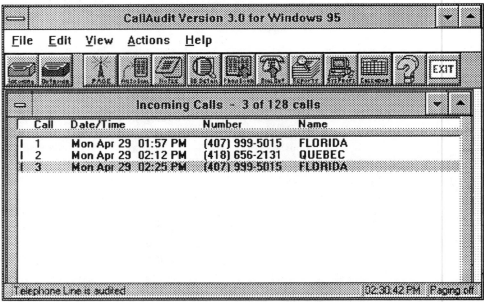

Figure 7.3 Integration of Caller ID ISDN services into workflow.

This application works with ISDN (either of the two BRI lines or all of the PRI lines), but it does not track the duration of call. I think this is useful information for tracking performance of sales representatives, customer support people, and generating compiled information about process costs. This screen does not show that the calls were contiguous. The gap between calls was the duration of each prior call. It is interesting to note that the area code for call number 3 is incorrect and should be in the (813) area code. I mention this because Caller ID is not perfect or thoroughly foolproof. The phone company states that the Caller ID is supplied by their switches and cannot be diddled with or faked. In any event, the matching Caller ID can be used to pop up a detailed screen about the caller. Figure 7.4 shows pop-up information for caller 3.

Incoming Call Details

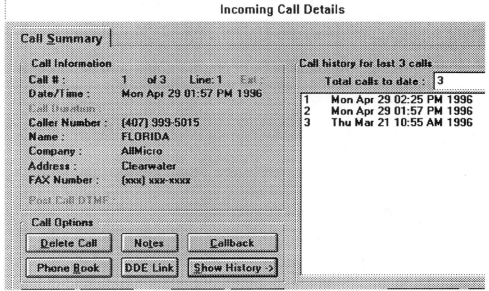

Figure 7.4 Caller ID supplemental information as delivered.

Note that this screen shows the history of the last calls. This can be useful, allowing you to be more friendly and personal with the caller. For example, I can ask if the product has been working well since the last call (on March 21) when someone solved an installation and reporting problem with a software product. This can be followed by "how can I help you today?" You might also note that the Caller ID can include just the phone number or supplemental information. Also, this capture does not get all the supplemental information in the correct fields. It does show what was sent, however. With any software product, you should be able to copy, paste, and augment this basic information, as shown in Figure 7.5.

If the status of the call showed that I missed it, or if I wanted to directly return that phone call, I would enable one of several screens. Notice that I could page a person directly by beeper or alphanumeric pager and even have the program automatically, through a Windows Dynamic Data Exchange (DDE) or Object Linking and Embedding (OLE), activate the page with an alpha message and the phone number part of the Caller ID. Figure 7.6 shows a call-back to the number from the Caller ID. However, in this instance, the area code is still incorrect and would yield the wrong results. The message is that the process should include some steps to minimize errors. In this example, a lookup at the information matching this Caller ID will provide the correct phone number. This problem is rare but not insurmountable as the prior pop-up screen demonstrated. I could also link the fax number to WinFax or NetFax to send back a technical support white paper or generate a return materials authorization for repair.

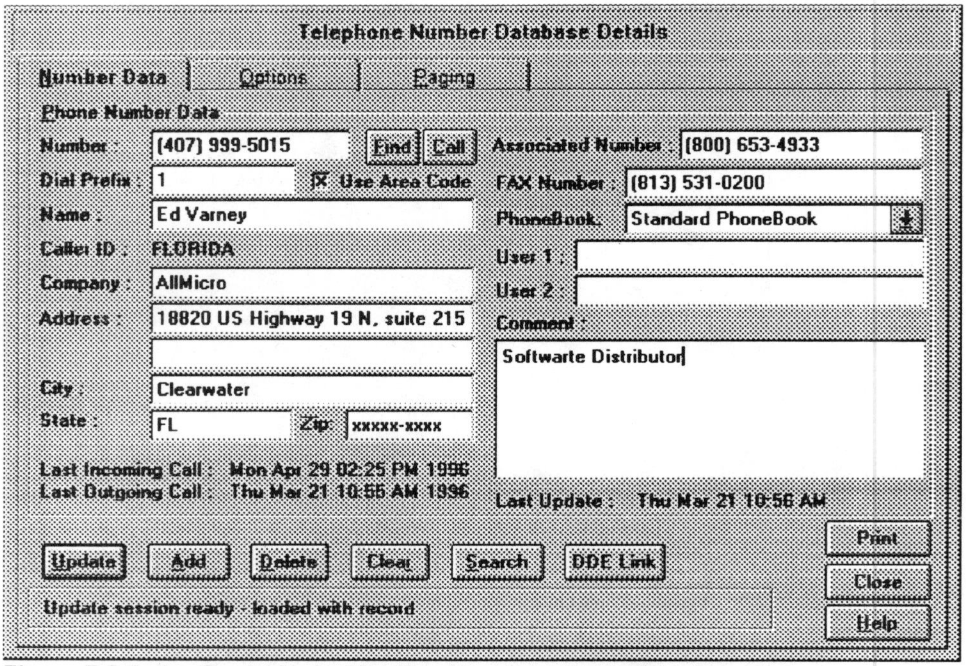

Figure 7.5 Caller ID matched with user-entered supplemental information.

Outbound telemarketing or call-backs based on voice mail can tap into the phone number capture from an accounting system. Although the voice mail stored on a hard disk may include a phone number, the other channel of information can be passed to the hard drive as well. This phone number can be passed to the telephony application and hence through the phone system as the calling number. The Caller ID is likely to be more accurate than the number in the voice mail message, certainly more intelligible. When there is a reason for security, any discrepancies between voice mail numbers and Caller ID numbers are useful pieces of information. It may point to a sloppy error that the caller may not have intended, or it may be someone's home phone number which they did not intend to provide. Although, personally, I am all for privacy and as much as possible, you need to know what services are available to your organization and how it can benefit from that. Whether you want to apply these services is a business decision with financial, legal, ethical, and moral ramifications.

You also need to realize that as you build databases on customers with account numbers, social security numbers, phone numbers, credit card account numbers, expiration dates, and personal histories, you become responsible for the integrity, use, and misuse of that information. Caller ID and big-brother databases are not "free" resources; with their use comes all the limitations and obligations of responsible ownership. In any event, outbound telemarketing can use phone numbers from inside a database to activate a phone call. Why have a person key in a phone number to the phone or into the database when that information can be captured and applied without minimal error? Although this process can be as easily implemented with analog as well as ISDN lines, the advantage is through process and workflow integration. A phone number represents both an entry key into the ongoing workflow and a starting point for new work. When you think of communications, both inbound and outbound, as an integral part of the process workflow, you will begin to see the efficiency to be gained from integrating telephony with computer processes into ISDN.

(CTI2.BMP)

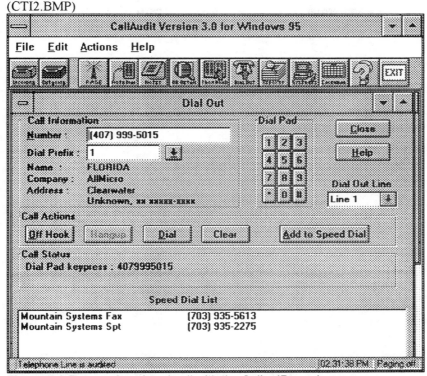

Figure 7.6 Return of a missed call with the Caller ID number.

ISDN Hub Systems

When you want to integrate ISDN services into operations so that call appearances can be routed to the correct person, you need to consider an ISDN PBX or hub. For example, suppose you rotate people who handle emergency calls on an hourly basis. This can be programmed into the PBX, and while the incoming phone number is always the same, who it gets forwarded to is a function of time and whether the primary person is already on the phone. That person can be flashed or the call can be directed to a backup person.

These services, while standard to some degree in most PBX systems such as Merlin, Rolm, and Honeywell, are a function of the software in the PBX or hub. Typical terminal functions in Electronic Key Telephone System (EKTS) include:

- Adjustable tone generator for custom ringing
- Group dialing
- Call forwarding and conferencing
- Call or data encryption
- On-hook dialing
- Call program monitor
- Call duration monitor

- Programmable functions
- Hold music and background music
- Call detail recording
- Bandwidth Allocation Control Protocol (BACP) or BRI breakout and distribution
- Voice mail
- Autoattendant (programmable scripts)
- Interactive voice response
- Connectivity and management for videoconferencing
- Traffic reporting
- 911 (emergency locator) pass-through
- Telephony Application Programming Interface (TAPI) and Telephone Servers API (TSAPI) compliance for computer workstations
- Automatic call distribution

There are other features that are supported as software-programmable functions, such as barge-in, camp-on, parking, night transfer, route selection, off-hook voice announce, station lock, toll restrictions, and privacy. The units typically look like wall-mounted PBXs. They are often flat units that mount inside a telecommunication closet on the wall. Increasingly, however, units are rack mounted because the wiring must link into LANs, file servers, and individual stations. Slot devices can connect to BRI, T-1, PRI, or other telephony services or host analog modems, a frame relay access device (FRAD), an X.25 packet assembler/disassembler (PAD), and terminal adapters. A standard rack-mounted unit is pictured in Figure 7.7.

The cards often support different services such as analog T-1 for compatibility with phased migrations, frame relay, PRI local loops, BRI local loops, and possibly even Broadband ISDN (B-ISDN) and Asynchronous Transfer Mode (ATM) services. Figure 7.8 points out that some ISDN PBX systems are scalable. Although it is obvious with the rack-mounted systems that you do not have to fully populate all the slots, the limits of this DICA-brand unit are 8 BRI lines, whereas the Cortelco system can support a mix of both BRI and PRI lines up to a maximum of 24 local loop (that is, ISDN circuits).

Notice the 25-pair D-plug connectors for the T-1 lines. ISDN PRI is often delivered on optical fiber and the individual channels are broken out with a 25-pair hydra, as described in Chapter 4. There simply isn't the space to handle the density of individual pairs for each channel. Even when such a hub creates a pass-through for an analog service T-1, the wiring is likely to be a four-pair multiplexed signal or the channeled signals on the hydra connection. When some of the slots are used for banks of modems (often 4 to 24 modems or terminal adapters per card), the complexity of external wiring is eliminated. An incoming or outgoing call is identified by type (that is, voice, digital data, analog data, or video) and routed from the end-user through the appropriate modem or terminal adapter through to the particular channel on the T-1 line. This bypasses any need for clumps of external wiring. As mentioned previously, a free B channel or several B channels are allocated as needed; there is no separation of analog from digital channels and this creates a unified, and thus simpler, wiring infrastructure.

Figure 7.7 Rack-mounted PBX and ISDN converter (courtesy of DICA).

Figure 7.8 Chassis-mounted PBX and ISDN converter (courtesy of Cortelco).

ISDN Terminal Equipment

ISDN phones are called *terminal equipment*. These are not to be confused with *terminal adapters*, which are the connection between the ISDN NT-1 and a computer device. Do not misunderstand. You can plug any analog handset (that is, a standard POTS telephone), even an old-fashioned rotary dial phone, into the phone port of most internal or external ISDN terminal adapters. In fact, the BitSurfr Pro even provides two ports so that you can simultaneously utilize each B channel with analog devices. You can string a fax, an analog data modem, an answering machine, and phones on daisy chain on each channel. The NT-1 handles the call appearance for each device based on unique directory numbers. However, you need phone equipment that is ISDN-compatible to provide the types of features described in the ISDN hub systems. Typically, ISDN terminals support these features:

- Speakerphone
- Display of Caller ID and Calling ID
- Voice message and message indicators
- Wireless handset compatibility
- Hearing aid or headset jack

However, AT&T and some other vendors provide home office and small office phones which plug into a standard ISDN jack without the need for the centralized ISDN hub or PBX. These phones provide standard phone service and some of the ISDN features. ISDN phones can even communicate with a PC for Caller ID, call routing, and pass-through of call information into call management software such as ACT. However, you cannot pool ISDN lines, create call appearances, or provide call forwarding and conferencing calls without actually creating an external connection through the local switch. This often means per-minute billing. These phones are illustrated in Figure 7.9.

More robust systems designed for somewhat larger organizations pack the hub electronics into a reception station (shown on the bottom left of the picture). This architecture does not typically support ISDN data services. The station actually terminates the PRI lines, and 25- or 100-pair wires lead back to a wiring closet. These lines are broken out to individual offices. Substation phones tie back to the receptionist and can replicate those functions when the receptionist is away from the station, overwhelmed, or forwarding a call to someone who might better know the disposition for that call. A substation is shown on the left of Figure 7.10.

The ISDN hub is essentially a switchboard that enables all the features when paired with compatible ISDN terminals. These Tadiran terminals, shown in Figure 7.11, include most of the features previously listed. However, full functionality is also a function of proper EKTS service provisioning, as described in Chapter 3.

New Tricks for Old PBXs

A number of products interface with existing PBX systems to provide ISDN services without replacing or augmenting the PBXs. AT&T Global Business Communications Systems builds a product called Definity Multimedia Link. Promptus Communications manufacturers a switched digital server (Oasis) that can connect into any PBX for ISDN pass-through. Of course, digital PBXs from Intecom, Siemens, Rolm, and Fujitsu have supported both BRI and PRI connections for several years. The connection for the

Multimedia Link and Oasis provides a B channel breakout for videoconferencing, data services, and other ISDN-specific services. This is illustrated in Figure 7.12.

Figure 7.9 ISDN "terminals" (courtesy of AT&T).

Figure 7.10 Hub-based ISDN terminals (courtesy of Tadiran Communications).

Figure 7.11 Full-featured ISDN terminal with call appearance, special service functions, and integrated serial port (courtesy of Tadiran Communications).

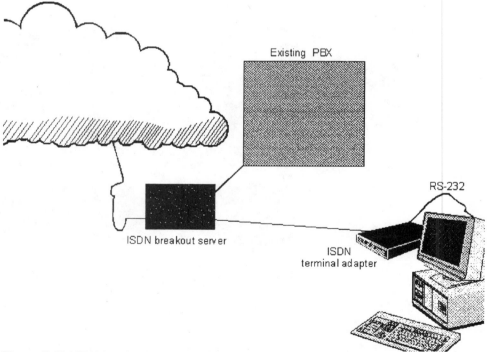

Figure 7.12 ISDN breakout for dedicated B channels and connection to an existing PBX.

However, realistically, this type of service is a method to extract some channels from what otherwise would be an analog T-1 for use in videoconferencing or ISDN dial-up services. The termination of each ISDN line still represents a dedicated wire through the facilities to the conference room or end-user. This is in addition to the lines for the PBX system. In no way does this breakout augment the PBX to support Caller ID, call forwarding, conference calls, flash messages, voice mail, or pooling of channels for voice, data, fax, or any mix of services as needed.

Conclusion

This chapter described the ISDN voice integration and the breakout of ISDN services for data applications and videoconferencing. In addition, it also described the transition from mixed PBX and analog phone services to a centralized focus for phone service through ISDN. The next chapter, Chapter 8, describes videoconferencing over ISDN and the integration of videoconferencing services with the LAN and networking.

8

Videoconferencing over ISDN

Introduction

This chapter describes videoconferencing over ISDN and the integration of videoconferencing services with the Local Area Network (LAN) and networking. I mention videoconferencing over the LAN in this chapter because the IsoEthernet standard supports ISDN and because the integration of Wide Area Network (WAN) and LAN facilities with telephony is available now. There are a number of products designed strictly to provide video telephone services over the WAN via ISDN. These single-purpose products, typically a computer box, camera, and monitor, are described in this chapter. In addition, there are a number of products that integrate into preexisting personal computer equipment. These are more robust and provide videoconferencing services that also support whiteboarding, desktop conferencing, desktop and application remote control, and collaboration.

Whiteboarding is the process of drawing on a whiteboard and transmitting that image in color or black and white to all videoconferencing participants. By the way, this can also be achieved with multiple-party (or *multiparty*) conferences by establishing a camera for the participants and another camera aimed at a blackboard, whiteboard, overhead screen, or other display medium. Although it is probably better to connect computer presentations directly into the conference signal, you can also aim a camera at the overhead screen for that type of presentation as well. Document annotation is another feature similar to whiteboarding and is sometimes called *shared whiteboard application*. It is more like a desktop conferencing. *Desktop conferencing* per se does not include sound and images of people but rather includes typed "chat" events and transmissions of images and files. *Desktop remote control* has been around a long time and allows a remote user to manipulate a local computer as though that person were local. When that is combined with collaboration, the remote control can be overridden or augmented by the local user. All these features are combined into some of the higher-end videoconferencing tools and also some of the more available tools. Some videoconferencing and collaboration tools run over LANs, WANs, and the Internet at the same time and are thus very flexible. When an Internet connection is coupled with an Internet Service Provider (ISP) providing ISDN dial-up services or dedicated lines (with a reasonable committed information rate), the ISDN Internet link-up is as viable a connection as a direct dial-up ISDN link.

The primary justification for videoconferencing is the savings in travel costs, and travel time and the increased communication possible with sight as well as sound. The improvement in employee quality of life is a more difficult incentive to validate with financial measurements but is nevertheless a very valid justification; most people prefer to be with their families at home instead of traveling. In addition, videoconferencing is about as developed as word processing was when I assembled an Apple (I) computer. Just as word processing has grown to include templates, mail merge, OLE integration with spreadsheets and databases, redlining, and multiple-person drafting, videoconferencing is becoming enabled with other applications and integrated with remote desktop control, whiteboarding, and collaborative procedures. Examples are automated transcription, multilingual transcription, and remote training. One major effort includes integrating Lotus RealTime Notes with Intel ProShare as shown in Figure 8.1.

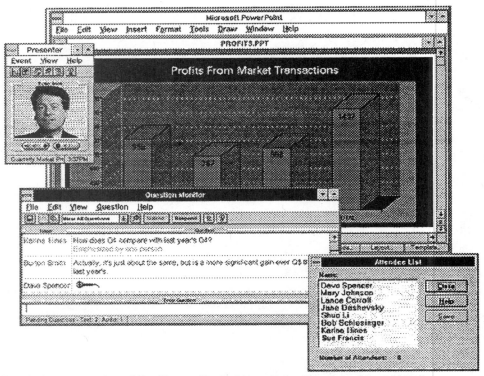

Figure 8.1 Integration of ProShare with desktop applications (courtesy of Intel).

Videoconferencing is a vertical application that can compensate for lack of personal contact with colleagues. It is suitable for most corporate activities and for banking, insurance, and government procedures. MCI and Sprint use it to prescreen employment candidates before on-site interviews. It provides medical diagnosis in remote or rural places that cannot afford specialists or even a single doctor.

Videoconferencing Technology

The personal computer has become our primary tool in the workplace. It is where you keep all your information: customer files, financial data, presentations, and all the other documents you need to conduct business on a daily basis. And conducting business means communicating. Whether with colleagues, customers, or vendors, communicating clearly is probably the most important business activity in which we engage. Communication today involves taking the information from our computers and communicating that information to others. The code phrase used to be *workgroup computing;* the new description is *collaboration*. This communication process usually consists of printing files from our computers, faxing, mailing, shipping, following up with a phone call, discussing, marking, and finally updating the files back in the computer. Each step takes time and introduces the possibility of miscommunication.

As more of the white-collar workplace becomes automated and computerized, and the technology becomes more robust to support face-to-face communication from the desktop, videoconferencing over the LAN and WAN becomes possible and desirable. Personal videoconferencing—with a videophone—allows you to communicate with others directly from your PC without all the intermediate steps. This segment is showing an annual compound growth rate of 33 percent with a current market size of $120 million, growing to $1.6 billion in 2000. With personal conferencing, you call your colleagues right from your PC; they see your documents as you see them, and you can share and mark up a whiteboard together. When remote control is enabled, you can make changes together or share a three-dimensional view with virtual reality modeling language (VRML). Various tools allow you to point to and mark up your information as you discuss it right in the application used to create the information in the first place.

With the addition of video capability, not only can you and your colleague both see the documents on which you are working, but you can see each other also. The subtleties of body language and facial expressions are things that no fax machine or telephone can communicate and forms the essence of why people like to communicate face to face. While you cannot smell and feel the other person, you can pick up the nuances of ticks, eye movements, and other subtle behaviors. It is the next best thing to actually being there. It has been the fantasy since *2001: A Space Odyssey* showed a glimpse of our possible technical future.

Each transport has its own unique characteristics, and when determining whether the transport is suitable for supporting videoconferencing within the entire corporation, the most important attributes to consider include:

- Bandwidth—the transport should have sufficient bandwidth to transmit video, audio and data at quality levels that allow you to see facial expressions and movements, hear comments, and work on a document.
- Latency—should not delay voice and video transmission from one end to the other so much that you cannot carry on an interactive session.
- Isochrony—description of the regularly timed delivery of data. The most suitable transports will support isochronous transmission of the video, audio, and data, ensuring they are delivered on time so audio and video do not break up.

- Resource contention—an ideal transport does not require the user to contend with other people on the transport for bandwidth. They are guaranteed a fixed amount of bandwidth at all times to transmit video, audio, and data.
- Availability—it should be widely available and accessible.

Videoconferencing is possible over dial-up lines and Internet connections from speeds above 9.6 kbps. The more bandwidth available the better the sound and the image. The trade-offs include quality of sound, quality and size of image, and speed of transmitting supplemental information such as screen images and collaborative whiteboards.

Videoconferencing and other allied tools require bandwidth. This means some of the WAN or LAN channel is reserved for call transmission management, session management, and conference management. You see or hear nothing with that bandwidth. It is all overhead. You also need bandwidth for sound, images, whiteboard, and desktop views. If the session includes desktop remote control, you need bandwidth for that as well. These bandwidth requirements are illustrated in Figure 8.2.

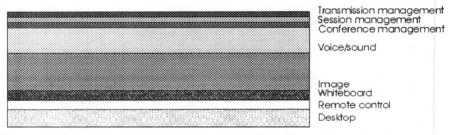

Figure 8.2 Relative bandwidths allocated within the videoconferencing stream.

However, this shows bandwidth requirements for just one side of a conversation. The other person wants to say something and be seen too. You need another channel for that as well, and typically a channel for each party in a multiparty conference session. This is indicated in Figure 8.3.

Sound is usually a priority item because the sound on videoconferencing is not tolerable if it is not at least as good as a plain old telephone service (POTS) call. Since the sound typically includes a person talking and also background noises (for example, other people at the conference table), you need to understand the dynamic ranges for sound. For example, Figure 8.4 illustrates the sound ranges for both compressed and filtered speech and concert-hall sound. You can compare half-duplex walkie-talkie sound to a good hi-fi sound system.

The dynamic range, really the width of the wavelength for the speech is between 4 and 8000 Hz (8 kHz) with a 4-kHz range, is comparable to a poor handset and most analog connections. The concert sound range is usually between 10 and 50,000 Hz (50 kHz), although many hi-fi systems provide ranges between 100 and 17,000 Hz (17 kHz). The deep rumbles occur at the greater wavelengths, between 10 and 500 Hz, whereas the piccolo sounds are above 10 kHz. Without compression, this is more bandwidth than can be supported on a slow analog modem linkage. Some products (under Windows) could not assign time slices for processes very well, with the result that compression created choppy sound. However, if you want more than just sound, you need to minimize the dynamic bandwidth to leave room for the pictures. Many products play games with the dynamic range by including the full range but chopping out portions of the range. For example, 501 to 560 Hz might be included, but 561 to 600 Hz might be removed. These dropouts are alternated for the full dynamic range since some people might consider these dropouts merely duplicate information.

The image is a function of image size, image depth or resolution, and frames per second. Size can range from the thumbnail to full screen. The next illustration shows image size changes given the same transmission. The images in Figure 8.5 are identical but are shown at different sizes.

The small image represents 100 X 70 pixels in 256-grayscale depth. You can also play with the color depth of the image from full color (24- to 36-bit) to 2-bit monochrome. This is illustrated in Figure 8.6.

True image size can also increase. For example, you can increase the information sent with each frame. This is shown by Figure 8.7.

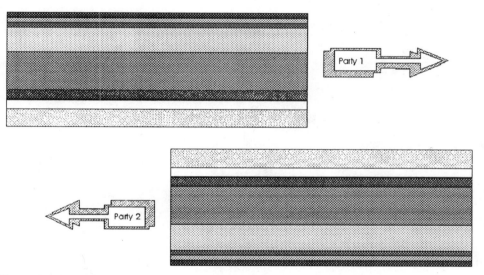

Figure 8.3 Relative bandwidths allocated within the videoconferencing stream.

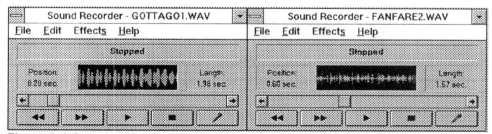

Figure 8.4 Speech which is compressed and cleaned on the right shows none of the range and subtleties of the music on the left.

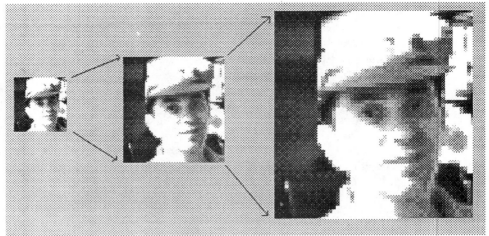

Figure 8.5 A frame from a videoconferencing image displayed at different frame sizes. Notice that the apparent resolution becomes worse as size increases.

Figure 8.6 These three images range from 1-bit depth on the right, 2 bits in the center, to 4-bits grayscale on right. It is also possible to provide color in 16 or 24, or use special algorithms for greater true color and tonal ranges.

Figure 8.7 Both images retain the same pixel depth and scale, but the right side image contains more information and requires a bigger transmission.

Another issue is the frequency of the transmission of the image. Typically, movies support anywhere from 12 to 32 frames/s. The National Television Systems Committee (NTSC) standard for video and TV transmission is 30 frames/s. Full motion requires at least 15 frames/s—the magic number for continuous motion—or it seems jerky, discontinuous, and nauseating. However, 10 frames/s is sufficient when people in head shots, like those shown above, do not jump around and *try* to stay within the frame. However, the smallest image shown here is at 100×70 pixels with 8 bits in depth. This corresponds to 56,000 bits, or a full second of an ISDN connection.

This is one of the reasons that analog connections are just not fully suitable for videoconferencing. That means—even at single channel ISDN speeds—you can send an image every 5 s or thereabouts because you need to provide sound and continue the overhead. The two B channels and particularly the D channel on ISDN are very convenient for call transmission, session, and conference management.

Sound and image compression goes a long way toward increasing the efficiency of videoconferencing. This is true for analog, LAN, and ISDN connections. Compression occurs differently for sound than for images. Silence can be removed from sound and not even transmitted. Since sound is an analog entity converted to a digital format, it can be compressed successfully. Since sound is also a sine wave about an axis, some of its information is duplicated and thus irrelevant. In addition, sound can be sampled at discrete frequencies so that the midranges can be ignored to reduce the information overhead. As I mentioned previously, the sound component of a videoconference should be at least as clear as a phone conversation, or people will be unhappy with it.

Still images can be compressed by 86 percent (expressed as 7:1 ratio) in general without any data loss. As an example, a 56-kbps image can be sent in 8 kbps, with the result that eight frames could be sent each second. However, images can be compressed even further with some loss. Loss could include tonal ranges, specs and artifacts, small objects, and color or motion depth. Compression ratios of 300:1 are possible with still images and minimal loss. Although the overhead for color is 3 times greater than for black and white, compression makes color very possible. In addition, moving images with videoconferencing can be compressed even more than still images because the updates between frames may represent a slight adjustment representing less information than an entire frame. This is a function of the algorithms, the time it takes to compress the image, and the desired frame rate. This all has to happen in real-time. PictureTel Live, CLI Eclipse, and RSI Systems ERIS can achieve 15 frames/s in some situations.

However, many products cannot achieve these rates, and in some cases, the site will not have ISDN availability. Videoconferencing is still available at the site with slower analog connections, albeit with reduced frame rates. Some chat-type products (for example Worlds Chat and AlphaWorld) substitute a three-dimensional representation, called an *avatar*, for real-time images. I think this is hokey and not really conducive to the face-to-face presentation possible with even very slow frame rates. Frame rate may be as slow as one frame every 10 s. Figure 8.8 shows the gap between frames.

A lot of motion and activity occurred between the frames (camera motion, person repositioned, and hat removed). It is not a blank screen between images; one image is displayed persistently until the next is received. This leads to jerky motion and some loss of comprehension as people process the new image in deference to the sound. Some of the newer products morph (that is, "transmorph") the old image into the new one to create a blended transition.

Figure 8.8 The image framing is a function of spacing and transmission rates.

However, the transitional effects require CPU time that might not be available on low-end systems. Intel 80486 is a minimal functional requirement and more than 16 MB of RAM (24 MB with Windows NT and Windows 95) is recommended for all products. The following table (Figure 8.9) describes the bandwidths required for different image sizes and frame rates.

Resolution	Frames/s	Compression	Bandwidth
640 × 480	30	None	2250
320 × 240	10	None	800
252 × 288	10	20:1	460
176 × 144	30	None	800
176 × 144	15	None	400
176 × 144	15	10:1	160
176 × 144	5	10:1	120

Figure 8.9 The video bandwidth requirements for bi-directional support.

H.261, the usual videocompression standard in use, is common across vendors and platforms, yields the p×64 framing and provides two common picture sizes, 252 × 288 and 176 × 144, which explains the sizes in the prior chart. I included 640 × 480 and 320 × 240 because these are standard PC display sizes corresponding to VGA and EGA. In effect, most connections between PCs will occur at less than full screen size, as the screen shot in Figure 8.10 shows.

The common and primary protocol for videoconferencing is the H.320 standard. Because most major vendors designed and created equipment that preceded any standards, the equipment is proprietary for the most part. The encoder/decoder boards and algorithms for supporting synchronized sound and video and providing support for desktop collaboration are not interoperable. However, most vendors support interoperability through the H.320 standard. Performance is very likely to be substandard, but at least there is some measure of interoperability.

Realistically, most of the systems on the market today provide from 2 to 8 frames/s. After all the issues of sound bandwidth, video bandwidth, and compression, there is also the issue of synchronizing the sound with the video. Obviously, when the frame rate is low, this is not an issue. However, as the frame rate approaches 15 frames/s and full-motion video, not fully cooked software and hardware can create the *dubbed movie effect* where the sound typically occurs after the image.

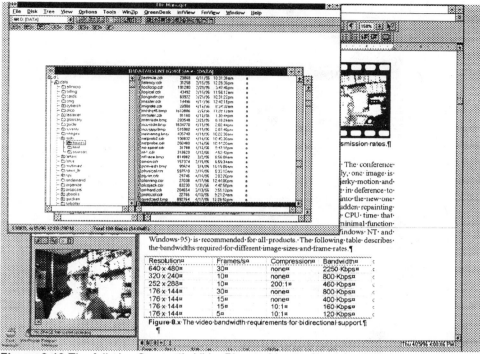

Figure 8.10 The full-size Connectix VideoPhone videoconferencing image (over the LAN) relative to the Windows screen size.

The high-end software matches the audio and video compression/decompression (codec) algorithms and interleaves the transmission of sound and video to resolve these problems. Interoperation of ProShare with PictureTel, for example, when using the older codecs which are common between them, creates this effect. If you can connect conference rooms at ISDN-H speeds (384 kbps) or have access to ISDN PRI and can bond six B channels (also 384 kbps), you are very likely to achieve the quality of a full screen at 30 frames/s. Also, you are likely to experience this with applications routed over a LAN or the Internet when the video stream is not matched to the available bandwidth.

Videoconferencing Equipment

The equipment required for videoconferencing ranges from a stand-alone conference room package (such as the PictureTel Concorde) with built-in ISDN adapters and a TV, camera with microphone to unbundled software packages. These cost from $20,000 to $60,000 per room. At the minimum, the hardware which you will require encompasses a microphone, a video camera, some connecting ports for microphone and camera, and the ISDN terminal adapter. The TV has speakers and provides sound. The stand-alone units, such as from PictureTel or RSI, are expensive but come with everything needed for each conference site. Figure 8.11 illustrates the basic videoconferencing hardware.

Connectix sells the QuickCam which is a black and white 256-bit grayscale camera that plugs into a parallel port. It will not coexist with a printer, but it can be attached to LPT2 or LPT3. You can add parallel port cards as needed to support the camera. A microphone typically connects into a sound card. The sound card provides output by way of speakers. These can be built into a multimedia monitor or can be discrete units. Headsets with a built-in microphone can be purchased for about $5 at discount outlets. Microphones cost from $5 to $100. In general, get one on a stand or stalk so that it can be positioned once and not moved. A hand-held microphone generates unnecessary and distracting background noise. Better units cost more than $60. You only need a unit as good as the quality of the connection.

You can insert a T-connector (or a "Y") *before* the P.A. system at presentations and route the signal from the speaker directly into the conferencing system. Do not route amplified sound, or you will saturate the sound at best and damage the equipment at worst. Consider inserting a filter on the microphone to filter out the clicks and hits to the microphone; that is helpful for everyone's ears, both locally and across the videoconferencing connection. If you are substantially compressing sound and video so that quality is compromised, an ordinary headset or microphone will do. Since most CD-ROM kits are bundled with a SoundBlaster-compatible sound board and a free microphone and speaker kit, consider this route as a solution for sound I/O.

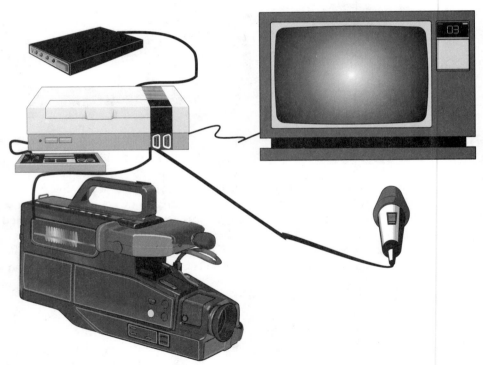

Figure 8.11 Basic videoconferencing equipment includes controller (attached ISDN terminal adapter if necessary), TV for sound and video, a video camera or camcorder for video and sound input, and a sound pickup or microphone.

Although it is possible to provide whiteboard and desktop document sharing as shown in Figure 8.12, the camera is essential to share real-time images of each other. It is possible to scan a photograph and create a fixed-image avatar, which is a real or imaginary visual representation, but most businesses will not tolerate such primitive technology. The incentive is to create a rapport between people to minimize the need for face-to-face meetings. Still images or, worse, disguised images (such as the avatar) typically decrease the rapport and make people more anxious for face-to-face events.

Figure 8.12 Whiteboard with document sharing (also with local video image) using Chameleon's InPerson applet (courtesy of NetManage).

Video cameras can be expensive. The Connectix QuickCam (black and white) is about $100. A new QuickCam color unit is about $230. The Intel and FlexCam cameras, as shown in Figure 8.13, cost about $600. Camcorders are available from $400 to $2000, and some of the digital units work very well for videoconferencing, although they start at $3200. Color video cameras start at $1200 and can cost as much as $4300. In addition, you will likely need a video capture board (also known as a video motion board, a digitizing board, or an encoder adapter) to mate with a camera or camcorder. This can add from $300 to $1500. At the present time, there are the following desktop solutions: the low-end, CU-SeeMe (free), enhanced CU-SeeMe ($150), VDOPhone, NetManage InPerson, or Connectix VideoPhone solution at about $250 per user, or the $1200 and up. Connectix costs about $150 alone, but do not overlook the cost of the sound board, microphone, and speakers. Unless you have specific use for cameras and camcorders, this route is not otherwise cost-effective.

When you run the Compaq or PictureTel desktop videoconferencing systems, the platform should be at least a high-end Intel 80486. A Pentium, given the negligible premium, makes a better platform. If you are building videoconferencing platforms from scratch or integrating videoconferencing into preexisting equipment, the codec used for imaging and the displaying images under Windows is very intensive for the CPU. If you off-load some of the processing to a specialized video display card (which is optimum), you will still see the best performance (faster than 75 Mhz) with Intel 80486. Some of the new video display cards support Motion Picture Experts Group (MPEG) playback, and when this is matched to the videoconferencing tools using the same codec, performance can approach 30 frames/s on the desktop.

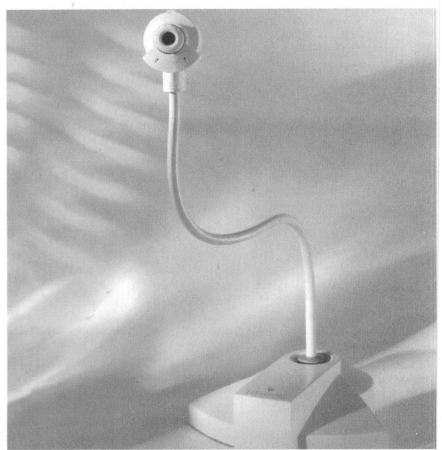

Figure 8.13 A convenient full-color video motion camera mounted on a flexible stand (courtesy of VideoLabs).

Enova Corporation (parent of San Diego Gas and Electric) and Philips Home Services are testing a screen phone in the San Diego metropolitan area. These units are expected to cost $299 in quantity. Although they do not provide full motion as most videoconferencing systems aim to do, they provide updated frames every few seconds. Ameritech and Energy Oasis are also testing this service in some markets.

Both VideoLAN VL 200 and CorelVideo represent another LAN/WAN option, but they require special wiring for the audio/video component. The audio and video is transmitted via the NTSC standard on a separate pair of coaxial cable or twisted pair. Similar in concept to IsoEthernet, but nevertheless different in implementation, hardware, and architecture, CorelVideo can interoperate with other H.320-standard systems.

Experimental products, those products at the low end, are actually quite useful. You have to work around the limitations since there are apt to be many more than with the higher-end products. On the other hand, these products are apt to be multiplatform, open, and very extensible. For example, CU-SeeMe runs on LANs, over the Internet, or any TCP/IP-based connection at low modem speed. It has the largest installed user base of any videoconferencing software and even supports your choice of audio codecs. This tool can support multiple-person conferences through the mechanism of a *reflector* site. This reflector provides conferencing and broadcasting through this server component for network and conference management (security with passwords). Platforms include Unix, Macintosh, and all versions of Windows. The White Pine version, a commercial release, has added whiteboarding, interoperability, and data and graphics sharing.

If you just want to try out videoconferencing, even when you have the ISDN circuits and want something more than the low-end experimental (but very serviceable) products, try out the conference rooms at Kinko's nationwide quick printing shops for $150 to $180 per hour for both ends of a conference. By the way, if you do have videoconferencing equipment and need to connect to a party who does not have comparable access—a prescreening for a job candidate is a good example—the cost for the conference is often cheaper than flying the candidate in for the initial interview.

Videoconferencing and ISDN

Videoconferencing capabilities can be point-to-point, much like a telephone call, or can be like a multipoint conference call. Multipoint video calls require that much more bandwidth and CPU capacity to handle each added participant. Since most videoconferencing systems allocate an ISDN B channel for each participant, ISDN is a very flexible and scalable transmission medium for this requirement. Because ISDN is a relatively high-bandwidth connection, it is very suitable for videoconferencing. It supports isochronous data transmission, and bandwidth is guaranteed once the connection is established. Thus there is a channel for each participant. Single-channel connections are possible but with reduced frame rates. Latency is manageable. ISDN service is becoming more available across the country and world, but availability can be an issue when deciding on this transport for wide-scale deployment of videoconferencing to the desktop.

LAN as a transport mechanism is widely available at most corporations today, and existing LAN installations have more than sufficient bandwidth to support videoconferencing connections (if the LAN load is low). Latency, even over several routers for a long-distance WAN link, is comparable to ISDN and within the limits for videoconferencing needs. Each caller must contend for this shared, asynchronous resource, so bandwidth is not secured once a connection is established. As you add more videoconferencing sessions, bandwidth and the management of its consumption on this medium does become an issue. Isochronous data transmission cannot be guaranteed because packets of information do not necessarily arrive at their destination at regular time intervals. This is the primary challenge of a LAN, and one that has been perceived as an insurmountable hurdle for real-time videoconferencing.

Similar to ISDN, Switched 56 traffic can travel over the same physical infrastructure that supports ISDN. As a result, the latency and isochronous nature of the transport is the same. Like ISDN, Switched 56 provides a two-channel call for videoconferencing, but the bandwidth is limited to 56 kbps per channel as opposed to 64 kbps per channel for ISDN. Switched 56 is an older technology with decreasing significance.

Considering the characteristics of the common transport protocols, we can focus on those infrastructures suitable for videoconferencing. Bandwidth on POTS is too low and the cost of ATM is too high. Switched 56 is technically feasible, but ISDN offers the same benefits with more bandwidth. The LAN has sufficient bandwidth, and wide availability, and its lack of isochronous data transmission support is a challenge that can be overcome. As a result, ISDN and LAN appear to be the transports of choice. Both are viable options, but each has associated strengths and weaknesses. These need to be considered carefully before deployment can begin. LAN latencies can vary widely and be very long when the LAN or internetwork is saturated from normal data traffic. ISDN requires a special in-house PBX or hub to handle site connectivity.

If you have over 50 people using videoconferencing, and you think they might be making a significant portion of the video calls within the building, create an ISDN infrastructure. Performance will be better than a LAN or internetwork, and it will interoperate with external local loops and provide other facilities as well. The PBX or hub must be ISDN-capable and requires at least single BRI lines from the PBX or hub directly to each desk. The local loop would be an ISDN PRI or several of them.

The connection works in this way. A call proceeds to the PBX or hub via the internal BRI line. From the PBX or hub, the call can be routed internally to another user. Calls can also be delivered to the LEC and then to the Interexchange Carrier, or the call can be routed directly to the IXC if there is dedicated access at the site. Once in the IXC's toll network, the call must traverse a similar path when exiting the network in the destination Local Access And Transport Area (LATA). If the call were to a videoconferencing unit in the same LATA (not behind the same PBX), the call would not be passed on to the IXC but may be passed through another LEC central office.

Even when you have many users and an ISDN-enabled PBX is financially unattractive, a portable ISDN hub can support a Basic Rate Interface line to the individual's desktop. Like the PBX connection, a call can originate from a BRI line provided by a portable ISDN hub. This hub serves the function of an ISDN-enabled PBX. Like the ISDN-enabled PBX, the portable ISDN hub can tie into the network via the Primary Rate Interface. ISDN transport for desktop videoconferencing requires an ISDN interface to your computer. BRI is the end-point ISDN interface required by most videoconferencing units; some do support PRI or ISDN-H when they are configured for multiparty conferences.

Videoconferencing Provisioning

Provisioning ISDN for videoconferencing is often slightly different from the setup for data and voice. However, if you anticipate full use of the BRI or PRI B channels, the provision should supplementally support that as well. Specifically, you will need to choose between NI-1 and Custom. You will also want to choose between one directory number or two. AT&T central office switches allow you to enter one number that is dialed twice in order to make the two B channel connections required for a typical ISDN dual-channel videoconference. In this environment, the Local Exchange Carrier (LEC) usually assigns the user one ISDN number, (thereby preserving numbers for future users). Northern Telecom switches require a directory number to be associated with each B channel. I recommend two separate directory numbers unless you anticipate high utilization of the ISDN circuit for videoconferencing and no other application for inbound calling (such as fax, data, or POTS).

Some local exchange carriers do not have data trunk groups to the Interexchange Carrier point of presence (POP). The POP is the IXC office where the call enters their toll network (the hand-off point between the LEC and IXC). In order for a long-distance data call to be completed, the LEC must have at least 56-kbps data-capable trunks between their central office and the POP. It is a function of clear channel.

ProShare on ISDN requires an ISDN BRI line and uses the entire 128 kbps provided to pass audio, video, and other conferencing data. Each B channel provides up to 64 kbps of capacity, and the 16 kbps available on the D channel is used only for signaling and call setup. The bandwidth is dynamically assigned with the only dedicated partition being 16 kbps for audio. This is also true for PictureTel, Vivo, TeleWork, and most other desktop conferencing systems. The conference room systems typically use fractional T-1 or ISDN-H for increased frame rates, picture size and depth, and multisite conferencing. ISDN provisioning must be mated to the particular vendor hardware requirements. When a large amount of data is being passed (such as when transferring a whiteboard, or a collaborative document or transmitting a file) the video bandwidth, and thus quality, is decreased to allow for rapid data transfer. Audio is not affected.

Provisioning for Intel Blue

Although the Intel ProShare system is not the most widely installed videoconferencing system, it was the first PC-based one. As such, many of the standards and interoperability issues have been resolved by Intel. If you anticipate connecting to or interoperating with ProShare, consider provisioning as per the Intel Blue definition. At the time of ISDN subscription, the user must request the phone company to provision the BRI line to match the type of local switch. The basic setting with the AT&T 5ESS switch is Custom with 2B1Q line code and 2B+D line multipoint. The line's configuration should be multipoint, with separate directory numbers for the ISDN phone set. The telephone company should supply you with different telephone numbers (DNs) and Service Profile Identifiers (SPIDs) for each channel. Also, each directory number should be programmed as follow:

- B1 circuit-switched voice/data
- B2 circuit-switched voice/data

The line must be set to allow both circuit-switched voice and data on both B channels. The D channel should be provisioned for signaling only. That is only User/Network Signaling is allowed on the D channel without D channel packet mode. Set the maximum number of terminals to 2. This tells the switch how many terminals are active on this line. Set the maximum number of B channels to 2, and set the Actual User flag to Yes to tell the switch that you are an actual user and may use both B channels simultaneously. Set the circuit-switch voice to 1, the circuit switch voice channel to Any, and the switch to allow only 1 B channel to actually be active for voice at a time. The *Any* tells the switch that it can use either B channel to deliver the call.

The switch for circuit-switched data should be set to 2 and the circuit-switched data channel to Any. This tells the switch that you may connect both B channels simultaneously. The *Any* tells the switch that either B channel may be used for data. Terminal type is A, the basic terminal. AT&T has defined the terminal types by letters (which have nothing to do with the various "simplified" provisioning or simplified ordering codes). This tells the switch that you are a basic Custom IS directory number terminal. Setting display to Yes tells the switch that you have display capabilities.

Set the call appearance quantity to default 1 for IS. This tells the switch that you have 1 voice call appearance per directory number. For the IS directory number phone set, the number of call appearances is flexible, usually 3. There are no additional charges for extra call appearances. Set the all appearance preferences to Idle. This tells the switch that your software will make a positive choice of which call appearance it will use to initiate a call. The telephone company will also need to know any additional voice features that you require on your IS directory number lines. Examples of these features are Caller ID and call forwarding, call hold, and flexible calling.

All settings for the Northern Telecom DMS100 BCS 36 which should be enabled for Custom setup are very similar to the 5ESS switch. The differences include that the circuit switch should be set to YES; and bearer restriction to no packet mode data (NOPMD). This tells the switch that you require circuit-switch ability on your B channel. The bearer restriction on your line means that you are not allowed to do packet data on your B channel. Set protocol to functional v1 (PVC 1) to tell the switch that the customer premise equipment (CPE) software is using a custom protocol.

Set the service profile identification suffix to 1. The SPID suffix for this switch and configuration can have from zero to eight digits and may vary with different telcos. Set the terminal end-point identifier (TEI) to Dynamic. This tells the switch that you can accept any TEI value from 64 to 126. The assignment of a dynamic TEI is the responsibility of the switch. Set ring to YES to send an alert to your CPE when there is an incoming call (or else you do not get notification of an incoming video call). Set the maximum keys to 10. This tells the switch how much memory to allocate for features. Set the key system (EKTS) option to No. This tells the switch that you are not a key system. A key system is where multiple telephone numbers are shared across terminals. If you are routing videoconferencing through a PRI hub, you will effectively need to provision that hub to allocate and initiate ISDN calls through the hub when it recognizes a videoconferencing call.

Set data option for PRoVLLC CMDATA (lower-layer compatibility). Place the lower-layer compatibility option for data on this B channel. This tells the switch that your CPE may utilize the lower-layer compatibility information element for compatibility checking with the called CPE. On a DMS-100, using the voice/data line configuration may require four different phone numbers (two for each B channel) unless the switch has been upgraded with a feature package to allow just one per B channel. If the switch still has a software generic lower than BCS32, it might be a problem. For all practical matters, support is nationwide, so that this should not be a problem.

Videoconferencing on the LAN

Other viable options for the deployment of videoconferencing are the existing LANs and WANs. These private networks facilitate data flow across the entire organization. Realize that videoconferencing may not work on existing network infrastructures without placing the current network traffic in jeopardy because of bandwidth overallocation.

According to Forrester Research Inc., 85 percent of the Fortune 1000's remote offices will be wired into headquarters over the next 2 years. Of the 2 million LANs that are not interconnected today, over 40 percent will forge WAN links in 3 to 4 years according to International Data Corp. Most major corporations not only have multiple LANs interconnected at various sites to create an internetwork, but these sites are also typically connected into one larger central network, with traffic passing over routers, bridges, and gateways. In order to be a compelling technology, LAN-based videoconferencing must offer connectivity to an entire corporate network.

The primary LAN technology is Ethernet, with Token Ring a compelling secondary option. Refer to my books, *Ethernet Management Guide, Token Ring Management Guide,* and the *Fast Ethernet Migration Strategies,* for management tips for these technologies. Only Fast Ethernet, Fiber Distributed Data InterChange (FDDI), and IsoEthernet really have the bandwidth for enterprisewide videoconferencing. FDDI is often used for high-speed corporate backbones between buildings or in high rises, as detailed in *FDDI Management Guide.* As networks overload, many corporations are replacing parts of the existing infrastructure with Ethernet switches to prolong the life of their 10-Mbps networks and are upgrading that backbone bandwidth to 100-Mbps technology such as FDDI and Fast Ethernet.

Networks at diverse geographical locations are linked together with WAN loops. As you no doubt understand from Chapter 2, these connections tend to be lower bandwidth than their LAN counterparts, usually a T-1 (1.54 Mbps) or some fraction of a T-1. WAN links of adequate bandwidth (128 kbps or greater) can support videoconferencing. However, these links tend to be the bandwidth bottleneck for intersite traffic in most corporations as companies begin to deploy LAN-based videoconferencing, and they may need to be upgraded to support significant numbers of users or separate video streams. Many companies are exploring higher-bandwidth options such as multiple T-1, T-3, or even ATM connections. When considering whether videoconferencing can be deployed on a LAN, there are several discrete issues:

- Bandwidth
- Isochrony
- Resource contention

These can be summed up in this way. The transport bandwidth should have sufficient bandwidth to transmit video, audio, and data at quality levels that allow you to see facial expressions and movements, hear comments, and work on a document. As previously mentioned, videoconferencing should provide the regularly timed delivery of data. The most suitable transports will support isochronous transmission of the video, audio, and data, ensuring that they are delivered on time. The ideal transport medium does not require the user to contend with other people on the transport for bandwidth. Each user is guaranteed a fixed amount of bandwidth at all times to transmit video, audio, and data. This is partly true with IsoEthernet, switched Token Ring, ATM, and Ethernet but not for any shared-media technology such as Fast Ethernet, FDDI, FireWire, SCSI III, or Fibre Channel.

Unlike the separated B channels in ISDN, the connection established between two points over the LAN is not dedicated, so the videoconferencing must compensate for the lack of a guaranteed channel. Each system must compete for the shared bandwidth on the LAN and attain real-time performance without isochronous data transmission. Bandwidth cannot correctly be resolved, so contention with other traffic will be an issue. Not using datagram protocols, packet identification, and buffering allows the isochrony hurdle to be overcome.

Reliable protocols such as SPX or TCP require acknowledgment of every packet sent. This provides inadequate performance for a real-time, interactive videoconferencing session over the LAN. Videoconferencing operates (independently by ITU standard

H.324) on top of the datagram components of those protocol stacks (such as UDP/IP). With the datagram protocol, packets do not have to be acknowledged by the receiver. The acknowledged packets would occur too late to be useful for real-time videoconferencing and only jam up the network with useless and outdated frames. This is one of the reasons filtering and spoofing were mentioned in Chapter 6. Spoofing becomes very important when videoconferencing runs over the LAN or internetwork.

Standards

Video telephony over narrowband ISDN is governed by a suite of ITU-T (formerly CCITT) interoperability standards. The overall video telephony suite is known informally as p x 64 (and pronounced *p star 64*) and formally as standard H.320. H.320 is an *umbrella* standard that specifies H.261 for video compression; H.221, H.230, and H.242 for communications, control, and indication; G.711, G.722, and G.728 for audio signals; and several others for specialized purposes. A common misconception, exploited by some equipment manufacturers, is that compliance with H.261 (the video compression standard) is enough to guarantee interoperability.

Bandwidth can be divided up among video, voice, and data in a bewildering variety of ways. Typically, 56 kbps might be allocated to voice, with 1.6 kbps to signaling (control and indication signals) and the balance allocated to video. This varies by vendor product and implementation and whether an interoperable connection is being established. An H.320-compatible terminal can support audio and video in one B channel using G.728 audio at 16 kbps. For a 64-kbps channel, this leaves 46.4 kbps for video (after subtracting 1.6 kbps for H.221 framing).

As previously mentioned, the resolution of an H.261 video image is either 352 × 288 (known as computer ionformation frame [CIF]) or 176 × 144 (known as quarter-CIF or QCIF). The frame rate can be anything up to 30 frames/s. Configurations typically use a 2B (BRI) or a 6B (switched-384 or 3xBRI with an inverse multiplexer) service, depending on the desired cost and video quality. In a 384-kbps call, a videoconferencing system can achieve 30 frames/s at CIF and looks comparable to a VHS videotape picture. In a 2B BRI call, a standard video phone can achieve 15 frames/s at CIF. Those who have seen the 1B video (or single Switched 56) call in operation generally agree that the quality is not sufficient for anything useful like computer-based training, only for the social aspect of being able to see the other party. A 2B picture, on the other hand, is for all practical purposes sufficient for remote education, presentations, and initial screening interviews. Rapidly changing scenes are still not handled well.

However, it should still be noted that 6XB and H0 do allow for dramatic improvement in picture quality compared to 2XB. In particular, H.320 video/audio applications will often allocate 56 kbps for audio, leaving only 68.8 kbps for video when using 2XB. On the other hand, using H0 allows 326.4 kbps for video with 56 kbps for audio (the concert-level sound). Alternative audio algorithms can improve picture quality over 2XB by not stealing as many bits. Note that 6B is not identical to H0; the latter is a single channel which will give you 80 kbps above that of six separate B channels. Inverse multiplexors can be used to combine B channels. The table in Figure 8.14 lists the common standards applied for videoconferencing based on the transmission carrier.

T.120 is the data-conferencing standard. It supports multipoint delivery (or "reflector sites" for conferences with many simultaneous users), interoperability over LANs, WANs. and different platforms and applications and reliable data delivery with error correction coding. In addition, T.120 is transparent to the transmission and network protocols. This means it is extensible from analog connections to ISDN and eventually to ATM or B-ISDN. It is scalable as well. H.320 is the main standard for videoconferencing. Primarily, H.320 supports multipoint delivery (for many users), interoperability over LANs, WANs, and diverse platforms and applications. Most importantly, H.320 is transparent to the transmission and network protocols so that it is extensible from ISDN to ATM and B-ISDN. These protocols are listed in the table in Figure 8.15.

Carrier	Overall standard	Video standard	Signal standard	Multiplexor standard	Comm interface
ISDN	H.320	H.261	H.242	H.221	I.400
B-ISDN	H.321	H.261	Q.2931	H.221	I.400
IsoEthernet	H.322	H.261	H.242	H.221	TCP/IP
LAN	H.323	H.261	H.24Z	H.22Z	TCP/IP
Mobile	H.324	H.263	H.245	H.223	Modem
PSTN	H.324	H.263	H.245	H.223	V.34

Figure 8.14 Videoconferencing standards for different carriers.

Standard	Application for videoconferencing
T.120	A collection of protocols for data conferencing and shared applications with whiteboard support over a base of network-specific protocols and communication links, including TCP/IP, NetWare, IPX, X.25, frame relay, POTS, and ISDN; associated with the Multipoint Control Unit (MCU), which is a multipoint router for managing the call sessions
T.124	Generic conferencing control
T.125	Multipoint communications control (sets up multiple connections and transfers messages)
T.126	Still image and image annotation protocol
T.127	File transfer protocol
G.711	Narrowband audio compression codec (3 kHz in 48 to 64 kbps)
G.722	Wideband audio compression codec (7 kHz in 48 to 64 kbps)
G.728	Narrowband audio compression codec (3 kHz in 16 kbps)
G.723	Audio transmission at 5.3 and 6.3 kbps with audio/video synchronization
H.245	Signal control
H.261	Audio compression codec allocated in p x 64 kbps
H.320	Video telephony at rates above 64 kbps (main definition)
H.321	ISDN circuit switching
H.322	Enhanced H.320 with a guaranteed quality of service
H.323	Conferencing standard for audio and video services; the RTP/RTCP real-time protocol for streaming data; connection between ISDN and LANs; and RSVP for reserving bandwidth over the Internet
H.324	Multimedia terminal for low-bit-rate visual telephone services over the general switched telephone network
H.263	Speech encoding in silicon

Figure 8.15 Important video and data-conferencing standards.

The Versit initiative is an implementation specification endorsed by Apple Computer, AT&T, IBM, Siemens Corporation, Intel, and other hardware vendors. This specification builds on the ITU H.320 and T.120 specifications describing videoconferencing over ISDN by adding support for interoperability between products from different vendors. Issues include call setup, phone books, secondary services, multiuser conferencing, and videoconference calls. An interim standard defines services such as data and whiteboard sharing building on the T.120 standard.

Videoconferencing Implementation

Test. Test. Test. If you want to provide a remote conference, you want assurances that everything will work. Test. Test. Test. Although you may foresee many technical problems, and there are likely to be many, there are also likely to be many cultural and organizational ones that seem salient to videoconferencing. In fact, the management and workflow problems are likely to be more troublesome than any ISDN connectivity problems. For example, a large bank set up a New York-to-Hong Kong one-day conference on strategy and marketing presentation. Rather than fly a large group from Hong Kong into New York (at $3200 per person), they decided to test videoconferencing. They tested it locally over ISDN lines and verified functionality of digital lines from North America to Asia. They tested performance of one channel, two channels, switchover of channels if the main channel failed, clear-channel support through various exchanges, linkages and fallback from 64 to 56 kbps, and overall reliability.

The Information Systems and the telecommunications groups had a lot of fun testing the video and sound features and found that they worked and the connections from ISDN to Switched 56 digital lines were reliable. They liked the system and began to use it for normal business. It worked well within this narrow framework.

Nevertheless, this was an expensive test since some equipment was bought for both sites; mostly, equipment was leased for the one-day trial. Some important people wanted to be in on the project and to be part of the presentation or participate some way in the one-day trial. Several large projection TVs were installed in the Hong Kong meeting hall so that everyone could see New York and a camera was setup so that New York could get feedback from Hong Kong on the mood of the crowd and the camera person could focus in on particular people who asked questions. There was a single video channel from Hong Kong. Some conferences dedicate one camera to the crowd and establish a second zoom and pan camera.

Everything worked well for the start of the conference. The opening presentation went flawlessly. The videoconferencing software mixed the talking head shot of the CEO into PowerPoint slides in the upper right corner where space had been reserved. Although this could be broadcast on two separate channels, the picture-in-picture minimizes the bandwidth requirements and general complexity of the presentation. This single composite image was shown on the three projection screens in Hong Kong. A second channel showed the New York audience. Then, everything went totally blank in the middle of the live presentation although the sound continued from New York to Hong Kong. Local diagnostics (from Hong Kong) showed that the lines were working correctly and no one was aware of the problem in New York. It took a few minutes for someone in Hong Kong to get the phone number for the right room where the presentation was originating; it wasn't in the IS offices anymore. Of course, the phone was forwarded so that the ringing would not disrupt the presentation. It took another few minutes for a person to actually barge into the room and explain the problem. The solution to this technical flaw was not immediately apparent.

The talking head shot could fill the screen, but then the participants could not see the slides at all so the continued presentation made little sense. Apparently, the size of the overall PowerPoint file was causing memory problems and the slides were not transmitting. Restarting the PowerPoint presentation was not a feasible solution because it worked correctly until the middle of the slide set, and the problem just reoccurred at the same place in the sequence. Jumping to the end of the set and scrolling backward merely caused the same problem somewhere else in the sequence. Since the presentation had been set up as a (compiled) runtime, there was no way to jump directly into the middle or show individual slides. Technical solutions were only eating up time and the conference was going nowhere. The conference was hastily aborted with continuation plans for the following week.

The presentation was resumed the following week where it went flawlessly—and we presume after all the bugs were worked out. At least it went well on the second live trial. Videoconferencing is software-intensive, and because it is complex software, it does have bugs. Furthermore, the software is not always robust enough to handle a platform problem or glitches with the D channel, which controls the audio and video transmission and synchronization.

So that you do not have to learn the same lesson the hard way, create a local echo. That means set up a local system to show how the presentation looks locally. Since the presentation is saving the cost of an airline ticket (and more), you can afford the extra hardware and a support person to monitor the quality of the presentation. You can install an ISDN BRI line emulator for this local loop or actually route the call out to the central office switch and back on separate local ISDN bearer lines. This is possible, if you recall, because ISDN supports multipoint drops, and this is part of basic provisioning. If you want actual feedback on the presentation hall and the quality of sound and video in real time, two video cameras and a video picture-in-picture (PIP) converter provide the basic videoconferencing signal. The combined two-image signal is only one frame as far as the videoconferencing software and hardware is concerned; this works with most H.120- and H.320-compatible systems. This local loop and feedback configuration is illustrated in Figure 8.16.

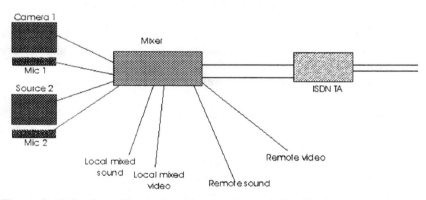

Figure 8.16 Configuration for local loop and remote feedback.

You can also provide feedback with a person running a camera at the presentation site to return any telltale signs of problems. There was a live autofocus video camera showing the audience in New York, but there was no one who was trained to anticipate and compensate for any problem. The Hong Kong audience politely nodded at the blank screen, listening to the words, and none of the technical people would swing the camera around or in any way interrupt the physical harmony of the presentation hall. This was partly a cultural problem. Although the technical people in Hong Kong and New York had a good rapport, the stricter hierarchy in Hong Kong prevented the technical people there from interrupting. Do not assume anything will work. The issue is not just technical, but also organizational or even mostly organizational.

The very public humiliation, partial success of the presentation, and waste of time for many important people focused resources on the retrial. Over 6 days, every aspect of the presentation was reviewed and tested, including a walk-through demo of the presentation to the entry presentation hall. Although PowerPoint had been installed and tested on the presenting machine, the full presentation file (combined with slides from all five speakers) had not been previously tested. The oversized runtime files were corrected so that the presentations were split into separate files. However, other flaws were discovered. Color, image, background templates, image sizes, and font usage displayed minor stylistic problems not seen at the initial presentation. Resolution and presentation templates differed, and the Corporate Communications group raised issues of stylistic conformity. At least, these errors were discovered and repaired before continuation.

For example, the video converter driving the three projection TVs was limited to 640 X 480 pixels, a common resolution limitation with overhead computer LCDs and projection equipment. The PowerPoint presentation segment was designed for 800 X 640 pixels and did not transmit at that depth over the videoconferencing equipment. When the presentation was tested on the smaller format, the bottom and left were not visible and the talking head shot obscured slide material. When rescaled to the smaller format, the images and fonts where not completely legible so that the segment needed to be redone. In addition, slight template differences (older versus newer corporate logos and older versus newer PowerPoint templates) created less than professional integration. Choice of fades, color backgrounds, and fonts, while consistent within each individual's presentation differed between presenters.

In several cases, choices of fonts used to create the initial PowerPoint segments were not encapsulated into the runtime and were not available on the presentation machine. Font substitutions were unprofessional (the substitution of Roman, the serif default for MS Windows 3.0, for ITC Leamington) and needed to be rationalized across the segments. In one case, Century Schoolbook mapped into Blippo Black, an unlikely image for any bank, rather than New Times Roman or even Ariel, which are standard in MS Windows 3.1x. The presenter's platform, Windows 3.11, was pared down to a plain vanilla system to minimize resources and lesson the chance of glitches. As a result, the corporate profile of Univers and Leamington fonts were not installed on this system and were not available because this problem was a new one.

While the presenters could not distinguish between Univers and Ariel or Leamington and Times Roman, it was a hot button for Corporate Communications, which had spent considerable money creating a font standard, a logo style and its placement position, and PMS colors. The cost for a corporate standard had included rationalizing letterheads, literature, and brochures in many languages in many sites. They had maintained this same consistency with commercials, training films, and promotional videos. They insisted that this cost not be eroded by the use of the wrong fonts, logos, and color scales in the live videoconferencing or in the desktop environment. Although this conference was only a test and a new technology, they made the valid point that tests tend to roll out to production unchanged. These monotype fonts were converted into Microsoft True-Type formats and distributed to the presenters so that the PowerPoint slides could be modified and not default to "comparable replacement faces" and loaded on the presentation machine. Also, Corporate Communications insisted that the color projection displays be adjusted to more accurately reflect the appropriate PMS colors.

Now that ISDN is widely available in Hong Kong, Singapore, Asia, Europe, and even in the United States, the uses for videoconferences have expanded throughout this bank. They have installed sites in all important offices and use the technology to minimize travel costs, increase rapport between remote locations, and resolve complex problems faster. The bank has also discovered that major accounts have the same videoconferencing equipment or compatible equipment so that conferences between these business partners have begun in earnest. It only takes a video phone call to start a two-person conference in some cases or a simple matter of setting a schedule for a meeting with more people.

When Things Go Wrong

Videoconferencing is the sum of TV recording and broadcasting and ISDN transmission. Everything that can go wrong with TV recording and broadcasting can go wrong with videoconferencing. Examples include people not fully dressed, hair not combed, presentation styles awkward or not framed within the area picked up by the camera, and sound out of synchronization with the picture. You could also experience a lack of sound or picture, a loss of one signal channel, or a problem with color depth. ISDN transmission problems were discussed in depth in Chapter 5, but since videoconferencing requires at least two B channels and the D channel for management, addition transmission problems include loss of a line, inability to bond the two channels together, inability to get two channels connected, and even alternate routes and mixed channel with 56 kbps.

Most video system software also provides a complete ISDN diagnostics package to assist in any ISDN line installation problems. Some of the more complex multisite, multiple-participant systems require a separate management station to control the sessions. For example, Intel sells LANDesk not only for LAN management but also for the videoconference session management.

National ISDN (NI-1) supports ProShare and Vivo with two directory numbers. However, you only need a single directory number. You want to match up ISDN to ISDN with clear channel or switched digital with clear channel (that is, the 64 kbps per channel). When your videoconferencing unit won't dial or receive calls and it is served by a 5ESS switch, your B channels may be configured for voice/data (both voice and data on each B channel), but you also need to have the line configured by specifying:

Data channels = 2

in order for both channels to work. The videoconferencing unit may not work at all unless both B channels are working. Try entering the phone number without any dashes or separators. The dashes work okay for modem calls but may confuse the ISDN terminal adapter. When the videoconferencing application requires you to dial two phone numbers to get both of the B channels active on a device served by a DMS-100 line, ask the telco to place the phone number for the second B channel in a hunt group with the phone number for the first B channel. When the first number is busy, the call automatically hunts to the second B channel, so the caller need only dial the first number.

Applications

There are a number of applications for videoconferencing. The following sections profile real-world applications and the users who benefit from this technology. Examples of such applications include conferencing, video mail, remote sales, distance learning, health care, and corporate presentations, as previously described with the multinational bank.

Conferencing and Video Mail

To support this growing thrust toward electronic meetings, a wide range of video and audio conferencing systems is becoming available, much of it specifically designed to exploit the speeds and channel flexibility of ISDN. In fact, using equipment available today, it is possible for even the smallest office or business to gain the advantages of dynamic, on-the-spot teleconferencing. The use of traditional full-motion, full-color video-conferences is virtually exploding as equipment costs come down and the availability of inexpensive dialed ISDN lines goes up. ISDN videoconferencing offers a range of equipment sizes and costs. High-end applications can combine, or bond, multiple ISDN channels for maximum video quality.

In these meetings, groups at two or more locations are photographed, usually by camera operators who add close-up zooms and other variety. The TV signals are transmitted through broadband digital connections, usually 384 kbps and above, and displayed at each of the other locations. Traditionally, speeds in this range required dedicated broadband lines set up especially for the purpose or leased virtual connections set up for specific daily or weekly time segments. With ISDN, it is possible to simply set aside an appropriate bundle of B channels through a corporate PRI, or for a smaller company to combine several BRI connections. When the meeting is over, usage charges stop, and lines can be returned to the normal pool of ISDN facilities.

A range of equipment manufacturers in the United States, Europe, and the Far East have demonstrated compact videophones that embody a range of voice, video, fax, and computer capabilities. In addition, just as automated voice messaging has become common today, we are going to see video mail. ISDN connections are beginning to link people in many different disciplines, offices, and locations into permanent or temporary virtual work-groups. Using dialed BRI lines, they communicate, share documents, files, and messages, interconnect to LANs or other networks, conduct video meetings, and in general interact with one another in much the same way they would if they shared an office.

Many who work in large organizations spend much of their working day in meetings. Meetings are, in fact, one of the primary ways companies plan, set schedules, solve problems, and make decisions. They are often pivotal events where the future of a product or even a company can be shaped. A growing range of organizations have discovered that carefully structured electronic meetings not only save money but are much easier to arrange with key people less likely to cancel. These meetings are frequently much more productive because participants can be better prepared to make or endorse decisions, commit to deadlines, and allocate resources, since they have access to the information, advisors, schedules, documents, and data they use every day. Electronic meetings also seem to promote a more systematic agenda, lessen the tendency to stray from a subject, and dramatically reduce outside and after-hours distractions.

ISDN virtual workgroups and team collaborations bring people together. Work can actually be more productive than in an ordinary meeting. At Jet Propulsion Laboratory, for example, Multipoint Bridge Conferences link scientists and engineers from many locations in a range of system-design, CAD/CAM, and other similar meetings. Worldwide group or team collaborations, in fact, are rapidly becoming practical in a number of disciplines. A architect in Silicon Valley uses as many as 12 ISDN B channels for regular CAD/CAM design collaborations with a number of engineers in Japan. The dialed ISDN connections link Sun workstations for a CAD/CAM application shared by and usable by all participants, two-way desktop video and voice connections to all participants, and a shared whiteboard where all can make annotations, gesture with on-screen markers, make rough drawings, and the like. The team can also share documents, faxes, and virtually any other form of electronically stored information.

Remote Sales

While many products are represented well with slicks or limited demos, perishable products can only be seen in real time. The Japanese are sizable consumers of fresh California produce, but they are also hesitant to buy without first seeing the product. Once, this meant shipping samples, packed in dry ice, by air. The cold sometimes altered the product, yet even when it didn't, prime picking times might have passed before a decision was made. Strawberries may seem like unlikely candidates for an international videoconference, but today strawberries, artichokes, and an entire market basket of fruits and vegetables are having their images beamed around the world through ISDN. Buyers throughout Japan see freshly picked products on live video transmissions using single BRI connections from sites throughout California.

Distance Learning

Distance learning refers to the linking of a teacher and students in several geographic locations via technology that allows for interaction. While initially designed to extend educational resources into geographically remote or rural areas, interest has grown in using the technology to share scarce resources in urban areas as well as to meet the needs of students who cannot reach traditional classrooms. Researchers have concluded that distance learning facilities overcome more than simply distance. According to a recent DRI study, the top three reasons that students report enrolling in television-based distance learning courses at the college level are time constraints, and work, and family responsibilities.

A broad range of educational institutions are beginning to use ISDN as the backbone for integrated voice, data, and video transmissions that bring expert teachers face to face with students in remote classrooms. In a more elaborate setup, classroom TV screens show the teacher and any visual being used, while a camera in the class lets the teacher see and interact directly with students.

Nevertheless, the prohibitive costs associated with many distance learning programs have kept these benefits from reaching a wider public. Distance learning programs are usually based on instructional programming delivered by satellite or through an Instructional Television Fixed Service (ITFS) network. Both methods require expensive equipment at the school and in the delivery system, and both allow for only limited interaction between teacher and student or among students.

Recent efforts to conduct distance learning over ISDN have been successful. According to a recent report commissioned and paid for by PacBell, "ISDN offers students and teachers in distance learning programs an interactivity level not available in ITFS or satellite programs without adding to the cost of the program." For these and other reasons, the researchers concluded that "ISDN has proved extremely efficient and effective technology for on-demand delivery of educational services to any region offering digitally switched phone service." ISDN is a viable, effective technology for distance learning, with shared workscreens, videoconferencing, and access to off-site resources in many settings.

A simpler setup might include a desktop video between a teacher and several students, a screen-sharing link for visuals and interchanged materials, and perhaps a remotely controlled computer screen showing documents, images, data, or almost anything else. ISDN connections, in fact, can effectively handle almost any combination of voice, visual, and data interchange. The Allegheny Health Education Research Foundation in Pittsburgh uses desktop videoconferencing from InSoft, called Communique. This technology increases collaboration, shares information, and trains medical students in remote locations. In addition to most office-type application, systems have been installed in operating rooms so that students can interact with the surgeons in real time. This provides a sterile environment, eliminates the nuisance of students under foot, and reduces the costs for transportation and travel time. In addition, specialized fiber optic surgery and microscopic surgery uses special TV cameras that already generate a signal transmitted to overhead monitors. This same video stream can be routed into the videoconferencing signal, too.

California State University, Chico, in partnership with AT&T and PacBell, has completed two successful trials of ISDN as a delivery system for distance learning. In May 1992, ISDN was used to link fifth-grade classes in three elementary schools in the Chico area. During the trial, students shared slide show presentations and participated in conference calls that included video, data, and image transmissions. Through this system, students and teachers were able to engage in real-time, interactive, two-way audio and video communication, view a laser disc video clip display, and send and annotate graphic images or text.

Appalachian State University, AT&T Network Systems, and BellSouth have built an ISDN-based distance learning network that delivers interactive voice, data, and video to three rural North Carolina schools. The 10-year project, called "Impact North Carolina: 21st Century Education," was touted as "one of the first in the nation to deliver interactive video instruction through existing copper phone lines." The system transmits interactive voice, data, and video at 112 kbps to two elementary schools and one high school in Watauga County. The Impact North Carolina system will give K–12 students access to remote lecturers, university libraries, and other distant resources and will also be used to improve teacher training, student teacher supervision, and continuing education at Reich College of Education, a major regional center for educating teachers.

Project Homeroom, an initiative "designed to improve student thinking, learning and computing skills," is a partnership among six Chicago area schools, AmeriTech, IBM, Illinois Bell, Prodigy, AT&T Network Systems, Central Telephone Co., and Eicon Technology Corp. a vendor of ISDN equipment. Over 550 students are participating in the project, which uses PCs, multimedia software with CD-ROM, video, voice, and text features. and on-line services over phone lines supplied by Illinois Bell. Students access on-line homework correction, instruction and tutoring, and computer communication with teachers, among other features. Seventy-six of the participating students from Stagg High School in Palos Hills, Ill., use the system over ISDN. It allows these students to exchange text, pictures, and calculations up to 8 times faster than other students can.

In Nashville, Tennessee, students at Carter Lawrence and Meigs Magnet Middle Schools can work together and with the Learning Technology Center at Vanderbilt University over ISDN. The pilot project, a joint effort of South Central Bell, Northern Telecom, Vanderbilt, and the Tennessee Public Service Commission, uses voice, video, and screen sharing technologies to enable students to see and talk with other students and with faculty at Vanderbilt, as well as share documents, graphics, and other information.

In the Research Triangle Park area of North Carolina, the North Carolina State University Center for Communications and Signal Processing, BellSouth, Southern Bell, GTE, IBM, Northern Telecom, and the Wake County Public School System are developing SCHOOLNET to demonstrate the enhancement of public education through advanced telecommunications technologies and are specifically testing ISDN. Among the functions that SCHOOLNET provides are video learning and distant instruction, electronic access to library materials, faculty support for exchange of curriculum materials and teaching aids, and administrative support for scheduling and staffing purposes.

ISDN could also enhance the availability and value of educational resources on the horizon. Congress has passed legislation calling for the creation of a National Research and Education Network (NREN) to link educational institutions, government, and industry. Among its purposes, Congress sought to promote the inclusion of high-performance computing into educational institutions at all levels. The investment needed to actually connect every one of the nation's 84,500 public schools and 24,000 private schools is far beyond the resources available in the NREN legislation. If ISDN was widely available, it could substantially leverage the value of the government's investment by enabling schools, especially at the K–12 level, to attach to the NREN and reap the benefits of this high-performance network. In combination with the World Wide Web (WWW) and ISDN, many students are surfing the web. Although many schools are without government funding, children and school administrators are soliciting vendors and parents to self-fund these projects.

Health Care

Many analysts have suggested that advanced telecommunications networks can have substantial benefits in improving the delivery and reducing the cost of health care services. While the publicity often focuses on medical consultation from home and remote diagnosis, other applications in the health care field include reducing administrative costs, providing health and medical information to help people take better care of themselves and make more informed decisions about their medical needs, providing health care in rural areas, and enabling doctors to consult with one another.

Medical images are especially dense with information, and the reliability of their transmission is paramount. A typical Computerized Axial Tomography (CAT) scan image contains about 5.2 MB of information, while a digitized x-ray requires 12 MB of information. Both CAT scan images and x-rays can be sent over phone lines today, but the process is slow: A single CAT scan takes 9 minutes, and an x-ray, 21 minutes. Using just one B channel of an ISDN line, those times can be reduced to 1.4 and 3 minutes, respectively. Of course, by combining the two B channels of an ISDN line, those times can be cut in half again.

The U.S. Public Health Service is facilitating the development of a multimedia telecommunication network for coordinating community health and human services and promoting shared group decision making for better case management. The Community Services Workstation will combine videoconferencing, document sharing among remote health care and social service workers, and access to databases with medical information, local services, and practical information that can be produced for clients such as maps and mass transit routes. Based on prototype research completed at Baylor University, the workstation now being tested at Howard University in Washington, D.C., is built on PCs connected via ISDN.

In Huntsville, Alabama, BellSouth with the U.S. Army have created an ISDN lab to develop voice, video, and data applications for the Army, including medical applications. Dr. Ira Denton, Jr., a neurosurgeon with The Alabama Back Institute, has demonstrated how ISDN can support remote consultation during surgery. Specifically, a remote specialist, linked with simplex video and full-duplex audio, views the operation through the operating microscope, getting the same view of the procedure as the surgeon on site. ISDN also enables postoperative follow-up exams of patients at remote locations. The exam can be performed by a nurse practitioner under the remote guidance of the surgeon, who has full video and audio contact with the exam site.

The General Computer Corporation, a company that processes claims for insurance and state benefit programs from pharmacies, doctors' offices, and hospitals, uses ISDN for claim processing and membership verification services in Pennsylvania and New Jersey. ISDN significantly reduces transaction time and the per-minute telecommunications charges. Currently, a routine authorization from a pharmacy or doctor's office that formerly required at least 30 s occurs in under 4 s with ISDN. General Computer projects that over 25 million pharmacy transactions are carried each year over the public network and that the switch to ISDN can cut response times on claims by up to 85 percent.

The U.S. Department of Veteran's Affairs (VA) Information System Center at Silver Spring, Maryland, is testing how ISDN can extend access to the VA's medical and document imaging system. The VA's integrated imaging system stores medical images, including pathology specimens, x-rays, cardiology studies, and endoscopy views in addition to the text-based patient information system available at all VA medical centers. The VA is using ISDN, configured to combine the two B channels for data transfer, to link the VA center at Silver Spring, the VA medical centers in Washington, D.C., and Baltimore, Maryland, and the NIST campus in Gaithersburg, Maryland. According to preliminary reports, the system can retrieve a 750-KB, 16 bit color image from the image server over ISDN in 60 s.

The University of Louisville, the State of Kentucky Cabinet for Economic Development, and South Central Bell have created the Telecommunications Research Center (TRC) at the University's Shelby Campus. The TRC recently demonstrated the transmission and reception of dental images using RadioVisioGraphy (RVG), a filmless dental x-ray system, over ISDN connections between the TRC and Washington, D.C. TRC forecasts uses for remote consultations for diagnosis, referrals, and follow-up.

Conclusion

This chapter presented the hardware, the applications for ISDN, and the techniques and troubleshooting issues for videoconferencing. As shown, videoconferencing is a parent service that also includes whiteboarding, desktop conferencing, desktop remote control, and collaboration. The next chapter shows how ISDN is scalable and provides many options for a planned, phased, or incremental migration from analog services to more capable digital connections. To this end, recognize that you can install ISDN in one of several migration paths to digital communication from analog POTS services.

9

ISDN Design Architecture

Introduction

Chapter 4 presented the technical and physical details for ISDN ordering, installation, provisioning, and hardware configuration, as well as some of the financial issues for selecting ISDN telephone services. This chapter presents a more cerebral look at ISDN from the point of view of a consultant, network designer, application integrator, or computer and network operations manager. Although the view is consistent with Chapter 4, this chapter pushes further into the issues of circuit selection, wiring, and connectivity options and practical information for wiring home offices, sites with a minimal number of ISDN users, and the corporate megalith that presumes to migrate all analog services into the integrated ISDN hub. ISDN is scalable and provides many options for a planned, phased, or incremental migration from analog services to more capable digital connections. To this end, recognize that you can install ISDN in one of several options. These are shown in Figure 9.1.

ISDN is specially designed to deliver digital connections through existing copper twisted-pair lines, and it is eminently suitable for small office and home office applications and systematic replacement of older Centrex and Private Branch Exchange (PBX) systems. ISDN can be the only line into a site, providing all telephone and data connectivity. In that scenario, you will need an NT-1 or NT-2, Electronic Key Telephone System (EKTS), or ISDN phone for the telephone and fax services. Data lines pass through the PBX or EKTS system. It is also feasible to install separate ISDN lines to each user as a primary or secondary line. At the financial crossover, it becomes desirable to replace Basic Rate Interface (BRI) lines with one or more Primary Rate Interface (PRI) lines. By the way, just because you have preexisting BRI lines does not mean you have to yank them out when you install a PRI circuit. You can mix and match BRI and PRI circuits through the same or different equipment. However, you probably want to arrange with the service provider for maintaining logical hunt groups, call forwarding options, and provisioning conformity so that lines and services are interchangeable. Figure 9.2 illustrates the interchangeability with ISDN.

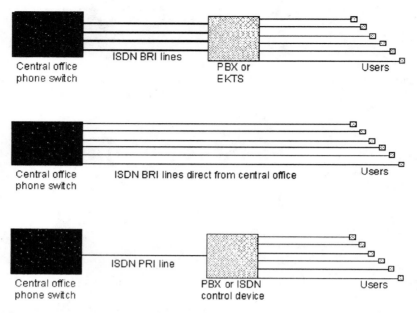

Figure 9.1 ISDN connection options.

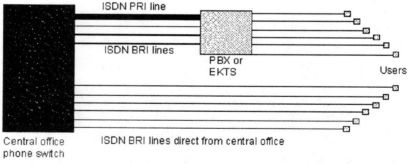

Figure 9.2 ISDN interconnection options.

You can route ISDN BRI into the same customer premises equipment (CPE) with one or many PRI lines or bypass the central facilities directly and route the BRI lines directly to users. An example of this architecture is a medium-sized business with PRI and supplemental BRI lines added as needed to handle standard PBX services. At some load, the costs for installing and maintaining another PRI line will be less than the BRI lines and they can be replaced. More likely, the telco will leave the existing BRI lines in place but will just deactivate them, assuming that growth will require additional BRI lines at some point in the future.

The direct BRI lines are typically routed to a Local Area Network (LAN) communication server for dial-in or dial-out data services. Ideally, the PBX or EKTS would be the termination point for all ISDN lines and the communication server would request lines as needed when needed. This pooling of ISDN channels increases the overall efficiency and utilization level of all the circuits. For example, if the communication designers assume 20 percent overcapacity on lines into the PBX or EKTS and similar overcapacity for the bank of lines to the communication server, there is more overcapacity than is truly needed. The consolidation of all lines typically means that a company will not need any direct ISDN lines at all. If or when voice or data callers get busy signals or cannot get a free outgoing channel, you add more lines and pool them.

While it may be possible to mix ISDN and standard T-1 or analog circuits into a PBX, EKTS, or hub device and thus create a less traumatic migration from analog to digital, you need to verify that you can integrate services and that you are not limiting the features available through ISDN. You may want to phase out analog services with the telemarketing group first so that they can use the Automatic Number Identification (ANI), Caller ID, automated attendant, conferencing features, and call hold features of ISDN that just are not necessary for outgoing analog modem lines. In addition, the telco may not be able to forward analog calls to ISDN channels or the other way around. You need to have a clear understanding of the technology and its limitations as implemented with your local telco carrier.

Also, ISDN BRI and ISDN PRI are not the end-all for digital communications. There are many other options that will integrate into the standard telecommunication infrastructure, as Chapter 2 discussed. Community Antenna Television (CATV) modems will not integrate with the phone system unless the cable TV companies install switching equipment that is similar to that of phone companies and provide the same penetration that the phone companies presently have. However, frame relay, X.25, Asynchronous Transfer Mode (ATM), Digital Subscriber Line (DSL), High-Data Rate Digital Subscriber Line (HDSL), MultiMegabit Data Service (MMDS), and Switched Megabit Data Service (SMDS) are part of the same telecommunication infrastructure. Realize that both cable modems and xDSL (such as ADSL, HDSL, and RDSL) represent shared media and unswitched technologies that still lack the flexibility of ISDN services. Many vendors are exploring ways to connect these services and Internet phone calls into the POTS matrix. While many of these protocols are not widely available or fully adapted to data, voice, and video transmission at the current time, they represent the migration from analog services to a unified digital totality. These services are shown in Figure 9.3.

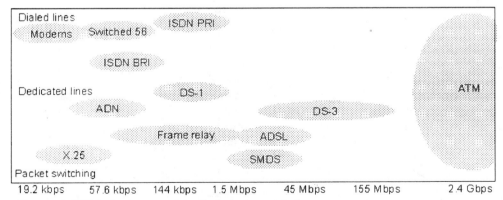

Figure 9.3 Digital telecommunication services spectrum.

Project Planning

Installation requires a good plan. Network blueprints should be detailed enough to represent all the network hardware components required for a complete installation. If the vendor or telephone service provider does not supply this or complete network parts and installation services, it is vital to coordinate the arrival of all the network components. You will need a complete bill of materials. The network will not work without the ISDN line, the ISDN line provisioned correctly, the ISDN terminal adapters, inside wiring, patch cords, or any number of other components, not the least of which is functioning network software. You need to corroborate that the network operating system or workstation can actually talk to the ISDN terminal adapter in a functional way. The installation can certainly be staged, because many installation procedures can be completed concurrently. Just about any missing part, no matter what its simplicity or general insignificance, can become a critically needed component. As I mentioned in Chapter 4, ISDN ordering and provisioning is not really a difficult process, perhaps requiring a maximum of 2 hours for a very difficult setup. However, those 2 hours could be spread over 10 days, and that elapsed duration could present a critical path.

Therefore, plan your installation timetable with an eye to the steps in network installation and coordinate delivery of components to ensure a smooth process. Figure 9.4 illustrates the major critical steps and a logical progression for network installation. Each step encompasses several individual stages not included in the figure. Inside wiring, for example, requires cutting cable to length, cutting a hole through an outside wall (do you have the masonry drill bits?), pulling the cable through the ceiling, applying end connectors, and terminating the actual cable at a patch panel or switch plate.

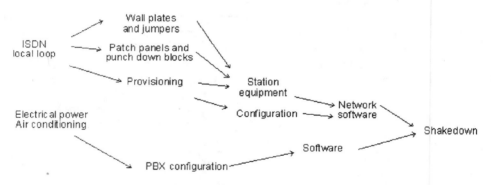

Figure 9.4 Critical paths for ISDN installation.

It is also relevant to create a project management time chart, sometimes called a Gantt chart, to show the relative sequence of events and the order in which they must occur. These charts also show the critical paths for network installation and what tasks can delay completion, as Figure 9.5 shows. By the way, just about any task or step can hold up installation completion, but critical steps (usually beyond your control) include arrival of materials and installation of wiring. Incompatibilities between software and hardware and version issues represent another significant stumbling block, one that is usually difficult to plot on a chart. It becomes a delay that you will want to build into your management planning model.

ISDN installation is a mechanically simple process, but a challenging experience if the technology is new to the organization. Sufficient skills may not have been developed to adequately identify and repair network problems. Therefore, consider the benefits of hiring the vendor or an experienced contractor to string and tap coaxial cable. Gather as much outside experience and expertise as possible to augment the internal network administrative team. The installation procedure requires a shakedown period. Premature acceptance of a new installation is unwise, and making or disseminating promises of network availability before most of the problems have been resolved will devalue the perception of the network administration team. Budget sufficient time for overruns, problems, and missing components. Be certain to negotiate preconditions for any final acceptance with the vendors. An installation may provide marginally acceptable operations but fail to perform the level of service desired. To forestall any future problems, define clearly what your expectations are. This is critical for successful vendor relations.

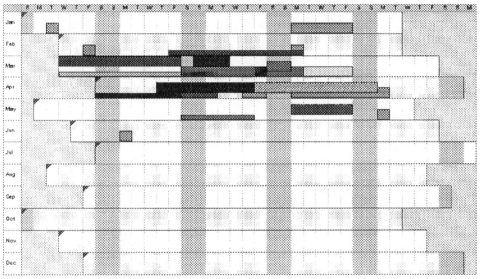

Figure 9.5 Project management time chart.

Site Planning

When you anticipate bringing ISDN service into a site, plan for the corresponding wiring, equipment availability, and switch provisioning. Factor in delivery times and the necessary lead times to pull wires and have those wires terminated and tested for functionality. Site planning is also a function of plotting the ISDN devices, functions, and services relative to a site wiring plan. You may discover you can consolidate or recycle existing wiring. You may also uncover some significant limitations.

If you are running BRI, you may be converting existing two-wire analog lines to ISDN or running new lines. If you are running PRI, this may come as fiber, or conversion of four-wire T-1 to PRI or you might run new lines. The issues for site planning are, thus, how to wire to the user termination from the local loop demarcation point. This wiring may include a telecommunication closet, NT-2 or line breakout equipment, and lines to individual users. If you are planning to recycle existing wiring, the only planning issues are wire compatibility for ISDN signals and jack compatibility and wiring for end-user or NT-1 devices. Since inside wiring is typically your responsibility, plan accordingly. You do not want to have the equipment in place, the local loop terminated outside the building, and a big gap between the users and the demarcation point. Refer to Chapter 4 for details of inside wiring.

In addition, you need to plan the wiring within the daisy-chain limitations of 280 m (991 ft) after the NT-2 and within 30 m (110 ft) of the NT-1. This is usually very generous for most installations. However, if you install an ISDN terminal adapter in one location and expect to connect the analog fax machine in another office to this, 30 m (110 ft) does not go a long way. You will need to factor in all heights and bends into this run. You may not be able to route the connection from the adapter back to the central wiring closet and then to another office since this will likely violate the crossconnect wiring length limitations. Although the length limitations for daisy-chained ISDN secondary connections are actually extensive, recall that Category 3 or Category 5 specifications limit horizontal runs to 100 m (330 ft); that should be the limiting factor in order to provide future wiring plant compatibility as discussed in Chapter 5 and described again later in this chapter with regards to wiring business sites.

Site Layouts

This section details wiring plans for various ISDN installations, including a residence, a small business, a large business, a typical office tower, and connections for remote sites, and building meshes for mixed digital and analog services.

Residence

The ISDN local loop runs in from a telephone pole or underground conduit to the demarcation point (usually a junction box) on the exterior of the house. You will need "inside wiring" from the junction box to the ISDN jack. In newer developments with underground wiring, the demarcation point is typically an underground junction box at the curbing or property line. A PVC pipe runs underground to an inlet in the house. You, or an electrician, will need to snake a line from the junction box to the house and connect it to the junction box and mount a jack where convenient. Wiring for a home office is shown in Figure 9.6.

The only complication is that pulling wiring through the ceiling, walls, or other cavities is typically inconvenient in most old or even new homes. The concept of modular wiring and smart houses is too expensive to be widely applied. You may typically run wiring inside the roof and drop it through the eaves outside to the approximate location of the home office. Drill a hole through the side of the building (first check for pipes, electrical wires, and other obstructions), and pull the wire through. There are various tools for checking for power, electrical signals, and metal before you drill a hole through a wall. Seal the holes to keep out water, dirt, and bugs, and retain any insulation inside the wall. Some local codes require that all wire is run inside a conduit; you will need to conform to those laws.

If you run wire outside the building and pull it through a window, use wire that is jacketed with a weather-resistant or weatherproof vinyl. Also, run it up high enough so that it doesn't sit in occasional snow or standing water.

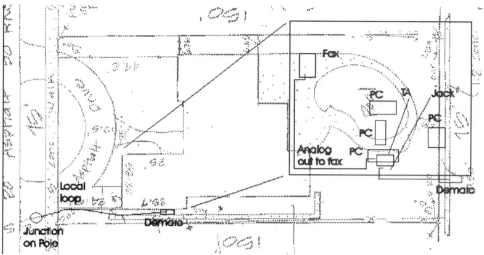

Figure 9.6 Typical home office ISDN wiring showing notations on a blueprint.

If the underground wiring if very old, you may need to dig a trench, line it with gravel, insert a new PVC pipe or shielded aluminum conduit (recommended and often required by local zoning and building codes), and string the ISDN connection from the junction box into the house. Installing the necessary wiring from the demarcation point to your ISDN hardware can be very inconvenient given the pressures of most other SOHO and work-at-home work loads, but nevertheless is a job for anyone who is reasonably handy. The difficult part is bypassing large rocks, tree roots, and buried debris. This trenching process is shown in Figure 9.7.

Figure 9.7 Digging a trench for underground wiring.

Small Business

The small business is often wired with an inside demarcation point. This is a factor of security. the fact that many businesses may be housed inside the same building, and a newer building may have been designed and prewired for the reality of phone systems. If the building is part of a strip mall, as is common in many parts of suburbia, phone access will be installed on the outside of the building in the back access alley. A friendly word of advise is to make certain the junction box in nondescript and securely latched with a padlock so that someone doesn't make after-hours calls. The wiring issues are not different from the home office scenario. Buildings tend to include false ceilings so that wiring can be pulled above the ceiling and dropped to jacks. ISDN wiring can be as simple as installing whatever runs are required for terminal adapters and new analog connections. as shown in Figure 9.8.

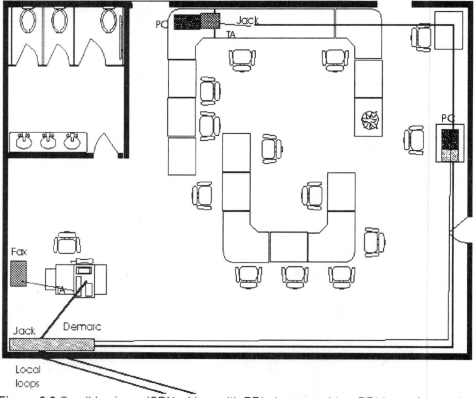

Figure 9.8 Small business ISDN wiring, with PRI phones and two BRI loops for terminal adapters providing Internet connectivity.

Large Business

You can wire a large business with spot termination of ISDN as shown for the small business. This is always an option and is in fact typically one of the ways to get clear channels past a corporate PBX system that doesn't provide dial tone for data, fax, or other services. Analogously, you would run ISDN lines past the PBX. If you wire an external NT-1, you can daisy-chain several ISDN terminal adapters, modems, or faxes to each BRI line. This will increase utilization on both channels. Since bonding is typically ineffective for increasing bandwidth for personal Internet connections, each person can get one B channel and better performance, in general, than with an analog modem. If you wire more than two users to each NT-1, you will increase line utilization but probably at the expense of immediate access for each user. More likely, ISDN at a large corporation will be implemented at a hub or PBX. This requires rewiring window offices and cubicles to support the new infrastructure. Cable paths are shown in Figure 9.9.

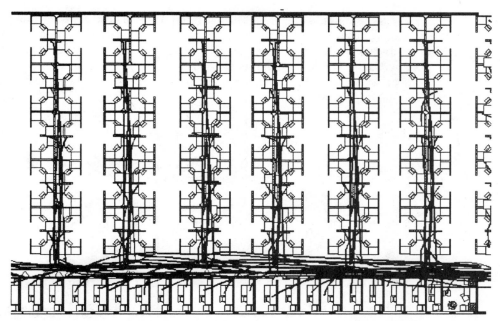

Figure 9.9 Large businesses are likely to rewire for ISDN to appreciate the full integration available.

If an organization is just replacing a PBX or key system and substituting a new digital PBX with ISDN facilities, existing wiring is probably sufficient for ISDN. Remember, ISDN only needs voice-grade or Category 1 wiring. It does need to test out for conductivity and proper impedance. However, when there are LANs and a desire for integration with client/server and desktop computing applications, typically each user will require several pairs for the phone system to support the forwarding and conference call features and another two pairs for LAN connections. Although pooling and consolidating modem and terminal adapter banks and the actual channels make more sense from a financial standpoint, some organizations include two more pairs for serial terminals, future videoconferencing support, and desktop modems or ISDN terminal adapters. When a large business or organization chooses this course of action, consider the suggestions in Chapter 4: Wire with Category 5, shielded foil, or Category 6 wiring, which is supposed to be completely suitable for ATM on copper to the desktop.

One of the issues which you should recognize with large sites is the need to maintain wire path lengths less than 100 m (330 ft) between end-user equipment and *each* telecommunication closet. The drive distances from PBXs and particularly with high-speed LANs are generally limited to that length. Although you can run the true ISDN signal from the telco up to 5.6 km (3.2 mi), realize that once the PRI or BRI channels have been demultiplexed, they lose the protective signal balance of all the channels. In addition, if you conform to the Electronics Industry Association/Telecommunication Industry Association (EIA/TIA) wiring recommendations, any new wiring infrastructure is more likely to support future LAN and telecommunication services.

Office Tower

The office tower is really no different from the large business site. However, each floor is more likely to include a single telecommunication closet within 100 m (330 ft) of every jack. In addition, you are likely to have multiple floors, each about the same as any other floor. Although wiring is distributed from the telecommunication closet on each floor, as shown in Figure 9.10, you need to be clear about where the demarcation point for ISDN lines will be.

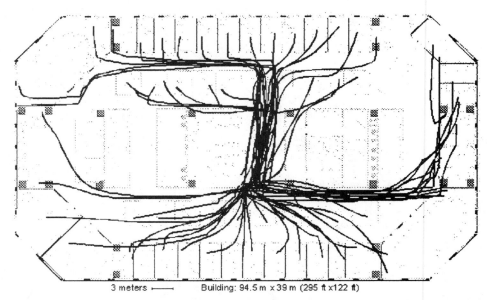

3 meters ⊢——⊣   Building: 94.5 m x 39 m (295 ft x122 ft)

Figure 9.10 A corporate tower is wired like any large site but also has the additional constraints of heights between floors on multiple floor sites.

There is an added height issue. Consider that a 45-story building represents 145 m (470 ft), and this can easily create signal drive distance problems for ISDN PBX systems. AT&T and PacBell solve problems for some organizations choosing ISDN by providing Centrex ISDN services. Although the organizations lose the flexibility provided by in-house call management and telephony integration, this solves the same vertical wiring problems because every line is effectively wired through the telco's central office switches. As previously stated, this can add to basic telephony costs because every call within the building becomes a call through the telco. However, the service provider should be willing to negotiate more favorable prices.

Conclusion

This chapter presented an overview of ISDN from the point of view of a consultant, network designer, application integrator, or MIS manager actually planning ISDN installation, provisioning, and usage requirements. The appendixes include supplemental references, a glossary of terms common to ISDN and the language specific to analog communications, and sources for additional information, Internet ISDN web sites, and contacts for ISDN vendors worldwide.

Sources and References

Sources

Bell, Jim, V.P., Sales and Marketing, Trend Communications, Inc., Reston, VA 22091.

Caswell, Sara, NIUF Secretariat, North American ISDN Users' Forum, Gaithersburg, MD 20899.

"ISDN Price Watch Residential and Business," *Computer Telephony*, October 1995, pp. 80-82.

Cwikla, John, Technical Support Analyst, US Robotics, Skokie, IL 60076.

Dabnor, John, V.P., Marketing and Planning, Telecommunications Division, Tadiran, Clearwater, FL 34620.

Davis, Trish, Marketing Communications, Elsa Inc., San Jose, CA 95131.

Emm, Gerry, Marketing Manager, Ascend Communications, Alameida, CA 94502.

Geble, Skip, IBM Product Support, Armonk, New York 01522.

Hamann, Elise, Director of Media Relations, American Power Conversion, West Kingston, RI 02892.

Haskell, Marcia, Technical Support Coordinator, Xircom Systems, Salem, NH 03079.

Lombardi, Michelle, Marketing Coordinator, Network Communications Corporation, Bloomington, MN 55437.

Piepiorra, Frank, President, Data TeleMark, McLean, VA 22102.

Rice, Dwight, BellSouth Business Systems, Birmingham, AL 35243.

Specht, Tracy, White Pine Software, Nashua, NH 03060.

Sturgess, James, Senior Marketing Manager, Motorola Systems Group, Toronto, Ontario, Canada M5C-2B8.

Thomas, Stacy, Public Relations Coordinator, ADTRAN, Huntsville, AL 35807.

Tillman, Alex, V.P. Marketing and Sales, Azure Technologies, Hopkingtin, MA 01749.

Winslow, Mike, National Accounts Manager, RADComm, Oak Moutain Side, NJ 07082.

Wood, Lori, Connectix Corporation, San Jose, CA 93027.

Provider Contacts

Ameritech: (800) 832-6328

  National ISDN Hotline (800) 543-ISDN ((800) 543-4736)

Bell Atlantic: (800) 570-ISDN

  Pricing, availability, tariffs, applications or ordering: (301) 236-8163

  ISDN Sales & Technology Center: (800) 570-ISDN or (800) 570-4736

Bellcore National ISDN Information Clearing House Hotline: (800) 992-4736

Bellcore's *ISDN Deployment Data*, Special Report (SR) 2102.

Bellcore document ordering: US: (800) 521-2673; other: (908) 699-5800

BellSouth: (800) 858-9413

Cincinatti Bell: (513) 566-DATA or (513) 566-3282

GTE: Menu-driven information service: (800) 4GTE-SW5.

  Florida, North Carolina, Virginia, and Kentucky: (800) 483-5200

  Illinois, Indiana, Ohio, and Pennsylvania: (800) 483-5600

  Oregon and Washington: (800) 483-5100

  California: (800) 483-5000

  Hawaii: (800) 643-4411

  Texas: (800) 483-5400

Nevada Bell: (702)-688-7124 (contact Lyle Walters)

New YorkNEX: (800) 438-4736, (800) GET-ISDN, (800) 698-0817, or (212) 626-7297.

 ISDN Information Hotline (800) GET-ISDN or (800) 438-4736

 or Roy Ray (914) 644-5152

Pacific Bell:

 ISDN Availability Hotline (automated audio response): (800) 995-0346

 ISDN Telemarketing (ordering information): (800) 662-0735

 ISDN service center: (800) 4PB-ISDN

 NI-1 Product Mgr.: (510) 823-5118

Rochester Telephone: (716) 777-1234

SNET: Donovan Dillon (203) 553-2369

Southwestern Bell: (800) SWB-ISDN or (800) 792-4736

 Austin: (512) 870-4064

 Dallas: (214) 268-1405

 Houston: (713) 567-4300

 San Antonio: (512) 351-8050

 Business Services (Cyd McInerney): (314) 235-1567

STENTOR (Canada): Steve Finlay (604) 432-3527

US West: (303)896-8370 (contact Julia Evans)

 Large businesses: Ron Woldeit 206-345-4759

 Small-Medium businesses: Mitch Glasser 602-351-6196

 Home Offices/Home-Based businesses: (800) 898-WORK (9675)

Commercial Contacts

Provider	Locations	Telephone Number
Ameritech	Illinois, Indiana, Michigan, Ohio, Wisconsin	(800) TEAMDATA
AT&T	Nationwide PRI, some BRI	(800) 367-7956
Bell Atlantic	Delaware, Maryland, New Jersey, Pennsylvania, Virginia, Washington D.C., West Virginia	(800) 570-ISDN
Bell South	Alabama, Florida, Georgia, Kentucky, Louisiana, Mississippi, North Carolina, South Carolina, Tennessee	(800) 428-ISDN
Cincinnati Bell	Kentucky, Ohio	(513) 566-DATA
MCI Communications	Nationwide (US) BRI and PRIq	(800) 727-2074
Nevada Bell	Nevada	(702) 333-4811
New YorkNEX	Maine, Massachusetts, New Hampshire, New York, Rhode Island, Vermont	(800) GET-ISDN
Pacific Bell (PacBell)	California	(800) 4PB-ISDN
Rochester Telephone	New York	(716) 777-1234
SBC Communications	85% availability of BRI and PRI of southwest	(314) 235-9553
So. New England Telephone	Connecticut	(800) 430-ISDN
Southwestern Bell	Missouri, Texas	(800) SWB-ISDN
Sprint Corporation	Nationwide (US) PRI, some BRI	(913) 624-4308
Stentor Alliance	Ottawa, Canada	(403) 944-8130
U.S. West	Arizona, Colorado, Idaho, Iowa, Minnesota, Montana, Nebraska, New Mexico, North Dakota, Oregon, South Dakota, Utah, Washington, Wyoming	(800) 898-WORK

International Contacts

IfN - Ingenierubuero fuer Nachrichtentechnik
Haidelmoosweg 52
D - 78467 Konstanz
Tel: +49 7531 97000-0
FAX: +49 7531 74998

United Kingdom

British Telecom ISDN Helpdesk
0800 181514 from within the UK,
+44 117 921 7764 from outside the UK

Mercury Data Communication
0500 424194 from within the UK,
+44 181 914 2335 from outside the UK

Internet WWW and FTP Addresses

Alliance for Competitive Communications: FTP://bell.com.pub

Ameritech: http://www.ameritech.com

Architectures for BISDN Networks: A Performance Study

http://www.niuf.nist.gov/sources/architec/broadban.html

AT&T (now Lucent Technologies): http://www.lucent.com

Bell Atlantic: http://bel-atl.com

Bell Atlantic ISDN support and information: http://www.ba.com/isdn.html

Belcore (now Lucent Technologies) http://www.belcore.com/ISDN

BellSouth: http://www.bellsouth.com

ISDN resources on CompuServe: GO ISDN

California ISDN User's Group: http://isdn.ciug.org

Combinet ISDN Availability Report: http://www.combinet.com/deploy.html

Dan Kegels' ISDN page: http://www.alumni.caltech.edu/~dank/isdn

David Ginsberg's ISDN Web Page: http://www.caprica.com/~dginsberg

Desktop videoconferencing resources:

http://www-cic2.lanl.gov/documentation/videocon.html

European ISDN User Forum:

    http://www.dcs.aber.ac.uk/Public/Research/Telematics/EIUF.index.html

Frequently asked questions (FAQs) on ISDN:

    ftp://rtfm.mit.edu/pub/usenet/news.answers/isdn-faq

ISDN News Group: comp.dcom.isdn

ITU (formerly CCITT) main page: http://info.itu.ch

ITU-T (formerly CCITT) standards: http://info.itu.ch/itudoc/3500/3734.html

National ISDN Users Forum (NIUF): www.niuf.nist.gov/misc/niuf.html

   ttp://www.niuf.nist.gov/misc/niuf.html#organization

Newspage ISDN: http://www.newspage.com

MCI: http://www.mci.com

Microsoft's TAPI SDK on CompuServe: GO WINEXT

ISDN to the Home: http://www.fc.net/~linas/isdn.html

Microsoft's ISDN Window support page: http://www.microsoft.com/windows/getisdn

National ISDN Users Forum (NIUF): www.niuf.nist.gov/misc/niuf.html

New YorkNEX: http://www.nynex.com

OCN ISDN Info Center: http://isdn.ocn.com

Pacific Bell: http://www.pacbell.com/isdn/book/toc.html

Pacific Bell Gopher Site: gopher://gopher.pacbell.com

SouthWestern Bell (SBC Communications): http://www.swbell.com

Sprint Corporation: http://www.sprint.com

Stentor Alliance: http://www.stentor.ca

US West: http://www.uswest.com

Contact Information for Public Commissions and Consumer Advocates

ISDN circuit tariffs vary widely by state and sometimes even within each state. The cost for providing the BRI circuit is not much different than from a standard POTS line. However, aspects of fixed-cost pricing and per-minute usage have been implemented different ways by the local exchange carriers (LECs). This in part accounts for some of the disparity. In addition, complaints from the LECs and Regional Bell Operating Companies (RBOCs) and local operating companies over the longer connection times with data calls (estimated at approximately 34 minutes) than with voice or fax calls (estimated on average at 4 minutes) makes ISDN circuits used for data applications more expensive to support. It is clear that the providers have underestimated the connection time for ISDN calls, but it is not clear that the providers incur additional service costs. However, it is clear that longer calls do create blocking patterns at central office switches that will increase the percentage of calls (POTS and ISDN) that cannot be completed on the first attempt. This will require increased central office switching capacity, but in reality it is more a function of providing more lines than in providing more voice lines.

Another serious concern to the service providers is that ISDN innately supports Caller ID, callback, conferencing, call-forwarding, and many other services not provided with POTS, not provided by electronic key telephone systems (EKTS), or most office Centrex switchboards. The providers earn extraordinary profits on these add-on services, which, for the most part with ISDN equipment and the newer ISDN-enabled software, are provided for very low costs. For example, Connectware's Cruiser (like many high-end communication packages for analog modems) provides small office/home office E-mail, Internet connectivity, voice mail, fax-on-demand, faxback, and BBS connectivity integrated into an ISDN environment. Even in Delaware or Florida where ISDN is expensive, it can be less expensive than comparable POTS lines with these Centrex add-on services. In states such as Tennessee or California where ISDN is a bargain and often less expensive than even a standard POTS line, ISDN can be a cost-effective bargain for voice (and analog fax) services only where users can bypass the expensive call-waiting, Caller ID, callback, E-mail, and call-forwarding services. This represents a fundamental shift in the economic basis for many service providers that will cause some to be less profitable or even unprofitable. They will eventually petition the tariff commissions to rebalance the rates for services (where a monopoly or oligopoly exists) and make ISDN more expensive relative to other services, and perhaps more expensive on a per-minute or per-byte unit than frame relay, SONET, ATM, dedicated lines, and the xDSL sevices. The reality, as I see it, is that ISDN is a service that seemed too little and too late to many, is more efficient than analog POTS, provides more services and bandwidth than POTS, but really costs no more to provide than POTS.

The listings of the public commissions are included here because an ISDN user (and every citizen for that matter) should have a voice in ISDN pricing. You should exercise that voice to prevent overpricing ISDN services. The listings of state public service commissions are listed beginning on the following page.

Alabama

Jim Sullivan, President
Alabama Public Service Commission
Office of Attorney
Box 991
Montgomery, Alabama 36101-4991
Voice: (205) 242-5207
Fax: (205) 240-3351

Mary Elizabeth Culberson, Acting Chief
Alabama Utilities Division
Alabama Statehouse
11 South Union Street
Montgomery, Alabama 36130
Voice: (334) 242-7414
Fax: (334) 242-74558

Alaska

Don Schroer, Chair
Alaska Public Utilities Commission
1016 W. 6th Avenue, Suite 400
Anchorage, Alaska 99501
Voice: (907) 276-6222
Fax: (907) 276-0160

Arizona

Greg Patterson
Arizona Corporation Commission
1200 W. Washington Street
Phoenix, Arizona 85007
Voice: (602) 542-2237
Fax: (602) 542-5560

Renz D. Jennings, Chair
Residential Utility Consumer
1501 West Washington Street, Suite 227
Phoenix, Arizona 85007
Voice: (602) 542-3733
Fax: (602) 542-3738

Arkansas

Shirley Guntharp
Office of the Attorney
P.O. Box 400
Little Rock, Arkansas 72203-0400
Voice: (501) 682-2051
Fax: (501) 682-1453

Sam I. Bratton Jr., Chair
Arkansas Public Service Commission Alabama
200 Tower Building
323 Center Street
Little Rock, Arkansas 72201
Voice: (501) 682-2007
Fax: (501) 682-8084

California

Daniel William Fessler, President
Toward Utility Rate Normalization
625 Polk Street, Suite 403
San Francisco, California 94102
Voice: (415) 929-8876
Fax: (415) 929-1132

Office of the Director
California Public Utilities Commission
California Consumer Affairs Branch
505 Van Ness Avenue
San Francisco, California 94102-3298
Voice: (415) 703-3703
Fax: (415) 703-1758

Michael Shames
Utility Consumers Action Network
1717 Kettner Boulevard, Suite 105
San Diego, California 92101-2532
Voice: (619) 696-6966
Fax: (619) 696-7477

Colorado

Robert J. Hix, Chair
Colorado Public Utilities Commission
1580 Logan Street, Office Level 2
Denver, Colorado 80203
Voice: (303) 894-2000 x 401
Fax: (303) 894-2065

Ronald J. Binz
Office of Consumer Council
1580 Logan Street, Suite 610
Denver, Colorado 80203
Voice: (303) 894-2121
Fax: (303) 894-2117

Connecticut

Reginald J. Smith, Chair
Connecticut Department of Public
Utility Control
Central Park Plaza
New Britain, Connecticut 06051
Voice: (203) 827-1553
Fax: (203) 827-2613

John F. Merchant
Office of Consumer Counsel
136 Main Street, Suite 501
New Britain, Connecticut 06051
Voice: (203) 827-7887
Fax: (203) 827-7760

Delaware

Robert J. McMahon, Chair
Delaware Public Service Commission
1560 South DuPont Highway
DE P.O. Box 457
Dover, Delaware 19903-0457
Voice: (302) 945-5212
Fax: (302) 739-4849

Patricia Stowell
Office of the Public Advocate
Carvel State Office Building
820 N. French Street, Fourth Floor
Wilmington, Delaware 19801
Voice: (302) 577-3087
Fax: (302) 577-3297

District of Columbia

Edward M. Meyers, Commissioner
District of Columbia Public Service
450 Fifth Street N.W.
Washington, DC 20001
Voice: (202) 626-5100
Fax: (202) 638-1785

Elizabeth A. Noel, Counsel
Office of the People's Commission
1133 15th Street N.W., Suite 500
Washington, DC 20005
Voice: (202) 727-3071
Fax: (202) 727-1014

Florida

Susan F. Clark, Chair
Florida Public Service Commission
2540 Shumard Oak Blvd.
Gearld Gunter Building
Tallahassee, Florida 32399-0850
Voice: (904) 413-6040
Fax: (904) 487-1716

Steve Shreve
Office of Public Counsel
812 Claude Pepper Building
111 West Madison Street
Tallahassee, Florida 32399-1400
Voice: (904) 488-9330
Fax: (904) 488-4491

Georgia

Nancy G. Gibson
Georgia Public Service Commission
244 Washington Street S.W
Atlanta, Georgia 30334-5701
Voice: (404) 656-4512
Fax: (404) 656-2341

Bob Durden, Chair
Office of Consumers' Utility
84 Peachtree Street N.W., Suite 201
Atlanta, Georgia 30303-2318
Voice: (404) 656-3982
Fax: (404) 651-9394

Hawaii

Yukio Naito, Chair
Hawaii Public Utilities Commission
465 South King Street
Kekuanao's Building No. 103
Honolulu, Hawaii 96813
Voice: (808) 586-2020
Fax: (808) 586-2066

Charles W. Totto
Division of Consumer Advocacy
P.O. Box 541
Honolulu, Hawaii 96809
Voice: (808) 586-2800
Fax: (808) 586-2780

Idaho

Ralph Nelson, President
Idaho Public Utilities Commission
P.O. Box 83720
Boise, Idaho 83720-0074
Voice: (208) 334-2898
Fax: (208) 334-3762

Illinois

Martin Cohen
Citizens Utility Board
208 S. LaSalle, Suite 1760
Chicago, Illinois 62604
Voice: (312) 263-4282
Fax: (312) 263-4329

Dan Miller, Chair
Illinois Commerce Commission
Leland Building, 527 East Capitol Avenue
P.O. Box 19280
Springfield, Illinois 62794-9280
Voice: (312) 814-2859
Fax: (217) 782-1042

Marie Spicuzza
Attorney's Office
Cook County State's Public Utility Division
28 N. Clark Street, Suite 400
Chicago, Illinois 60602
Voice: (312) 345-2436
Fax: (312) 345-2401

Indiana

John F. Mortell, Chair
Indiana Utility Regulatory Commission
Indiana Government Center South, Room N501
302 West Washington Street
Indianapolis, Indiana 46204
Voice: (312) 232-2702
Fax: (312) 232-6758

Ann Becker
Office of Utility Consumer
100 N. Senate Avenue, Suite E306
Indianapolis, Indiana 46204-2208
Voice: (317) 232-2494
Fax: (317) 232-5923

Iowa

Allan T. Thoms, Chair
Iowa Utilities Board
Lucas State Office Building
State Capitol
Des Moines, Iowa 50319
Voice: (515) 281-5167
Fax: (515)281-8821

James Maret
Office of Consumer Advocate
State Capitol
Lucas State Office Building, 4th Floor
Des Moines, Iowa 50319
Voice: (515) 281-5984
Fax: (515) 242-6564

Kansas

Susan M. Seltsam, Chair
Kansas Corporation Commission
1500 S.W. Arrowhead Road
Topeka, Kansas 66604-4027
Voice: (913) 271-3166
Fax: (913) 271-3354

Citizens' Utility Ratepayer Board
1500 S.W. Arrowhead Road
Topeka, Kansas 60604-4027
Voice: (913) 271-3200
Fax: (913) 271-3116

Kentucky

George Edward Overbey Jr.,Commissioner
Kentucky Public Service Commission General
730 Schenkel Lane
P.O. Box 615
Frankfort, Kentucky 40602
Voice: (502) 564-3940
Fax: (502) 564-3460

Dennis Howard
Office of the Attorney
Public Service Litigation Branch
P.O. Box 2000
Frankfort, Kentucky 40602
Voice: (505) 573-4994
Fax: (505) 573-8315

Louisiana

John F. Schwegmann, Chair
Louisiana Public Service Commission
P.O. Box 91154
Baton Rouge, Louisiana 70821-9154
Voice: (504) 838-5250
Fax: (504) 838-5252

Maine

Thomas L. Welch, Chair
Maine Public Utilities Commission
242 State Street, State House Station 18
Augusta, Maine 04333
Voice: (207) 287-3831
Fax: (207) 287-4317

Stephen Ward, Public Advocate
State House Station 112
Augusta, Maine 04333
Voice: (207) 287-2445
Fax: (207) 287-1039

Maryland

H. Russell Frisby Jr., Chair
Maryland Public Service
Commission Office of People's Counsel
6 Saint Paul Street, 16th Floor
Baltimore, Maryland 21202-6806
Voice: (410) 767-8072
Fax: (410) 333-6495

Michael Travieso
6 St. Paul Street, Suite 2102
Baltimore. Maryland 21202
Voice: (410) 767-8150
Fax: (410) 333-3616

Massachusetts

Kenneth Gordon, Chair
Massachusetts Department of Public Utilities
Public Protection Bureau
100 Cambridge Street
Boston. Massachusetts 02202
Voice: (617) 727-3517
Fax: (617) 723-8812

George Dean
Office of Attorney General
131 Tremont Street
Boston. Massachusetts 02202
Voice: (617)727-2200
Fax: (617) 727-1047

Michigan

John G. Strand. Chair
Michigan Public Service Commission, Special Litigation Division
Mercantile Building
6545 Mercantile Way
P.O. Box 30221
Lansing. Michigan 48909-7721
Voice: (517) 334-6370
Fax: (517) 882-5170

Paul Lark
Office of Attorney General
P.O. Box 30212
Lansing. Michigan 48909
Voice: (517) 373-1123
Fax: (517) 373-9860

Minnesota

Donald A. Storm, Chair
Minnesota Public Utilities Commission Division
121 7th Place East. Suite 350
St. Paul. Minnesota 55101-2147
Voice: (612) 296-0621
Fax: (612) 297-7073

Joan Peterson
Residential Utilities
Office of Attorney General
NCL Tower, Suite 1200
445 Minnesota Street
St. Paul, Minnesota 55101-2130
Voice: (612) 296-9412
Fax: (612) 296-7438

Mississippi

Curt Herbert Jr., Chair
Mississippi Public Service Commission
Walter Sillers State Office Building, 19th Floor
P.O. Box 1174
Jackson, Mississippi 39215-1174
Voice: (601) 961-5440
Fax: (601) 961-5469

Frank Spencer
Public Advocacy Division
Attorney General's Office
P.O. Box 220
Jackson, Mississippi 39205
Voice: (601) 359-4209
Fax: (601) 359-4273

Missouri

Allan G. Mueller, Chair
Missouri Public Service Commission
P.O. Box 360
Truman State Office Building
Jefferson City, Missouri 65102
Voice: (314) 751-3234
Fax: (314) 526-1847

Martha S. Hogerty
Office of the Public Counsel
P.O. Box 7800
Jefferson City, Missouri 65102
Voice: (314) 751-4857
Fax: (314) 751-5562

Montana

Nancy McCaffree, Chair
Montana Public Service Commission
1701 Prospect Avenue
P.O. Box 202601
Helena, MT 59620-2601
Voice: (406) 444-6165
Fax: (406) 444-7618

Nebraska

Daniel G. Urwiller, Chair
Nebraska Public Service Commission
300 The Atrium
1200 N. Street
P.O. Box 94927
Lincoln, NE 68509-4927
Voice: (402) 471-0216
Fax: (402) 471-0254

Nevada

John F. Mendoza, Chair
Nevada Public Service Commission
727 Fairview Drive
Carson City, Nevada 89710
Voice: (702) 486-2600
Fax: (702) 687-6110

Fred Schmidt
Advocate for Customers of Public Utilities
1000 East Williams Street, Suite 200
Carson City, Nevada 89710
Voice: (702) 687-6300
Fax: (702) 687-6304

New Hampshire

Douglas L. Patch, Chair
New Hampshire Public Utilities Commission
8 Old Suncook Road, Building No. 1
Concord, New Hampshire 03301-5185
Voice: (603) 271-2442
Fax: (603) 271-3878

Michael W. Holmes
Office of the Consumer Advocate
8 Old Suncook Road, Building #1
Concord, New Hampshire 03301-5185
Voice: (603) 271-1172
Fax: (603) 271-1177

New Jersey

Herbert H. Tate, President
New Jersey Board of Public Utilities
Two Gateway Center
Newark, New Jersey 07102
Voice: (201) 648-2013
Fax: (201) 777-3330

Blossom Peretz
Division of Ratepayer Advocate
P.O. Box 46005
Newark, New Jersey 07101
Voice: (201) 271-2690
Fax: (201) 624-1047

New Mexico

Lawrence B. Ingram, Chairman
New Mexico Public Utility Commission
Marian Hall
224 East Palace Avenue
Santa Fe, NH 87501-2013
Voice: (505) 827-6942
Fax: (505) 827-6973

Charles F. Noble
Consumer Protection Division
Office of Attorney General
Drawer 1508
Santa Fe, NH 87504-1508
Voice: (505) 827-6010
Fax: (505) 827-5826

New York

Richard Kessler
New York State Consumer Protection Board
Utility Intervention Office
99 Washington Avenue, 10th Floor
Albany, New York 12210
Voice: (518) 474-3514
Fax: (518) 474-2474

Harold A. Jerry Jr., Chair
New York Public Service Commission
Three Empire State Plaza
Albany, New York 12223
Voice: (518) 474-2523
Fax: (518) 473-2838

Robert Piller
Public Utility Law
39 Columbia Street
Albany, New York 12207
Voice: (518) 449-3375
Fax: (518) 432-6178

Robert Piller
Project of New York
One Penn Plaza, 8th Floor
New York, New York 10019
Voice: (212) 290-4416
Fax: (212) 290-4435

Robert Ceisler
Citizens' Utility Board
Ellicott Square Building
295 Main Street
Buffalo, New York 14203
Voice: (716) 847-3400
Fax: (716) 847-3426

Gordian Raacke
Citizen's Advisory Panel
1767 Veterans Highway, Suite 46
Central Islip, New York 11722-1538
Voice: (516) 234-0232
Fax: (516) 234-2403

North Carolina

Jo Anne Sanford
Office of Attorney General
P. O. Box 629
Raleigh, North Carolina 27602
Voice: (919) 733-7214
Fax: (919) 715-0599

Hugh A. Wells, Chair
North Carolina Utilities Commission
430 North Salisbury Street
Dobbs Building
Raleigh, North Carolina 27603

Utilities, Energy & Insurance Section
P.O. Box 29510
Raleigh, North Carolina 27626-0510
Voice: (919) 733-4249
Fax: (919) 733-7300

Robert P. Gruber
Public Staff
P.O. Box 29520
Raleigh, North Carolina 27626-0520
Voice: (919) 733-2435
Fax: (919) 733-9365

North Dakota

Leo M. Reinbold, President
North Dakota Public Service Commission
600 East Boulevard, 12th Floor
Bismarck, North Dakota 58505
Voice: (701) 328-2400
Fax: (701) 328-2410

Ohio

Robert Tongren
Office of the Consumers' Counsel
77 South High Street, 15th Floor
Columbus, Ohio 43266-0550
Voice: (614) 466-8574
Fax: (614) 466-9475

Craig A. Glazer, Chairman
Ohio Public Utilities Commission
180 East Broad Street
Columbus, Ohio 43215-3793
Voice: (614) 466-3204
Fax: (614) 466-7366

Oklahoma

Cody L. Graves, Chair
Jim Thorpe Office Building
P.O. Box 52000-2000
Oklahoma City, Oklahoma 73152
Voice: (405) 521-2267
Fax: (405) 521-6045

Oregon

Joan H. Smith, Chair
Oregon Public Utility Commission
550 Capitol Street N.E.
Salem, Oregon 97310
Voice: (503) 378-6611
Fax: (503) 378-5505

Bob Jenks
Oregon Citizens Utility Board
921 Southwest Morrison, Suite 550
Portland, Oregon 97205
Voice: (503) 227-1984
Fax: (503) 274-6847

Pennsylvania

Irwin A. Popowsky
Office of Consumer Advocate
1425 Strawberry Square
Harrisburg, Pennsylvania, 17120-1102
Voice: (717) 783-5048
Fax: (717) 783-7152

John M. Quain, Chair
Pennsylvania Public Utility Commission
P.O. Box 3265
Harrisburg, Pennsylvania 17105-3265
Voice: (717) 783-7349
Fax: (717) 783-7230

Bernard A. Ryan, Jr.
Office of Small Business Advocate
Commerce Building Suite
300 North Second Street
Harrisburg, Pennsylvania 17101
Voice: (717) 783-2525
Fax: (717) 783-2831

Rhode Island

James J. Malachowski, Chair
Rhode Island Public Utilities Commission
100 Orange Street
Providence, Rhode Island 02903
Voice: (401) 277-3500 x 108
Fax: (401) 277-6805

South Carolina

Nancy Vaughn Combs
Division of Consumer Advocacy
Department of Consumer Affairs
2801 Devine Street, 2nd Floor
P.O. Box 5757
Columbia, South Carolina 29250
Voice: (803) 734-9464
Fax: (803) 734-9365

Rudolph Mitchell, Chair
South Carolina Public Service Commission
P.O. Drawer 11649
Columbia, South Carolina 29211
Voice: (803)737-5250
Fax: (803) 737-5199

South Dakota

Kenneth D. Stofferhn, Chair
South Dakota Public Utilities Commission
State Capitol
500 East Capitol Street
Pierre, South Dakota 57501-5070
Voice: (605) 773-3201
Fax: (605) 773-3809

Tennessee

Vincent Williams
Consumer Advocate Division
Office of the Attorney General
405 James Robertson Parkway
Nashville, Tennessee 37243
Voice: (615) 741-8700
Fax: (615) 741-8724

Keith Bissell, Chair
Tennessee Public Service Commission
460 James Robertson Parkway
Nashville, Tennessee 37243-0505
Voice: (615) 741-3668
Fax: (615) 741-2336

Texas

Patrick H. Wood III, Chair
Texas Public Utility Commission
7800 Shoal Creek Blvd.
Austin, Texas 78757
Voice: (512) 458-0100
Fax: (512) 453-8271

Suzi Ray McClellan, Counsel
Office of Public Utility
7800 Shoal Creek Blvd.
Austin, Texas 78757
Voice: (615) 741-8700
Fax: (615) 741-8724

Utah

Sandy Mooy
Committee of Consumer Services
P.O. Box 45809
Salt Lake City, Utah 84145-0809
Voice: (801) 530-6645
Fax: (801) 530-7655

Stephen F. Mecham, Chair
160 East 300 South
P.O. Box 45585
Salt Lake City, Utah 84145
Voice: (801) 530-6716
Fax: (801) 530-6796

Vermont

Richard H. Cowart, Chair
Vermont Public Service Board
Chittenden Bank Building
112 State Street
Drawer 20
Montpelier, Vermont 05620-2701
Voice: (802) 828-2358
Fax: (802) 828-3351

James Volz
Division of Public Advocacy
State Office Building
120 State Street
Montpelier, Vermont 05620
Voice: (802) 828-2811
Fax: (802) 828-2342

Virginia

Preston C. Shannon, Chair
Virginia State Corporation Commission
Tyler Building
P.O. Box 1197
Richmond, Virginia 23209
Voice: (804) 371-9608
Fax: (804) 371-9376

Edward L. Petrini
Office of Attorney General
Insurance & Utilities Regulatory Section
900 E. Main Street
Richmond, Virginia 23219
Voice: (804) 786-3433
Fax: (804) 371-2086

Washington

Rob Manifold
Public Counsel Section
Bank of California Building
Suite 2000
900 4th Avenue
Seattle, Washington 98164
Voice: (206) 464-6595
Fax: (206) 389-3058

Sharon L. Nelson, Chair
Washington Utilities and Transportation Commission
Chandler Plaza Building
P.O. Box 47250
Olympia, Washington 98504
Voice: (206) 753-6430
Fax: (206) 586-1150

Evergreen Legal Services
King County Office
401 Second Avenue South, Suite 401
Seattle. Washington 98104
Voice: (206) 464-1422
Fax: (206) 382-3386

West Virginia

Boyce Griffith. Chair
West Virginia Public Service Commission
700 Union Building
201 Brooks Street
P.O. Box 812
Charleston. West Virgina 25323
Voice: (304) 340-0306
Fax: (304) 340-0325

Billy Jack Greeg
Consumer Advocate Division
723 Kanawha Boulevard East
Charleston. West Virgina 25301
Voice: (304) 558-0526
Fax: (304) 558-3610

Wisconsin

Cheryl L. Parrino. Chair
Wisconsin Public Service Commission
610 North Whitney Way
Madison. Wisconsin 53705-2729
Voice: (608) 251-3322
Fax: (608) 251-7609

David Merritt
Citizens Utility Board
16 N. Carroll Street. Suite 300
P.O. Box 7854
Madison. Wisconsin 53703
Voice: (608) 266-7897
Fax: (608) 266-1401

Wyoming

Steve Ellenbecker. Chairman
Wyoming Public Service Commission
700 West 21st Street
Cheyenne. Wyoming 82002
Voice: (307) 777-7427
Fax: (307) 777-5700

B

Glossary of Terms

0B+D

ISDN BRI service (D channel only) used solely for packet data transmission.

100Base-T

The IEEE 802.3u extension for providing Ethernet transmission at 100 Mbps on twisted pair and powered signal-regenerative hubs.

10Base-T

A reference to the Ethernet IEEE 802.3 standard supplemental definition, specifically to twisted pair wiring and connectors, twisted pair variations, and signal-regenerative powered hubs. The number scheme designates that these are baseband networks with transmission rates of 10 Mbps. The maximum contiguous cable segment length is usually limited to 100 m (324 ft) due to the extreme signal interference on the unshielded cabling.

24×7

A reference to non-stop data processing operations that run 24 hours per day, 7 days per week. Also 24-by-7.

2B+D

The Basic Rate Interface (BRI) in ISDN. A single ISDN circuit is divided into two 64 kbps digital channels for voice or data and one 16 kbps channel for low-speed data and signaling.

66 Block
The most common family of connecting blocks. Typical applications are in large buildings. but occasionally they are used on smaller installations.

A&B bit signaling
Procedure used in most T-1 transmission facilities where 1 bit from every sixth frame of each of 24 T-1 subchannels is used for carrying supervisory signaling information.

AAL
Acronym for ATM Adaptation Layer (q.v.).

ABM
Acronym for Asynchronous Balanced Mode (q.v.).

ABR
Acronym for Available Bit Rate (q.v.).

Abnormal Preamble
A packet error that occurs when the preamble doesn't match the legal 8-byte Ethernet synchronization pattern.

AC
Abbreviation for Alternating Current (q.v.).

Access Loop Diversity
A service offered by a Regional Bell Operating Company consisting of multipath communication connections to a customer site.

Access Number
The phone number that must be dialed by someone calling you.

ACD
Acronym for Automatic Call Distribution (q.v.).

ACK
Acronym for Acknowledgment (q.v.).

Acknowledgment
A message typically sent from one network device to another to acknowledge that some event (for example. receipt of a message) has occurred. Referred to as ACK.

Adapter
A PC board, usually installed inside a computer system, that provides network communication capabilities to and from that computer system. The term adapter often is used interchangeably with a Network Interface Card (q.v.).

Adapter Board
A PC board that plugs into any computer, including mainframe, minicomputer, personal computer, or workstation. Within the networking context it often refers to the network access unit, network interface card, or a network controller adapter.

Address
(1.) Data structure used to identify a unique entity, such as a particular process or network location. (2.) A reference to a source or destination station on a network.

Address Error
A packet is improperly labeled with either source or destination information.

Agent
(1.) The managed nodes on a network. (2.) A network management entity. (3.) Generally, software that processes queries and returns replies on behalf of an application, gathers information about a network device, and executes commands in response to a management console's requests. (4). In network management systems, agents reside in all managed devices and report the values of specified variables to management stations. In the Simple Network Management Protocol, the agent's capabilities are determined by the Management Information Base.

AIS
Acronym for Alarm Indication Signal (q.v.).

Alarm
A message notifying an operator or administrator of a network problem.

Alarm Indication Signal
In T-1, an all-ones signal transmitted in lieu of the normal signal to maintain transmission continuity and to indicate to the receiving terminal that there is a transmission fault that is located either at, or upstream from, the transmitting terminal. Abbreviated as AIS.

Alert
A message which warrants action at the control point.

Alternate Network Service Agreement
Customers who reside in areas where the central office switch does not support ISDN can be serviced from a neighboring central office at no additional charge. From the

customer's perspective, ISDN is readily available and affordable, but the customer *must* agree to migrate to the local central office if and when service becomes available. In most cases this will involve a change in phone number. This agreement pertains to Bell South customers only. Abbreviated as ANSA.

Alternate Serving Wire Center Support
A service offered by a Regional Bell Operating Company consisting of multipath communication connections from multiple central office or point-of-presence facilities to a customer site.

Alternating Current
Typically refers to the 120 volt (220 volt in Europe, Asia, and Africa) electricity delivered by your local power utility to the three-pin power outlet in your wall. Called *alternating current* because the polarity of the current alternates between ±60 (±50 in Europe. Asia. and Africa) times a second. Abbreviated as AC.

American National Standards Institute
The coordinating body for voluntary standards groups within the United States and a member of the International Organization for Standards. A governmental agency that maintains standards for science and commerce, including a list of acceptable standards for computer languages, character sets, connection compatibility, and many other aspects of the computer and data communications industries. Better known by the ANSI acronym.

America Standard Code for Information Interchange
A code used extensively in data transmission. 128 numerals, letters, symbols and control codes are each represented by 7-bit binary numbers. Abbreviated as ASCII.

American Wire Gauge
A measurement system for electrical wire where larger numbers represent thinner wires (based upon the extrusion of a fixed amount of metal over longer distances). Abbreviated as AWG.

Analog
A data-transfer method that uses continuously variable physical quantities for transmitting voice and data signals over conventional telephone lines. Analog transmission speed is limited by the bandwidth of the human voice.

Analog Signal
A transmission in which information is represented as physical magnitudes of electrical signals.

Analog Transmission
Signal transmission over wires in which information is conveyed through a variation of some combination of signal amplitude, frequency, and phase. Analog cellular systems transmit calls on electromagnetic radio waves that mimic the wave pattern of the voice. Analog is a duplex transmission requiring a channel to receive and a channel to transmit.

ANI
Acronym for Automatic Number Identification (q.v.).

ANSA
Acronym for Alternate Network Service Agreement (q.v.).

ANSI
Acronym for American National Standards Institute (q.v.).

API
Acronym for Application Programming Interface (q.v.).

Applet
An application started from the control panel or a hot key sequence. Control panel applets each configure a particular system feature; for example, printers, video drivers, or system sounds.

Application Programming Interface
A specification call convention that defines an interface to a service. Abbreviated as API.

Architecture
The way hardware or software is structured, usually based on a specific design philosophy. Architecture affects both a computers abilities and its limitations.

ARM
Acronym for Asynchronous Response Mode (q.v.).

ARQ
Acronym for Automatic Repeat Request (q.v.).

ASCII
Acronym for American Standard Code for Information Interchange (q.v.).

Asymmetric Digital Subscriber Line
A TV *set-top* box data communication delivery protocol supporting 6.2-Mbps downstream link with a 64-kbps upstream (and control) link supported on POTS two-wire UTP. A useful protocol for downloads but ineffective for peer-to-peer data delivery or remote LAN access. Abbreviated as ADSL.

Asynchronous
A process where overlapping operations can occur independently.

Asynchronous Balanced Mode
An HDLC (and derivative protocol) communication mode supporting peer-oriented point-to-point communications between two stations, where either station can initiate transmission. Abbreviated as ABM.

Asynchronous Response Mode
An HDLC communication mode involving one primary and at least one secondary, where either the primary or one of the secondaries can initiate transmissions. Abbreviated as ARM.

Asynchronous Transfer Mode
(1.) A cell relay packet network providing from 25 Mbps to Gigabits per second from central offices to central offices. (2.) A form of packet switching; a subset of cell relay that uses 53-byte cells (5 bytes of overhead, and another 4 bytes for LAN sequencing adaptation) as the basic transport unit. In concept, circuits of different signaling speeds can move data from desktop to desktop and across long-distance services without major changes in data format. Abbreviated as ATM.

Asynchronous Transmission
Operation of a network system wherein events occur without precise clocking. In such systems, individual characters are usually encapsulated in control bits called *start* and *stop* bits, which designate the beginning and ending of characters.

AT
The Hayes sequence for introducing commands to operate a widely used modem interface and with many V.35-compatible ISDN terminal adapters.

ATM
Acronym for Asynchronous Transfer Mode (q.v.).

ATM Adaptation Layer
The protocols which translates LAN packets into ATM cells for transport across the ATM switch. A set of four standard protocols that translate user traffic from the higher layers of the protocol stack into a size and format that can be contained in the payload of an ATM cell and returned it to its original form at the destination. Each protocol con-

sists of two sub layers: the segmentation and reassembly (SAR) sublayer and the convergence sublayer . Each is geared to a particular class of traffic, with specific characteristics concerning delay and cell loss. All protocol functions occur at the ATM end station rather than at the switch. (2.) The layer of the ATM protocol stack that handles most of the processing and routing activities. These include building the ATM header, cell multiplexing/demultiplexing, cell reception and header validation, cell routing using VPIs/VCIs, payload identification, quality of service specification, flow control, and prioritization. Abbreviated AAL.

Attenuation
Loss of communication signal energy.

Automatic Call Delivery
When roaming, incoming calls ring on your phone just as in your home area. Automated inbound call distribution system.

Automatic Call Distributor
A method of routing calls based upon caller input to the appropriate person or device. Abbreviated as ACD.

Automatic Call Reconnect
Feature permitting automatic call rerouting away from a failed trunk line.

Automatic Number Identification
A feature, often associated with ISDN, that passes a caller's telephone number over the network to the receiver so that the caller can be identified. Also known as *Caller ID*. Abbreviated as ANI.

Automatic Repeat Request
Communication method where the receiver detects errors and requests retransmissions. Abbreviated as ARQ.

Available Bit Rate
The congestion and flow control proposals for asynchronous transfer mode. This is the logic which prioritizes traffic by type, source, burstiness, importance, and criticality. *See also* Constant Bit Rate, Available Bit Rate, and Virtual Circuit Flow control. Abbreviated as ABR.

Available Channel
A channel is available if it is an allocated channel that is not currently in use.

AWG
Acronym for American Wire Gauge (q.v.).

B Channel
See Bearer channel.

B-ISDN
Acronym for Broadband Integrated Services Digital Network (q.v.).

Baby Bells
See Regional Bell Operating Company.

Backhaul
Network of connecting telephone trunks (leased lines and/or microwave) to link the sites to the central office.

BACP
Acronym for Bandwidth Allocation Control Protocol (q.v.).

Balanced Cable
Twisted-pair wiring.

Band
A portion of the radio frequency.

Bandwidth
(1.) The range (band) of frequencies that are transmitted on a channel. The difference between the highest and lowest frequencies is expressed in hertz (Hz) or millions of hertz (MHz). (2.) The wire speed of the transmission channel. (3.) The range of frequencies on the electromagnetic spectrum allocated for wireless transmission.

Bandwidth Allocation Control Protocol
An IETF standard for ISDN for dynamically adding and dropping channels on a MLPPP connection. Primary authors of this standard included Ascend Communications, Microsoft Corporation, Bay Networks, Cisco Systems, and Shiva Corporation. Abbreviated as BACP.

Bandwidth on Demand
(1.) An inverse-multiplexing method used to combine multiple channels into one single channel. (2.) A feature of inverse multiplexers which can allocate bandwidth to bursty traffic as it exceeds the currently available bandwidth. (3.) The ability to regulate channel capacity, traffic composition, and bandwidth based upon normal and burst loads. This service is primarily available with Frame relay and ATM, but also with ISDN and standard dial-up modem or digital lines. *See also* Committed Information Rate. Abbreviated as BOND.

Bandwidth Reservation
A feature of SVC where call bandwidth is allocated for high-priority or high-bandwidth.

Baseband
A transmission channel which carries a single communications channel, on which only one signal can transmit at a given time.

Basic Input/Output System
The MS-DOS library of calls for access to data. Shields applications from different types of computer hardware. Abbreviated as BIOS.

Basic Rate Interface
The ISDN interface composed of two B channels and one D channel for circuit-switched communication of voice, data and video. The interface for connecting a desktop terminal (data) and telephones (voice) to the ISDN switch. The BRI includes two 64-kbps B channels for simultaneous voice and data service and one 6-kbps D control channel for call information and customer data. *See also* Broadband Integrated Digital Services Network and Integrated Digital Services Network. Abbreviated as BRI.

Basic Rate Interface Terminal Extender
An ISDN local loop wiring extension technology to provide for central office to user demarcation lengths greater than 5 km (18,000 ft). Abbreviated as BRITE.

Baud
(1.) A unit of signaling speed represented by code elements (often bits) per second. (2.) A French-language term which represents the transfer of 1 bit.

Bearer Channel
A circuit-switched digital channel that sends and receives voice and data signals at speeds of 64 kbps. Referred to as the B channel.

Bearer Services
Services provided by the B channel, including digital telephony, toll-free inbound and outbound calls. circuit-switched data at 64 kbps, packet-switched (X.25) data, and frame relay data.

Bell Operating Companies
The local telephone companies that existed (prior to deregulation, under which AT&T was ordered by the courts to divest itself in each of the seven U.S. regions. *See also* Regional Bell Operating Company. Abbreviated as BOCs.

Bend Radius
The minimum bend allowable for Ethernet coaxial cable. PVC cable is 24.5 cm (10 in). The minimum bend for Teflon® cable is 45.7 cm (18 in).

BER
Acronym for Bit Error Rate (q.v.).

BICI
Acronym for Broadband Intercarrier Interface (q.v.).

Binary Synchronous Communications
A character-oriented data link protocol for half-duplex applications. Usually referred to as *Bisync*.

BIOS
Acronym for Basic Input/Output System (q.v.).

Bipolar
Electrical characteristic denoting a circuit with both negative and positive polarity.

BISDN
Acronym for Broadband Integrated Digital Services Network (q.v.).

Bisync
Acronym for Binary Synchronous Communication (q.v.).

Bit
Binary digit. Unit used in the binary numbering system. Can be 0 or (1.)

Bit-Oriented Protocol
Class of link layer communication protocols that can transmit frames without regard for frame content. Bit-oriented protocols are more efficient and more reliable than byte-oriented protocols, and they provide full-duplex operation.

Bit Error Rate
Percentage of transmitted bits received in error. Abbreviated as BER.

Bit Rate
The speed at which bits are transmitted, usually expressed in bits per second (bps).

Bit Stream
The sequence of bits transmitted on the forward or reverse channel of a channel stream.

Bit Time
The time it takes for a network to transport 1 bit of data. It is equivalent to the reciprocal of the network transmission rate and is a convenient measurement in performing network performance calculations.

Bit Transmission Rate
The number of bits transmitted in 1 second.

Bits per Second
A rate at which data is transmitted over a communications channel. Abbreviated as bps.

Blackout
Total loss of all electrical power.

BOC
Acronym for Bell Operating Company (q.v.).

BOND
Acronym for Bandwidth on Demand (q.v.).

Bonding
A reference to multiple channels providing concurrent communication using the protocol, Bandwidth on Demand (q.v.).

Bottleneck
(1.) Any obstruction that impedes completion of a task. (2.) The critical path in task completion; the path without slack time. (3.) The task or process in a larger system that has no slack time or the task which delays subsequent and sequential tasks, so that system completion is delayed.

BPS
Acronym for Bits Per Second (q.v.).

Break
A physical break (electrical or optical) in the network media that prohibits passage of the transmission signal.

Breakout Bundle
A single package containing multiple optical fibers, wires, and cables supporting extraction of single wires at intermediate places along the full length of the wiring package.

BRI
Acronym for Basic Rate Interface (q.v.).

Bridge
A device that interconnects networks using *similar* protocols. *See also* Router.

BRITE
Acronym for Basic Rate Interface Terminal Extender (q.v.).

Broadband
As contrasted with baseband, a transmission system that multiplexes multiple independent signals onto one cable. In telecommunications terminology, any channel having a bandwidth greater than a voice-grade channel (4 kHz). In LAN terminology, a coaxial cable on which analog signaling is used. Also called *wideband*.

Broadband Integrated Services Digital Network
A digital data network for transmitting large amounts of data, voice, and video over long distances. SONET or ATM. Communication standards developed by the ITU/TTS to handle high-bandwidth applications such as video BISDN will use ATM technology over SONET-based transmission circuits to provide data rates of 155 Mbps to 622 Mbps and beyond. *See also* Basic Rate Interface, Integrated Services Digital Network, and Primary Rate Interface. Abbreviated as B-ISDN.

Broadband Intercarrier Interface
A carrier-to-carrier interface similar to private network-to-network interface (PNNI) but lacking some of the detailed information offered by the PNNI because carriers are less likely to let their switches share routing information or detailed network maps with their competitors' gear. This supports PVC and SVCs . Abbreviated as BICI.

Broadcast
(1.) A transmission in one or more channel streams that is intended to be processed by all the channel stream. (2.) The transmission from one station on a network, to two or more stations on the network or to a frame with a destination broadcast address. (3.) A message sent to all network destinations.

Brownout
When voltage drops more than 10 percent below what is normal.

BSC
Acronym for Binary Synchronous Communication (q.v.).

Buffer
A temporary storage area that is used as a staging area when moving memory between RAM, applications, and storage disks. Usually refers to a memory structure for contain-

ing a block read from or written to a hard disk. A storage area used for handling data in transit. Buffers often are used to compensate for differences in processing speed between network devices. Bursts of data can be stored in buffers until they can be handled by slower processing devices.

Burst
A directed transmission from one network device to another that represents a significant portion of network transmission bandwidth.

Bus-Mastering Controller
A specialized device controller which bypasses the normal CPU access process and bus access controller to move segments of memory between RAM and the device which it controls. The alternative is an ISDN slave controller.

Busy/Idle Flag
An indicator that is transmitted periodically to indicate whether the reverse channel is current in the busy state or the idle state.

Busy State
A channel stream is in a busy state if signals in the busy/idle flag indicates it.

Byte
Generic term used to refer to a series of consecutive binary digits that are operated upon as a unit; for example, an 8-bit byte.

Byte-Oriented Protocol
Class of data link communications protocols that use a specific character from the user character set to delimit frames.

Byte Reversal
The process of storing numeric data with the least-significant byte first.

Cable
A transmission medium of wires or optical fibers wrapped in a protective cover.

Cable Scanner
A testing tool which verifies the integrity and performance of network wiring and cable. It tests for electrical breaks, shorts, impedance, capacitance, as well as for signal cross-talk and signal attenuation. These tools are sometimes called *pair* or *ring scanners* when designed for Token-Ring or FDDI.

Cable Television
Formerly called Community Antenna Television, a communication system where multiple channels of programming material are transmitted to homes using broadband coaxial cable. Abbreviated as CATV.

Cable Tester
See Cable Scanner.

CAC
Acronym for Connection Admission Control (q.v.).

CACH
Acronym for Call Appearance Handling (q.v.).

Call Appearance Handling
An ISDN supplemental service for handling incoming analog or digital calls. Abbreviated as CACH.

Call Attempts
An attempt by a user to place or receive a call by pressing the send key.

Call Priority
Priority assigned to each origination port in circuit-switched systems. This priority defines the order in which calls are reconnected. Call priority also defines which calls can or cannot be placed during a bandwidth reservation.

Call Reference Number
Abbreviated as CRN.

Call Setup Time
The time required to establish a switched call between DTE devices.

Caller ID
The header with the source phone number identification that can be captured by various telephony equipment. Abbreviated as CID.

Canadian Standards Association
Agency within Canada that certifies products to Canadian national safety standards.

CAP
Acronym for Carrier Amplitude/Phase Modulation Technique (q.v.).

Capacitance
The electrical properties of the network cable and hardware.

Carrier
(1.) A company that provides telephone (or another communications) service. Also, an unmodulated radio signal. *See also* wireline. (2.) A signal suitable for modulation by another signal containing information to be transmitted.

Carrier Amplitude/Phase Modulation Technique
Paradyne signal encoding method for ADSL and ISDN services.

Category 1
The EIA/TIA recommendation for two-pair twisted (TP) to support data transmission rates to 1 Mbps. This is voice-grade wire suitable for analog telephone, facsimile, and modem connections. It is not typically used for digital lines although it is suitable for ISDN BRI and PRI. It is not in any way similar to IBM Type 1 cable.

Category 2
The EIA/TIA recommendation for two-pair twisted pair (TP) to support data transmission rates to 4 Mbps. This grade of wire is suitable for 4-Mbps Token-Ring.

Category 3
The EIA/TIA recommendation for two-pair twisted pair (TP) to support data transmission rates to 10 Mbps. This is suitable for Ethernet and 4-kbps Token-Ring.

Category 4
The EIA/TIA recommendation for two-pair twisted pair (TP) to support data transmission rates to 16 Mbps. This is suitable for Ethernet and 16-Mbps Token-Ring, but is not necessarily suitable for switched or full-duplex data transmission because of the near-end crosstalk or impedance. IBM Type 1 cable does correspond to these performance recommendations.

Category 5
The EIA/TIA recommendation for two-pair twisted pair (TP) to support data transmission rates to 155 Mbps. This is suitable for Fast Ethernet, FDDI, and ATM. IBM Type 1 (coaxial) cable might correspond to these performance recommendations.

CATV
Acronym for Cable Television (q.v.).

CBR
Acronym for Constant Bit Rate (q.v.).

CCITT
Acronym for Consultative Committee for International Telephone and Telegraph (q.v.). *See also* International Telecommunications Union.

Central Office
The telephone-switching station nearest the customer's location. A local telephone company office to which all local loops in a given area connect and in which circuit switching of subscriber lines occurs. The central office serves the businesses, organizations, and residences connected to its local loop lines. Abbreviated as CO.

Centrex
An improved PBX that also provides direct inward dialing and automatic number identification of the calling PBX. Refers to a specific AT&T telephone system product.

CEPT
Acronym for Conference Europeéne des Postes et Telecommunications (q.v.).

Challenge Handshake Authentication Protocol
A security feature that prevents unauthorized access to devices running the feature supported only on lines using PPP encapsulation. Abbreviated as CHAP.

Channel
(1.) An individual communication path that carries signals. (2.) The path between a sender and receiver (two PCs, for example) that carries one stream of data. (3.) Also describes the specific path between large computers and attached peripherals.

Channel Acquisition
The process of acquisition, up to the point of establishing digital synchronization.

Channel Attached
Pertaining to attachment of devices directly by data channels (I/O channels) to a computer.

Channel Bandwidth
The frequency range of an ISDN channel.

Channel Capacity
The amount of data that might be transferred across a channel stream per unit time. The number of users that might be supported by a channel stream.

Channel Congestion
The phenomenon whereby the amount of data that might be transferred across a channel stream decreases as the number of users increases.

Channel Seizure
The event when a channel is assigned to a phone and a path is opened up for the call.

Channel Service Unit
A digital interface device that connects end-user equipment to the local digital telephone loop. The piece of equipment that terminates the long-distance circuit in the customer's location. It is often paired with a Digital service unit. Abbreviated as CSU.

Channel Service Unit/Data Service Unit
The hardware interface on the customer side of a circuit from a central office often needed for T-1, T-3, switched digital services, and ISDN. The ISDN Terminal Adapter is the Channel Service Unit/Data Service Unit. Sometimes called *Customer Service Unit/Data Service Unit*. Abbreviated as CSU/DSU.

CHAP
Acronym for Challenge Handshake Authentication Protocol (q.v.).

CID
Acronym for Caller ID (q.v.).

Circuit
(1.) A communications link between two or more points. (2.) The physical medium on which transmission signals are carried.

Circuit Switch
A device using an open-pipe technique to establish a temporary dedicated connection between two points for the duration of the call. Used heavily in the phone-company network for ISDN, circuit switching often is contrasted with contention and token passing as a channel-access method, and with message switching and packet switching as a switching technique.

Circuit Switching
An open-pipe technique that establishes a temporary dedicated connection between two points for the duration of the call. It is a switching system in which a dedicated physical circuit path must exist between sender and receiver for the duration of the 'call.' Used heavily in phone company networks, circuit switching often is contrasted with contention and token passing as a channel-access method, and with message switching and packet switching as a switching technique.

Circuit-Switched
A communications method used by the phone company to establish a dedicated path between two devices.

Clamping Level
The amount of voltage a line protector will let through before cutting off anything higher. *See* Surge Suppresser.

Clear Channel
A reference to switched digital connections in general, ISDN in particular, applying the SS7 protocol for 64 kbps transmission capability (rather than only 56 kbps).

Clear To Send
A circuit in the RS-232 specification that is activated when the DCE is ready to accept data from the DTE.

Client
(1.) A network user or process, often a device or workstation. (2.) A secondary processor which relies upon a primary processor or server.

Client Layer
The collective term that is used to refer to the data link and physical layers of the OSI reference model.

Client/Server Computing
Term used to describe distributed processing (computing) network systems in which transaction responsibilities are divided into two parts: client (front end) and server (back end). Both terms (client and server) can be applied to both software programs or actual computing devices.

Client/Server Database
A database product with a central database engine (the server) and numerous front ends (the clients). The engine performs the database manipulations and sends the results to the clients. *See also* Client/Server.

Client/Server Processing
The establishment of a host computer (Server) to provide end-user (Client) services.

Clipping
However, when a modem signal exceeds the harmonic distortion when a signal exceeds a connection's transmission frequency range.

Cloud
The logical representation of the physical circuits required for point-to-point ISDN connectivity through private virtual networks.

CO
Acronym for Central Office (q.v.).

Coaxial Cable
An insulated, tinned copper conducting wire surrounded by foamed PVC or Teflon, and shielded by tinned copper braid or an aluminum sleeve. It can carry data transmissions at very high data rates with little loss of information and with a high immunity to outside interference.

CODEC
Acronym for Coder/Decoder (q.v.).

Coder/Decoder
A device that typically uses pulse code modulation to transform analog voice into a digital bit stream and vice versa. Abbreviated as CODEC.

Coding
Electrical techniques used to convey binary signals.

Common Carrier
Companies which provide communication networks (such as AT&T, Sprint, and PTT). A licensed, private utility company that supplies communication services to the public at regulated prices.

Communication
Transmission of information.

Composite
To combine. This is the generation of a single image from multiple image elements. It forms the basis of cast-based animation.

Compressed SLIP
A variation on SLIP in which the IP header information is compressed. Abbreviated as CSLIP.

Computer Telephone Integration
The connection of telephony services into computer services. Abbreviated as CTI.

Conference Europeéne des Postes et Telecommunications
European Common Market subcommittee that defined internetworking and communication standards for member countries. Abbreviated as CEPT.

Confidentiality

A security process that ensures that the content of a transmitted message cannot be determined except by the possessor of a key associated with the message, even if the transmitted bits are observed.

Congestion

A slowdown in a network due to a bottleneck or excessive network traffic.

Excessive network traffic.

(1.) An indication that a station does not have sufficient buffer space to copy a cell, frame, or packet addressed to it. (2.) A packet sent to the file server that cannot buffer the frame, indicates that the server NIC is considered to be congested. The board might require reconfiguration to allocate additional incoming data buffers.

Connecting Block

An insulating base with binding posts upon which connections to phones, etc. can be made. Often, insulation-splicing clips are used instead of screw terminals. Also called *terminal blocks.*

Connection Admission Control

Two mechanisms used to control the set up of virtual circuits. Overbooking, which allows one connection to exceed permissible traffic limits, assumes that other active connections are not using the maximum available resources. Full booking limits network access once maximum resources are committed and only adds connections that specify acceptable traffic parameters. Abbreviated as CAC.

Constant Bit Rate

A user network interface congestion control method designed for ATM to limit traffic to a guaranteed level of service where nodes negotiate with switches for bandwidth and reserve a fixed rate for the virtual connection. Nodes which exceed negotiated bandwidth are monitored by a *Leaky Bucket Algorithm.* Abbreviated as CBR.

Consultative Committee for International Telephone and Telegraph

Outdated name for the International Telecommunications Union (q.v.).

Contention

Access method in which network devices compete for the right to access the physical medium. *See also* Circuit Switching.

Continuous DTMF

Sends dual-tone multi-frequency (DTMF) tones—also called touchtones—for as long as the key is held down, allowing access to special services such as voice mail and answering machines that require long-duration tones.

Control Channel
The shared return channel used for transmission of digital control information.

Convergence
The ability of (and speed with which) a group of internetworking devices running a specific routing protocol to agree on the topology and routing path through the various segments.

CPE
Acronym for Customer Premises Equipment (q.v.).

CRN
Acronym for Call Reference Number (q.v.).

CSA
Acronym for Canadian Standards Association (q.v.).

CSLIP
Acronym for Compressed SLIP (q.v.).

CSU
Acronym for Channel Service Unit (q.v.).

CSU/DSU
Acronym for Channel Server Unit/Data Service Unit (q.v.).

CTI
Acronym for Computer Telephone Integration (q.v.).

Current
A measure of how much electricity passes a point on a wire in a given time frame. Current is measured in amperes or amps.

Customer Premises Equipment
Terminating equipment, such as terminals, phones, and modems, supplied by the phone company, installed at customer sites, and connected to the phone company network. Abbreviated as CPE.

Cutover
A term which defines the time when a new telecommunications system will be placed in service. Or, the act of transferring service from an existing system to a new system.

Data
When used in the context of communications refers to transmitted information, particularly information that is not interpreted by a particular protocol entity but merely delivered to a higher-level entity, possibly after some processing.

Data Communications Equipment
EIA term for devices and connections of a communications network that connect the communication circuit with the end device. Abbreviated as DCE.

Data Compression
(1.) A method of reducing the space required to represent data, as either bits, characters, replacement codes, or graphic images. (2.) A reduction in the size of data by exploiting redundancy. Many modems incorporate MNP5 or V.42 bis protocols to compress data before it is sent over the phone line. Dual compression (at the file and modem levels) actually increases the transmitted size. For compression to work, data must be sent over a clean, noise-free telephone line. Since common error-correction protocols are synchronous, there is usually a throughput gain there as well.

Data Processing
Computer operations which are geared to the entry, manipulation, and dissemination of information. Abbreviated as DP.

Data Service Unit
A device used in digital transmission for connecting data terminal equipment (DTE), such as a router, to a digital transmission circuit (DTC) or service. Abbreviated as DSU.

Data Stream
All data transmitted through a communications line in a single read or write operation.

Data Terminal Equipment
A computer terminal which connects to a host computer. It might also be a software session on a workstation or personal computer attaching to a host computer. Abbreviated as DTE.

Data-Over-Voice
A technology that allows users to converse, send faxes, receive data, and exchange mail during the same phone call and over the same analog connection. Abbreviated as DOV.

Datagrade Twisted Pair
Telephone-type wire twisted over its length to preserve signal strength and minimize crosstalk and electromagnetic interference for high-speed data communications networks. Abbreviated as DTP or as Datagrade TP.

dB
Abbreviation for Decibel (q.v.).

DC
Acronym for Direct Current (q.v.).

DCE
Acronym for Data Communications Equipment.

Decibel
A value expressed in decibels is determined as 10 times the logarithm of the value taken base 10. Abbreviated as dB.

Decompression
The restoration of redundant data that was removed through compression.

Dedicated Bandwidth
(1.) The allocation of the full protocol-and media-dependent bandwidth to two communication network devices through a packet switch. (2.) The allocation of transmission segments to devices on a network.

Dedicated Channel
An RF channel that is allocated solely for the use of CDPD.

Dedicated Line
A communications line that is not switched. When the line is not owned by the user, the term *leased line* is more common.

Delay
The time between the initiation of a transaction by a sender and the first response received by the sender. Also, the time required to move a packet from source to destination over a given path.

Demarc
Demarcation point between carrier equipment and private telephone equipment (CPE).

Demarcation Point
The position where the local telephone company wiring and their service responsibility typically ends. This is typically a wiring closet, an exterior service box, or a panel inside a building in a telecommunications closet. Sometimes referred to as just the *demarc* or the *demarcation*.

Demodulation
Process of returning a modulated signal to its original form. Modems perform demodulation by taking an analog signal and returning it to its original (digital) form.

Demultiplexing
The process of distributing data received in a shared data stream to the several entities that share the data stream. Demultiplexing can occur at several layers of a protocol stack.

DDR
Acronym for Dial-on-Demand Routing (q.v.).

DHCP
Acronym for Dynamic Host Configuration Protocol (q.v.).

Dial-on-Demand Routing
(1.) A technique whereby a router can automatically initiate and close a circuit-switched session. It permits routing over ISDN or phone lines using an external ISDN terminal adapter (TA) or modem. The router communicates to the TA what numbers to dial using V.25 bis protocol. (2.) Routing feature that provides on-demand network connections in an environment using the Public Switched Telephone Network (PSTN). Abbreviated as DDR.

Dial-Up Line
Communications circuit that is established by a switched circuit connection using the telephone network.

DID
Acronym for Direct Inward Dialing (q.v.).

Digital
A representation of information by a unit of length of the signal inversion or size of the signal strength.

Digital Service Unit
A synchronous serial data interface that buffers and controls the flow of data between a network portal, such as a bridge or router, and the channel service unit. *See also* Data Service Unit. Abbreviated as DSU.

Digital Simultaneous Voice and Data
Analog modem which supports voice communication over data communication in the same phone call. Provides 19.2 kbps with voice and 28.8 kbps with data only. Abbreviated as DSVD.

Digital Subscriber Line
A data delivery service from a telco providing 384 Mbps to T-1 rates. Abbreviated as DSL.

Digital Subscriber Loop
The physical connection from the central office to the customer site. Abbreviated as DSL.

Digital Transmission
(1.) A transmission of information represented by electrical units. (2.) Digital conversion of a conversation into the binary computer language of 0's and 1's, and transmitting the signal as a pattern of radio pulses. At the receiving end the conversation is reconstructed. Digital transmission consumes less power and the compaction of the signal allows for greater use of the radio spectrum.

Direct Inward Dialing
A feature allowing an incoming call from the public network to reach a specific telephone (i.e. an office extension) without the assistance of an attendant. Abbreviated as DID.

Directory Number
The logical phone number assigned to a bearer ISDN channel. Abbreviated as DN.

Directory Service
It is like a white pages service in which the user supplies the name of an entity, and receives information about that entity, such as address, title, and description. The directory can also be used as a yellow pages. Even if the user does not know the name of an entity, the user can supply some characteristic by which a search can be initiated and can receive one or more directory entries that match that search. A directory user can request the address of another entity in order to establish communication. Abbreviated as DS.

DN
Acronym for Directory Number (q.v.).

DOV
Acronym for Data-over-Voice (q.v.).

Downlink
The process of receiving information from a source computer. *See also* Download.

Download
The process of connecting a local computer to a remote one to acquire information from that remote computer.

DP
Acronym for Data Processing (q.v.).

DS
Acronym for Directory Service (q.v.).

DS-0
(1.) A single 64-kbps channel of a DS-1 digital facility. (2.) Data Switching class 0; a 56 kbps data channel.

DS-1
(1.) Digital (transmission) System 1, or Digital Signal level 1. (2.) Term used to refer to the 1.544 Mbps (U.S.) or 2.108-Mbps (Europe) digital signal carried on a T-1 facility.

DS11/DTI
Domestic trunk interface circuit to be used for DS-1 applications with 24 trunks.

DSL
Acronym for Digital Subscriber Line (q.v.) and Digital Subscriber Line (q.v.).

DSU
Acronym for Data Service Unit (q.v.) and Digital Service Unit (q.v.).

DSVD
Acronym for Digital Simultaneous Voice and Data (q.v.).

DTE
Acronym for Data Terminal Equipment (q.v.).; a computer terminal.

DTP
Acronym for Datagrade Twisted Pair (q.v.).

Dynamic Host Configuration Protocol
IETF specification which enables clients to borrow a uniquely assigned IP address for a short period. Abbreviated as DHCP.

E channel
64-kbps ISDN circuit-switching control channel.

E-1
The provisional transmission rate in the European digital infrastructure providing 2.048 Mbps.

E-2
The transmission rate generally available in the European digital infrastructure (8.488 Mbps). Consists of four E-1 circuits.

E-3
The highest transmission rate generally available in the European digital infrastructure (34.368 Mbps). Consists of 16 E-1 circuits.

E.164
ITU/TTS recommendation for international telecommunication numbering, especially ISDN, B-ISDN, and SMDS. An evolution of normal telephone numbers.

E-mail
Electronic mail. Computer network by which messages can be sent to others in the network.

Edge Routing
An early routing algorithm developed for Internet connectivity that is now superseded by Source Bridge Routing or the Interior Gateway Protocol.

EDI
Acronym for Electronic Data Interchange (q.v.).

EIA-232-D
Interface between data terminal equipment and data communications equipment employing serial binary data interchange.

EIA/TIA
Acronym for Electronics Industry Association/Telecommunication Industry Association (q.v.).

EIA/TIA 568/569
Premise wiring recommendations for telecommunications and data communications in commercial and high-rise buildings and campus facilities. These are not standards or specifications, but represent rather a concept. Conformance testing to wiring category definitions is based upon an industry-based general adaptation and consensus to these wiring concepts.

Electronic Data Interchange
The electronic communication of operational data such as orders and invoices between organizations. Abbreviated as EDI.

Electronics Industry Association/Telecommunication Industry Association

Two industry groups merging telephonic and communications experience which have created a series of joint standards for consolidating premise wiring and measuring quality and performance. EIA was established in 1924 by common carrier radio interest and the electronic industry community to respond to governmental regulatory matters. Abbreviated as EIA/TIA.

Encryption

The processing of data under a secret key in such a way that the original data can only be determined by a recipient in possession of the key. The application of a specific algorithm to data so as to alter the appearance of the data to make it incomprehensible to those who might attempt to steal the information. The process of decryption applies the algorithm in reverse to restore the data to its original appearance.

Ethernet

A popular baseband local area network from which the IEEE 802.3 standard was derived. Ethernet applies the IEEE 802.2 MAC protocols and uses the persistent CSMA/CD protocol on a bus topology based upon original specifications invented by Xerox Corporation and developed jointly by Xerox, Intel, and Digital Equipment Corporation. Ethernet networks operate at 10 Mbps using CSMA/CD to run over coaxial cable. Ethernet is similar to a series of standards produced by IEEE, referred to as IEEE 802.3. Ethernet is a persistent transmission protocol. 100Base-T and 100BaseVG-AnyLAN are developing standards that provide a maximum transmission rate of 100 Mbps. Sometimes abbreviated as *Enet*.

Explorer Frame

The message transmitted in a source route bridging environment (such as NetWare) to determine the optimal route to another networked device.

Facility

A wide area telecommunication connection provided by a common carrier (such as Sprint, MCI, or AT&T).

Fade

A temporary reduction in received signal strength.

Far-End Crosstalk

Signal interference on a twisted pair usually due to improper far-end termination at a modular plug or cross-connect block. Abbreviated as FEXT.

Facsimile

The device or process for transmitting a document as a line-scanned image over standard telephone lines. Abbreviated as fax.

Fast Ethernet
A reference to Ethernet running at 100 Mbps based upon either the 100Base-T or 100BaseVG-AnyLAN IEEE standards.

fax
Abbreviation for Facsimile (q.v.).

FDDI
Acronym for Fiber Data Distributed Interchange (q.v.).

FEP
Acronym for Fluorinated Ethylene Propylene (q.v.).

FEXT
Acronym for Far-End Crosstalk (q.v.).

Fiber Data Distributed Interchange
Optical fiber network based upon the ANSI X3.139, X3.148, X3.166, X3.184, X3.186, or X3T9.5 specifications. FDDI provides 125 Mbps signal rate with 4 bits encoded into 5-bit format for a 100-Mbps transmission rate. It functions on single or dual ring and star network with a maximum circumference of 250 km, although copper-based hardware is an option. See also CDDI, SDDI, TP-DDI, and TP PMD. Abbreviated as FDDI.

Fibre Channel
A high-end networking media and protocol based upon parallel and synchronized transmission of light through optical fibers for point-to-point connectivity.

File Transfer Protocol
IP application protocol for transferring files between network nodes. Abbreviated as FTP.

Filter
Generally, a process or device that screens incoming information for certain characteristics, allowing a subset of that information to pass through.

Firewall
(1.) A mechanism to protect network stations, subnetworks, and channels from complete failure caused by a single point. (2.) A device, mechanism, bridge, router, or gateway which prevents unauthorized access by hackers, crackers, vandals, and employees from private network services and data. (3.) A moat between public data networks (i.e., CompuServe, Internet, and public data carrier networks) and the enterprise network.

Flow Control
(1.) A procedure which allows an entity to cause a remote entity to suspend or resume the transmission of data. (2.) The protocol that ensures that the modem or computer is not supplied with more data than it can cope with (based on a signal to the data source to stop sending). Ideally, every link in the communication path should have a way to manage flow control with its peers, otherwise data can be lost. (3.) In IBM networks, this technique is called *pacing*.

Fluorinated Ethylene Propylene
Better known as Teflon®. This high-temperature insulator is used in cable coatings and foam compositions where building codes specify fire-resistant, high-temperature applications. Abbreviated as FEP.

FR-SSCS
Acronym for Frame Relay Service-Specific Convergence Sublayer (q.v.).

Fractional T-1
One or more of the 24 separate 64-kbps circuits in T-1. Abbreviated as FT-1.

FRAD
Acronym for Frame Relay Access Device (q.v.).

Frame
(1.) A self-contained group of bits representing data and control information. The control information usually includes source and destination addressing, sequencing, flow control, preamble, delay, and error control information at different protocol levels. (2.) A packet with framing bits for preamble and delay. (3.) Data transmission units for FDDI. The terms *packet, datagram, segment,* and *message* are also used to describe logical information groupings at various layers of the OSI reference model and in various technology circles.

Frame Relay
(1.) A packet switching device. *See* Packet Switching Network. (2.) A switching interface to get frames or packets over parts of the network as quickly as possible. *See* Packet Switching Network. (3.) A protocol used across the interface between user devices (for example, hosts and routers) and network equipment (for example, switching nodes). Frame Relay is more efficient than X.25, the protocol for which it is generally considered a replacement. *See also* X.25.

Frame Relay Access Device
The digital service unit required to split the frame relay signal into packets. Abbreviated as FRAD.

Frame Relay Interface
Very fast and efficient interface that unlike conventional systems allows bandwidth on demand.

Frame Relay Service—Specific Convergence Layer
The frame relay sublayer portion of the convergence sublayer that is dependent upon the type of traffic that is being converted. Abbreviated as FR-SSCS.

Framing
The process of assigning data bits into the network time slot.

Frequency
Measured in hertz (Hz), the number of cycles of an alternating current signal per unit time.

ft
Abbreviation for a foot.

FT-1
Acronym for Fractional T-1 (q.v.).

FTP
Acronym for File Transfer Protocol (q.v.).

Full Duplex
A two-way transmission method that echoes characters to ensure proper reception. Also, the ability for information to flow in both directions simultaneously over a communications link.

Fuse
An electrical device, typically a wire or strip, of a material that melts to the internal circuit when current exceeds the rated level of the fuse. The idea is that on an electrical circuit the fuse should be the weakest point—thus the point that heats up when things go wrong and melts.

Gateway
1. A device that routes information from one network to another. It often provides an interface between dissimilar networks and provides protocol translation between the networks, such as SNA and TCP/IP or IPX/SPX. A gateway is also a software connection between different networks; this meaning is not implied in this book. The gateway provides service at levels 1 through 7 of the OSI reference model. See also Bridge and Router. 2. In the IP community, an older term referring to a routing device. Today, the term router is used to describe nodes that perform this function, and gateway refers to a

special-purpose device that performs a Layer 7 conversion of information from one protocol stack to another.

Ground
The actual connection or the process of connecting equipment by some conductor (wire) to an earth (ground) for electrical purposes. One purpose of a *ground* wire is to carry spurious voltage (lightning strikes, spikes, and surges) away from the electrical and electronic circuit.

Ground Loop
A condition that occurs when a circuit is grounded at one or more points.

H Channel
Full-duplex ISDN primary rate channel operating at 384 kbps.

H-0
Primary rate interface for Integrated Services Digital Network.

Handset
The part of a phone you pick up and put to your ear. It contains a miniature speaker to reproduce sound and a microphone to pick it up. The handsets of cellular phones also usually contain a keypad for entering phone numbers and commands and a means of displaying those numbers and responses from the phone.

Handshake
Sequence of messages exchanged between two or more network devices to ensure transmission synchronization.

Handshaking
The exchange of signals between transmitting and receiving devices or their associated modems to establish that each is working and ready to communicate, and to synchronize timing.

Hardware Address
Also called *physical address* or *MAC-layer address*, a data link layer address associated with a particular network device. Contrasts with network or protocol address, which is a network layer address.

HDLC
Acronym for High-Level Data Link Control (q.v.).

HDSL
Acronym for High-Bit-Rate Digital Subscriber Line (q.v.).

Hertz
Signal frequency used for voice, data, TV, and other forms of electronic communications represented by the number of cycles per second. Abbreviated as Hz.

High-Bit-Rate Digital Subscriber Line
A data delivery service from a telco providing bi-directional 2 Mbps. Abbreviated as HDSL.

High-Level Data Link Control
Popular ISO standard bit-oriented, link-layer protocol derived from SDLC. It specifies an encapsulation method of data on synchronous serial data links. Abbreviated as HDLC.

HTML
Acronym for HyperText Markup Language (q.v.).

HTTP
Acronym for HyperText Transport Protocol (q.v.).

Hub
(1.) A network interface that provides star connectivity. *See* MAU. (2.) A wiring concentrator. (3.) Generally, a term used to describe a device that serves as the center of a star-topology network. In Ethernet/IEEE 802.3 terminology, a hub is an Ethernet multiport repeater, which is sometimes referred to as a *concentrator*. The term is also used to refer to a hardware/software device that contains multiple independent but connected modules of network and internetwork equipment. Hubs can be active (where they repeat signals sent through them) or passive (where they do not repeat, but merely split, signals sent through them).

Hub Adapter
A network interface board that provides access for additional network nodes. It essentially doubles as a wiring concentrator inside another device such as a PC.

Hunt Group
A series of telephone numbers in sequence that allows the calling party to connect with the first available line.

HyperText Markup Language
A hypertext markup language used in WWW clients to format and present information. Abbreviated as HTML.

HyperText Transport Protocol
An application-level protocol supporting the typing and negotiation of data representation independent of the systems for platforms. Abbreviated as HTTP.

Hz
Acronym for hertz (q.v.).

I.500
ISO series includes recommendations for ISDN network interfaces.

I.600
ISO series includes recommendations for ISDN maintenance.

ICX
Acronym for Intercarrier Exchange (q.v.).

IE
Acronym for Information Element (q.v.).

IEC
Acronym for Interexchange Carrier (q.v.).

IEEE
Acronym for the Institute for Electrical and Electronic Engineers (q.v.).

IEEE 488
Parallel interface for data collection.

IEEE 802.3
An Ethernet specification derived from the original Xerox Ethernet specifications. It describes the collision detection protocol on a bus topology using baseband transmissions.

IETF
Acronym for Internet Engineering Task Force (q.v.).

Impedance
The mathematical combination of resistance and capacitance that is used as a measurement to describe the electrical properties of the coaxial cable and network hardware.

In-Band
The usage of actual channel bandwidth for measurement, control, and management of that same transmission channel.

In-Band Control
Control information that is provided in the same channel as data.

In-Band Signaling
Transmission within a frequency range normally used for information transmission. Contrasted with out-of-band signaling, which uses frequencies outside the normal range of information-transfer frequencies.

Inductance
The property of electrical fields to induce a voltage to flow on the coaxial cable and network hardware. It is usually a disruptive signal that interferes with normal network transmissions.

Information Element
The ISDN network switch setup response message. Abbreviated as IE.

Infrastructure
(1.) The premise wiring plant supporting data communications; the risers, jumpers, patch panels, and wiring closets; the hubs, PBXs, and switches; the computer operations providing services to network devices; the network division; the test equipment; the design and support staff maintaining operations. (2.) The physical and logical components of a network. Typically, this includes wiring, wiring connections, attachment devices, network nodes and stations, interconnectivity devices (such as hubs, routers, gateways, and Switches), operating environment software, and software applications.

Input/Output
The data flows into and out of a computer device. Abbreviated as I/O.

Inside Wiring
Wiring that is done from the point of demarcation to the jack in the wall where the line terminates.

Institute for Electrical and Electronic Engineers
A membership-based organization based in New York City that creates and publishes technical specifications and scientific publications. Abbreviated as IEEE.

Integrated Services Digital Network
A limited set of standard interfaces to a digital communications network defined by the ITU (formerly the CCITT). A network for carrying data, voice, and video on digital circuits. It provides transmission in real time for interactive and multimedia applications and includes graphical communications interfaces. Communication protocols proposed by telephone companies to permit telephone networks to carry data, voice, and other source material. *See also* BRI, B-ISDN, and PRI. A dial-up common digital carrier service supporting switched service, basic rate interface, two directional channels at 56 kbps (64 kbps with SS7 clear channel) each, and a separate out-of-band bearer control channel at 16 Mbps-and multirate services. Abbreviated as ISDN.

Intel Blue
The specifications required to provision the ISDN line to meet the needs of Intel ISDN based products. This is the only information you need to give to the phone company when asked how the line needs to be provisioned.

Intercarrier Exchange
The long-distance telephone service provider. Abbreviated as ICX.

Interconnection
Junction (telecommunication) connecting two communication carriers such as cellular and land line networks, allowing mutual access by customers to each carrier's network. The interface between a cellular carrier and the local land line network, allowing calls to originate in one and terminate in the other.

Interconnectivity
The process where different network protocols, hardware, and host mainframe systems can attach to each other for transferring data to each other.

Interexchange Carrier
Essentially, a long-distance telephone company. Interexchange carriers might offer circuit-switched, lease-line, or packet-switched service or some combination of the three. Abbreviated as IEC.

Interface
A device that connects equipment of different types for mutual access. Generally, this refers to computer software and hardware that enable disks and other storage devices to communicate with a computer. In networking, an interface translates different protocols so that different types of computers can communicate together. In the OSI model, the interface is the method of passing data between layers on one device.

Interface Error
A condition indicative of hardware or software incompatibilities.

Interference
Unwanted communication channel noise.

InterLATA
Telephone call origination and termination between two different LATA regions.

International Standards Organization
The standards-making body responsible for OSI, a set of communications standards aimed at global interoperability. The United States is one of 75 member countries. Abbreviated as ISO.

International Telecommunications Union
Previously named the CCITT (Comite Consultatif International Telegraphique et Telephonique). Leading group to develop telecommunications standards. Abbreviates as ITU or ITU/TTS.

Internet Address
A unique domain name or numerical address for any device on the Internet. Abbreviated as IP Address.

Internet Engineering Task Force
The standards-making body for the Internet; the arbiter of TCP/IP standards, including SNMP, and the IAB task force consisting of over 40 groups responsible for addressing short-term Internet engineering issues. Abbreviated as IETF.

Internet Packet Exchange
A common network-level protocol proposed and applied by vendors, including Novell. IPX is based upon the original Xerox XNS frame specifications. Abbreviated as IPX.

Internet Protocol
A layer 3 (network layer) protocol that contains addressing information and some control information that allows packets to be routed. Documented in RFC 791. *See* Transaction Control Protocol/Internet Protocol. Abbreviated as IP.

Internet Service Provider
A business that provides Internet tie line connectivity, dial-up service, or website hosting services. Abbreviated as ISP.

Internetwork
A collection of networks interconnected by routers that functions (generally) as a single network. Sometimes called an *internet*, which is not to be confused with the Internet.

Internetworking
General term used to refer to the industry that has arisen around the problem of connecting networks together. The term can refer to products, procedures, and technologies.

Interrupt
A request for an attention signal sent by either hardware or software to the CPU that causes the CPU to suspend some operations and transfer control to an interrupt handler. *See also* Interrupt Request Line.

Interrupt Request Line
Hardware communication channels over which I/O devices can send interrupts to the CPU. Priority levels are prearranged to each line. Abbreviated as IRQ.

IntraLATA
Within LATA boundaries.

Inverse Multiplexing
The combination of several lower-speed circuits into one broadband circuit to give the bandwidth needed for high-speed applications. An inverse multiplexor also pulls together and synchronizes multiple channels at the receiving end of a transmission.

I/O
Abbreviation for input/output. Refers to all memory movement with the bus, data moved to and from disks, user and screen presentations, and data frames placed to and from the network channel.

IOC
Acronym for ISDN Ordering Codes (q.v.).

I/O Channel
See Bus.

IP
Acronym for Internet Protocol (q.v.).

IP address
See Internet Address.

IPX
Acronym for Internet Packet Exchange (q.v.).

IRQ
Acronym for Interrupt Request Line (q.v.).

ISDN
Acronym for Integrated Services Digital Network (q.v.).

ISDN Ordering Codes
Specific purchasing and provisioning information for ISDN customer premise equipment and carrier services defining a set of switch translations and interfaces. Bellcore has final responsibility for administering these codes, testing procedures, and carrier conformity. Abbreviates as IOC.

ISDN Phone
A telephone set which is able to communicate according to the ISDN protocol, for both voice and data.

ISO
Acronym for the International Standards Organization (q.v.).

ISO 11801
ISO cable testing specifications functionally equivalent to EIA/TIA Category 5.

ISO 2593
Connector pin allocations for use with high-speed data terminal equipment.

ISO 4903
Fifteen pin data communication DTE/DCE interface connector and pin assignments used to connect the serial port to an ISDN terminal adapter.

Iso-100Base-T
A combination of 100Base-T and six ISDN BRI lines as defined by the IEEE 802.3u and IEEE 802.9a standards.

ISO/EIC 8877
Media interface connector with eight connector pins. Commonly referred to as RJ-45. *See also* RJ-45.

Isochronous Communication
The consistent and uninterrupted stream of data across the network that is the necessary form of communication for video sessions.

Isochronous Transmission
Asynchronous (start-stop) transmission over a synchronous data link. In telephony, isochronous implies constant bit-rate sampling and is referred to as the inverse of asynchronous transmission.

IsoEthernet
A combination of 10Base-T and six ISDN BRI lines as defined by the IEEE 802.9a standards.

ISP
Acronym for Internet Service Provider (q.v.).

ITU
Acronym for International Telecommunications Union (q.v.).

ITU/TSS
Acronym for International Telecommunications Union-Telecommunications Standards Sector (q.v.).

Jack
A socket or receptacle into which a plug is inserted for purposes of making an electrical connection. Also called *outlets* and *plugs*.

Jack Type
Different types of jacks (RJ-11, RJ-45, or RJ-48) can be used for an ISDN line. The RJ-11 is the most common and is most often used for analog phones, modems, and fax machines. RJ-48 and RJ-45 are essentially the same, since they both have the same eight-pin configuration, except in Europe where the RJ-45 is likely to have six pins. An RJ-11 jack can fit into an RJ-45/RJ-48 connector; however, an RJ-45/RJ-48 cannot fit into an RJ-11 connector.

k
Abbreviation for kilobytes (q.v.).

KB
Acronym for kilobytes.

kbps
Abbreviation for kilobits (1000) per second.

kft
Abbreviation for kilofeet (q.v.).

kHz
Abbreviate for kilohertz (q.v.).

Kilobytes
A term meaning 1034 bytes of memory. Abbreviated as KB or k.

Kilofeet
An odd RBOC measurement equal to 1000 ft (308.6 m) used by RBOC for measuring local loop distances from branch offices to user sites. Abbreviated as kft.

Kilohertz
A measure of audio and radio frequency (a thousand cycles per second). Abbreviated as kHz. The human ear can hear frequencies up to about 20 kHz. There are 1000 kHz in a megahertz (MHz).

Kilometer
A unit representing a 1000 m, or approximately 3200 ft. It is abbreviated as km.

km
Acronym a kilometer (q.v.).

LAN
Acronym for Local Area Network (q.v.).

LAN Emulation
A cell-based implementation of LAN connectivity through private virtual network clouds, such as X.25, frame relay, or ATM.

LAN Network Manager
Source-bridge and Token-Ring management package provided by IBM. Typically running on a PC, it monitors source route bridges and Token-Ring devices and can pass alerts up to Net View.

LAN Operating System
See Network Operating System.

Land Telephone Company
A conventional telephone company, local exchange carrier, Regional Bell Operating company, local operating company, or a telecom company (telco).

LAPD
Acronym for Link Access Protocol for the D Channel (q.v.).

LAPM
Acronym for Link Access Procedure for Modems (q.v.).

LAT
Acronym for Local Area Transport (q.v.).

LATA
Acronym for Local Area Telephone Access (q.v.) and Local Access And Transport Area (q.v.).

Latency
(1.) The waiting time for a station desiring to transmit on the network before it gets access or permission to transmit. (2.) The delay or process time that prevents completion of a task. Latency usually refers to the lag between a request for delivery of data over the network and when it is actually received. (3.) The period of time after a request has been made for service before it is fulfilled. (4.) The amount of time between when a device requests access to a network and when it is granted permission to transmit.

Leased Line
(1.) A dedicated common carrier circuit providing point-to-point or multipoint network connection. Also called a *private line*. (2.) An open-pipe communications circuit reserved for the permanent and private use of a customer.

LEC
Acronym for Local Exchange Carrier (q.v.).

LED
Acronym for Light-Emitting Diode (q.v.).

Light-Emitting Diode
An transistor junction which emits light when forward-biased. May be used individually as a pilot light. or combined into a dot matrix array for display of alpha-numeric and other symbols. Abbreviated as LED.

Line
The individual telephone line number within a central office. Thus, a full 10-digit telephone number can be represented as NPA or NXX-LINE.

Line Conditioning
The use of equipment on leased voice-grade channels to improve analog characteristics, thereby allowing higher transmission rates.

Link Access Protocol for the D Channel
The ISDN link-layer protocol for the D channel derived from the ITU/TTS X.25 protocol. It is designed primarily to satisfy the signaling requirements of ISDN Basic Access. Defined by ITU/TTS Recommendations Q.920 and Q.921. Abbreviated as LAPD.

Link Access Procedures for a Modem
The handshaking process between computer and modem to establish a communication path. Abbreviated as LAPM.

Local Area Network.
A network limited in size to a floor, building, or city block. This usually services from 2 to 100 users. Abbreviated as LAN.

Local Echo
A host implementation for capturing and displaying individual keystrokes on terminal or serial connections.

Local Exchange Carrier
A provider of local telephone services. Abbreviated as LEC.

Local Loop
The line from a telephone subscriber's premises to the telephone company central office (CO).

Logical Channel
A nondedicated, packet-switched communications path between two or more network nodes. Through packet switching, many logical channels can exist simultaneously on a single physical channel.

Logical Device
(1.) Any addressable node on a network (2.) A description that lists how the network references physical devices.

Loop
Route where packets never reach the destination but simply cycle repeatedly through a constant series of network nodes.

Loopback Test
A test for faults over a transmission medium where received data is returned to the sending point (thus traveling a loop) and compared with the data sent.

Loop Qualification
A test done by the phone company to make sure the customer is within the maximum distance of 5555 m (18,000 ft) from the central office that services that customer.

Loopback Test
A test for faults over a transmission medium where received data is returned to the sending point (thus traveling a loop) and compared with the data sent.

m
Abbreviation for meter (q.v.).

MAC
Acronym for Media Access Control (q.v.).

MAN
Acronym for Metropolitan Area Network (q.v.).

Management Information Base
A database that defines what information can be gathered and what aspects can be controlled for a network device. Referred to as MIB.

Management Protocols
Procedures concerned with configuration, administration, and similar functions.

MAU
Acronym for Multistation Access Unit (q.v.).

MB
Abbreviation for a megabyte (q.v.).

Mbps
Abbreviation for Megabytes per second (q.v.).

Media Access Control
A hardware-level protocol for networking corresponding to ISO level 1. It is a best effort datagram delivery service with low delay useful for bursty LAN and WAN traffic. Abbreviated as MAC.

Megabits Per Second
The number of bits (in units of 1024×1024) transferred per second. Abbreviated as Mbps.

Megabyte
A term meaning 1024 kilobytes of memory or disk space. Abbreviated as MB.

Megabytes Per Second
The number of millions of bytes transferred per second. Abbreviated as Mbps.

Megahertz
Signal frequency used for voice, data, TV, and other forms of electronic communications in the millions of cycles per second. Abbreviated as MHz.

Message
(1.) Any cell, frame, or packet containing a response to a LAN-type network request, process, activity, or network management operation. (2.) A packet data unit (PDU) of any defined format and purpose. (3.) An application-layer logical grouping of information. *See also* Packet, Frame, and Payload.

Message Handling System
The software application defined by various IETF specifications for the interchange of electronic mail over the Internet and private networks.

Metal Oxide Varister

Typical electrical device included in backup power supplies, surge suppressers, and power filters to protect electronic equipment from electrical surges. It loses potency with usage and can become a fire hazard when it finally fails. Abbreviated as MOV.

Meter

Unit of measurement equivalent to 39.25 in, or 3.27 ft. Abbreviated as m.

Metric

A formal measuring standard or benchmark. Network performance metrics include Mbps, throughput, error rates, and other less formal definitions.

Metropolitan Area Network

A network that spans buildings, or city blocks, or a college or corporate campus. Optical fiber repeaters, bridges, routers, packet Switches, and PBX services usually supply the network links. Abbreviated as MAN.

MHS

Acronym for Message Handling System (q.v.).

MHz

Acronym for Megahertz (q.v.).

MIB

Acronym for Management Information Base (q.v.).

Microcom Network Protocols

A set of modem-to-modem protocols that provide error correction and compression. Abbreviated as MNP.

Microsecond

1×10^{-6} second. Abbreviated as µs.

Millisecond

1×10^{-3} second. Abbreviated ms.

Milliwatt

One-thousandth of a watt. Hand-held cellular phones usually have a maximum output power of 100 milliwatts. Abbreviated as mW.

MLPPP

Acronym for Multilink Point-to-Point Protocol (q.v.).

MLT-3
Acronym for Multilevel Transmission-3 (q.v.).

MNP
Acronym for Microcom Network Protocols (q.v.).

MNP2
Microcom Network Protocols with error correction using asynchronous transmission.

MNP3
Microcom Network Protocols with error correction using synchronous transmission between the modems (the DTE interface is still asynchronous). Since each 8-bit byte takes 8 rather than 10 bits to transmit, a 20 percent increase in throughput is possible. Unfortunately the MNP3 protocol overhead is rather high, so this increase is not realized.

MNP4
Microcom Network Protocols using data phase optimization, which improves on the rather inefficient protocol design of MNP2 and MNP3. Synchronous MNP4 comes closer to achieving 20 percent throughput.

MNP5
Microcom Network Protocols with simple data compression. Dynamically arranges for commonly occurring characters to be transmitted with fewer bits than rare characters. It takes into account changing character frequencies as data flows. Also encodes long runs of the same character. Typically compresses text by 35 percent.

MNP10
Microcom Network Protocols for cellular or wireless transmission applying compression. error detection, error correction, data rate fallback, and readjustment.

Modem
Acronym for Modulator-Demodulator (q.v.).

Modular Wiring
See Premise Wiring and EIA/TIA 568/569.

Modulation
The analog waveform of a transmitted digital signal. Process by which signal characteristics are transformed to represent information. Types of modulation include frequency modulation (FM), in which signals of different frequencies represent different data values, and amplitude modulation (AM), in which signal amplitude is varied to represent different data values.

Modulator-Demodulator
A device which transforms two-level serial computer output into a form suitable for telephonic transmission and vice-versa. Normally converts digital information to a change in amplitude, frequency or phase of an audio tone, or a shift between tone frequencies. Abbreviated and better known as a Modem.

MP
Acronym for Multilink Point-to-Point Protocol (q.v.).

MP+
Acronym for Multilink Point-to-Point Protocol Plus (q.v.).

ms
Abbreviation for Millisecond (q.v.).

µs
Abbreviation for Microsecond (q.v.).

Multicast
(1.) The ability to broadcast to a select subset of nodes. (2.) Single packets copied to a specific subset of network addresses. These addresses are specified in the destination-address field. In contrast, in a broadcast, packets are sent to all devices in a network.

MultiLATA
Through more than one LATA. *See* LATA.

Multilink Point-to-Point Protocol
A multiple drop series of connections extending the functionally of the single-user connection afforded by Point-to-Point Protocol. Abbreviated as MLPPP or MP.

Multilink Point-to-Point Protocol Plus
PPP enhanced with multiple line support, number configuration, remote management, and multiple simultaneous data links. Abbreviated as MP+.

Multimeter
A test tool that measures electrical voltages and resistance. Also called a *multitester*. Sometimes called an *ohmmeter* or *Ωmeter*.

Multiplexing
The process of mixing data originating from several entities into a single shared data stream. Multiplexing can occur at several layers of a protocol stack.

Multirate ISDN
Provides multiple ISDN channels at the same time. The multiple channels can provide dedicated or switched service for individual uses, or can be multiplexed into a single pipeline for demanding data communication requirements, such as videoconferencing.

Multistation Access Unit
A device which connects directly to a lobe wire and broadcasts and receives information over that cable and switches the signals to the next active downstream station. It might be abbreviated as MSAU, but more often as MAU.

mw
Abbreviation for milliwatt.

N-ISDN
Acronym for Narrowband ISDN. *See also* B-ISDN.

Nanosecond
1×10^{-9} second. Abbreviated as vs or ns.

NANP
Abbreviation for North American Numbering Plan (q.v.).

Narrowband ISDN
Voice-grade transmission at 2.4 Mbps or less, or the sub-voice-grade transmission from 5 Mbps to 15 Mbps. Abbreviated as N-ISDN.

National Bureau of Standards
See National Institute of Standards and Technology.

National Institute of Standards and Technology
Formerly NBS, this U.S. government organization supports and catalogs a variety of standards. Abbreviated as NIST.

National ISDN 1
A specification for a standard ISDN phone line. The goal is for National ISDN 1 to become a set of standards which every manufacturer can conform to. For example, ISDN phones that conform to the National ISDN 1 standard will work, regardless of the central office the customer is connected to. Future standards, denoted as NI-2 and NI-3, are currently being developed. Abbreviated as NI-1.

National Television Systems Committee
NTSC is the video standard for North America and many countries. Abbreviated as NTSC.

NAU
Acronym for Network Access Unit (q.v.).

NCOS
Acronym for Network Class of Service (q.v.).

NDIS
Acronym for Network Device Interface Specification (q.v.).

NetBEUI
Acronym for NetBIOS Extended User Interface (q.v.).

NetBIOS
Acronym for Network Basic Input Output System (q.v.).

NetBIOS Extended User Interface
A simple (nonroutable) network protocol introduced by IBM PC LAN and used extensively by Microsoft (LAN Manager and NTAS). It is not suitable for the enterprise network so it must be translated or encapsulated. Abbreviated as NetBEUI.

Network
(1.) Hardware and software that allow computers to transmit data over both local and long distances. (2.) An area comprising two or more MSAs (Metropolitan Statistical Area) and portions of any RSA (Rural Service Area) or RSAs located between such MSAs, in which the company provides or plans to provide uninterrupted cellular telephone service to system users traveling in and between such MSAs. The creation of a network typically leaves intact the ownership interests in the respective systems which comprise the network but allows the systems to be combined for operational and marketing purposes. (3.) A collection of computers and other devices that are able to communicate with each other over some network medium.

Network Access Unit
The network controller. Abbreviated as NAU.

Network Address
Also called a *protocol address*, a network layer address refers to a logical, rather than a physical, network device.

Network Administrator
Person who helps maintain a network.

Network Analyzer
A hardware/software device offering various network troubleshooting features, including protocol-specific packet decodes, specific preprogrammed troubleshooting tests, packet filtering, and packet transmission. *See also* Protocol Analyzer.

Network Basic Input/Output System
The first level of network software which controls network hardware. A Novell NetWare specification for OSI level 1 data exchange. Abbreviated as NetBIOS.

Network Class of Service
AT&T 5ESS ISDN switch and line setting defining the interconnection of ISDN, switched digital, frame relay, X.25, X.31, and ATM services. Abbreviated as NCOS.

Network Device Interface Specification
A Microsoft network interface specification for operating system and protocol independent device drivers. An effort to create a standard for bridging different types of network adapter cards and multiple protocol stacks. This network-level protocol is supported by IBM LAN Manager and new Microsoft networking products, such as MS Windows for Workgroups and NTAS. Abbreviated as NDIS.

Network Interface
The boundary between a carrier's network and a local installation.

Network Interface Card
The network access unit which contains the hardware, software, and specialized PROM information necessary for a station to communicate across the network. Usually referenced as Network Interface Controller. Abbreviated as NIC.

Network Interface Controller
The network access unit which contains the hardware, software, and specialized address information necessary for a station to communicate across the network. Abbreviated as NIC.

Network Interface Unit
See Network Interface Controller.

Network Layer
The third layer of the OSI reference model which activates the routing with network address resolution and flow control in terms of segmentation and blocking. Also, this layer provides service selection, connection resets, and expedited data transfers. IP or IPX, both common network software, run at this level.

Network Node
Any device attached to a local or wide area network, including PCs attached to the Internet with an ISDN connection.

Network Operating System
(1.) A platform for networking services that combines operating system software with network access. This is typically not application software but rather an integrated operating system. (2.) The software required to control and connect stations into a functioning network conforming to a protocol and providing a logical platform for sharing resources. Abbreviated as NOS.

Network Termination
The description of ISDN equipment required at the customer site of a digital subscriber loop. Abbreviated as NT-1. *See also* Channel Service Unit/Data Service Unit.

Network Termination 1
A device that is required to connect ISDN terminal equipment to an ISDN line. The NT-1 connects to the two-wire line (twisted pair copper wiring) that your telephone company has assigned for your ISDN service. Abbreviated as NT-1.

NI-1
Acronym for National ISDN 1 (q.v.).

NIC
Acronym for Network Interface Card (q.v.) and Network Interface Controller (q.v.).

NIST
Acronym for National Institute of Standards and Technology (q.v.).

Node
(1.) A logical, nonphysical interconnection to the network that supports computer workstations or other types of physical devices on a network that participates in communication. (2.) Alternatively, a node might connect to a fan-out unit providing network access for many devices. A device might be a terminal server, or a shared peripheral such as a file server, printer, or plotter.

Noise
(1.) Undesirable communications channel signals. (2.) Electrical signal interference on a communications channel that can distort or disrupt data signals. Generally, this refers to electromagnetic interference or radio-frequency interference.

North American Numbering Plan

A reference to a new plan to include the numbers 0 and 1 in area codes and local exchanges so as to increase the availability of unique phone numbers. Abbreviated as NANP.

NOS

Acronym for Network Operating System (q.v.).

NPA

Acronym for Number Plan Area (q.v.).

ns

Abbreviation for nanosecond (q.v.).

NT

Acronym for Network Termination (q.v.).

NT-1

Network termination type 1 for ISDN BRI. The connector at either end of an ISDN link. It converts the two-wire ISDN circuit interface to four wires, for up to eight terminal devices.

NT-2

Network interface/terminator for ISDN PRI and ISDN-based PBXs.

NTSC

Acronym for National Television Systems Committee (q.v.).

Number Plan Area

The area code. Another name for your local cellular calling and billing area. This might include neighboring area codes in addition to the one assigned to your phone. The first three digits of a full ten-digit telephone number. Abbreviated as NPA.

One-Touch Dialing

A phone feature that allows you to press and hold one key to dial preprogrammed numbers.

Open-Pipe

A description of how data moves on a circuit-switched or lease-line digital data connection. In an open-pipe connection, data moves in a steady stream. It is not in packets, and all of the data follows the same path from the sending LAN to the receiving LAN.

Open Systems Interconnection
The suite of communications standards mandated by ISO and the U.S. government. CMIP is part of this suite. Abbreviated as OSI.

OSI Reference Model
Open Systems Interconnection reference model defined by the International Standards Organization (ISO), which has determined a data communication architectural model for networking.

Out-of-Band
The use of other frequencies or other channels (than the transmission channel) for measurement, control, and management of that transmission channel. Also called *sideband*.

Out-of-Band Signaling
Transmission using frequencies or channels outside the normal frequencies or channels used for information transfer. Out-of-band signaling is often used for error reporting in situations in which in-band signaling can be affected by whatever problems the network might be experiencing.

Outlet
Any ISDN device or POTS phone that can connect into the ISDN BRI or PRI ports.

Overflow
Amount of traffic directed to another cell site when all channels are busy. Expressed as a percentage of offered load.

Overhead
CPU, disk processing, and/or network channel bandwidth allocated to the processing and/or packaging of network data.

Overhead Message
The message sent on the control channel as a part of the paging data stream to give the mobile equipment certain descriptive information about the local system.

Packet
A self-contained group of bits representing data and control information. A logical grouping of information that includes a header and (usually) user data. The control information usually includes source and destination addressing, sequencing, flow control, and error control information at different protocol levels. Generally refers to data transmission units for Ethernet, Token-Ring, and switching. *See also* Cell, Frame, Message, and payload.

Packet Assembler/Disassembler

The software that assembles data inside ISDN, X.25, and other frames and packets, and then extracts that data at the delivery destination.

Packet Switched Node

An Internet packet switch. Also, a switching node in the X.25 architecture. Usually, the PSN is data communication equipment (DCE) and allows for connection to data terminal equipment (DTE). *See also* X.25. This acronym is also commonly used as an expansion for packet-switched network.

Packet-Switched Public Data Network

A public network providing packet data delivery. Examples include X.25, frame relay, and X.31. Abbreviated as PSPDN.

Packet Switching

A network transmission methodology that uses data to define a start and length of a transmission for digital communications. A process of sending data in discrete blocks. Network on which nodes share bandwidth with each other by intermittently sending logical information units (packets). In contrast, a circuit-switching network dedicates one circuit at a time to data transmission. *See also* Circuit Switching and Message Switching.

Packet-Switching Network

A network transmission methodology that uses data to define a start and length of a transmission for digital communications. A process of sending data in discrete blocks. A network consisting of a series of interconnected switches that route individual packets of data over one of several redundant routes. Most commonly, packet-switching networks refer to X.25, ATM, or frame relay. Packet-switched networks offer flexibility for multipoint connections, high reliability, and flexible pricing.

PAD

Acronym for Packet Assembler/Disassembler (q.v.).

Pair Scanner

A testing tool which verifies the integrity and performance of network wiring and cable. It tests for electrical breaks, shorts, impedance, capacitance, signal crosstalk and signal attenuation.

Parity

A method of checking the accuracy of binary numbers. An extra bit (the parity bit) is added to the number. If even parity is used, the sum of all 0s in the number and its parity bit is even. If odd parity is used, the sum of all 1s and the parity bit is odd.

Parity Check
A process for checking the integrity of a character. A parity check involves appending a bit that makes the total number of binary 1 digits in a character or word (excluding the parity bit).

Payload
The actual data, information, or message contents carried within a cell, frame, or packet.

PBX
Acronym for Public Branch Exchange (q.v.).

PC Card
Sixty-eight-pin, integrated circuit and rotating media cards designed to add functionality to portable computers, while using the smallest feasible packaging. Also called a *PC credit card.*

PC Credit Card
See PC Card.

PCI
Acronym for Peripheral Component Interface (q.v.).

PCMCIA
Acronym for Personal Computer Memory Card International Association (q.v.).

PCWS
Acronym for Public Corporate Web Service (PCWS).

PDN
Acronym for Public Data Network (q.v.).

Peripheral Component Interface
The Intel bus standard providing bus throughputs up to 120 Mbps. Abbreviated as PCI.

Permanent Virtual Circuit
A logical, nonphysical service location. This compares to the open-pipe link, which is a physical link. Abbreviated as PVC. *See also* Committed Information Rate and Frame Allocated Bandwidth.

Personal Computer Memory Card International Association
A standard for a computer plug-in, credit card-sized card which provides about 90 percent compatibility across various platforms, BIOS, and application software. Abbreviated as PCMCIA. *See* PC Card.

Physical Address
The unique address associated with each workstation on a network. A physical address is devised to be distinct from all other physical addresses on interconnected networks. A worldwide designation unique to each unit.

Physical Device
Any item of hardware on the network.

Physical Layer
Level 1 of the OSI reference model which insulates the data link layer from the medium-dependent physical characteristics.

Plain Old Telephone Service
The capability afforded by a rotary dial telephone and an electromechanical central office. Standard analog telephone service used by many telephone companies throughout the United States. Abbreviated as POTS.

Point of Demarcation
The physical point where the phone company ends its responsibility with the wiring of the phone line. Also known as simply the *Demarcation, Demarcation Point,* or *Demarc.*

Point-of-Presence
The physical access point for a telecommunication service, usually a central office, for T-1. ISDN. frame relay, ATM, and other digital services. The issue of the point-of-presence is that many services are not available in all markets, and the connection might not be economically feasible or locally available. Abbreviated as POP.

Point-to-Multipoint Delivery
Delivery of data from a single source to several destinations.

Point-to-Point Protocol
A TCP/IP serial interface designed and implemented to overcome the shortcomings in SLIP. Supports datagrams, multiple protocols, link control negotiation, authentication, encapsulation. compression, LLC, NCP, and other common network protocols. *See also* SLIP. Abbreviated as PPP.

Polling
An access method involving a central node asking each node in a predetermined order if it has data to send. This is often used in mainframe environments, and the order is often determined as a function of priority.

Polyvinyl Chloride
An extensively used insulator in cable coatings and coaxial cable foam compositions. Abbreviated as PVC.

POP
Acronym for Point-Of-Presence (q.v.).

POPS
One unit of population (i.e., one person). The POPS concept is used to measure relative market sizes.

Postal Telephone and Telegraph
A government-sponsored agency that provides telephone services existing in most areas outside North America. Abbreviated as PTT.

POTS
Acronym for Plain Old Telephone Service (q.v.).

PPP
Acronym for Point-to-Point Protocol (q.v.).

Premise Wiring
A telecommunications and data communications wiring infrastructure which embodies the concept of flexible, recyclable, reconfigurable, and reusable modular components centered about wiring closet.

Presentation Layer
This is the sixth layer of the OSI reference model which transfers information from the application software to the network session layer of the operating system. At this level, software performs data transformations, data formatting, syntax selection (including ASCII, EBCDIC, or other numeric or graphic formats), device selection and control, and data compression or encryption.

PRI
Acronym for Primary Rate Interface (q.v.).

Primary Rate Interface
ISDN interface to primary rate access. Primary rate access consists of the single 64-kbps D channel plus 23 (in the case of 1.544 Mbps) or 30 (in the case of 2.048 Mbps) B channels for voice and/or data-the equivalent of one European E-1 link. The interface at each end of the high-volume trunks linking PBX and central office facilities or connecting network switches to each other PRI transmits using 24 64-kbps B channels (with maximum speeds of 1.544 Mbps) and one 64-kbps D channel. Abbreviated as PRI.

Protocol Analyzer
Test equipment that transmits, receives, and captures Ethernet packets to verify proper network operation.

Provisioning
The process of selecting the exact switch parameters at the central office for an ISDN circuit.

PSN
Acronym for Packet Switch Node (q.v.).

PSPDN
Acronym for Packet-Switched Public Data Network (q.v.).

PSTN
Acronym for Packet-Switched Telephone Network (q.v.).

PTT
Acronym for Postal Telephone and Telegraph (q.v.).

Public Branch Exchange
A telephone switchboard or switching service typically installed locally to an organization to route calls to the appropriate people and devices. Abbreviated as PBX.

Public Corporate Web Service
The device providing web services to the world. This is better known as a *host ISP* service or a *web server*. Abbreviated as PCWS.

Public Data Network
A network operated either by a government (as in Europe) or by a private concern to provide computer communications to the public, usually for a fee. These enable small organizations to create a WAN without all the equipment costs of long-distance circuits. Abbreviated as PDN.

Public Switched Telephone Network
The telecommunications network traditionally encompassing local and long-distance land-line carriers and now also including cellular carriers. Refers to the telephone network. Abbreviated as PSTN.

Public Utilities Commission
State organizations which have jurisdiction over telecommunications. Abbreviated as PUC.

PUC
Acronym for Public Utilities Commission (q.v.).

PVC
Acronym for Permanent Virtual Circuit (q.v.) and Polyvinyl Chloride (q.v.).

Q.920
ISDN specifications for the user-network interface (UNI) data link layer. *See also* UNL.

Q.921
ISDN specifications for the user-network interface (UNI) data link layer. *See also* UNL.

Q.931
ITU recommendation. The standard for signaling to set up ISDN connections.

Q.93B
Standard for signaling to set up ATM virtual connections. An evolution of ITU/TTS Recommendation Q.931.

RBHC
Acronym for Regional Bell Holding Company (q.v.).

Radio Frequency Interference
Electronically propagated noise from radar, radio, or electronic sources. Abbreviated as RFI.

RBOC
Acronym for Regional Bell Operating Company (q.v.).

Receiver Terminal Identification
A 32-character alphanumeric destination identification field which most stand-alone facsimile machines exchange as part of the current protocol. These are user-supplied fields which must be programmed into the machines. It really is part of the CIS field. Abbreviated as RTI.

Regional Bell Holding Company
One of seven telephone companies created after the AT&T divestiture in 1984 owning and operating the Regional Bell Operating Company and other businesses. The RHBC crosses state lines. Abbreviated as RBHC.

Regional Bell Operating Company
One of the seven independent organizations created by the 1974 AT&T consent decree dividing that company into the so-called Baby Bells. Abbreviated as RBOC.

Reliability

Ratio of expected to received keepalives from a link. If the ratio is high, the line is reliable. Used as a routing metric.

Remote Monitoring

The software, processes, or equipment used to collect and store information from remote network devices. Abbreviated as RMON.

Remote Monitoring MIB

An extended MIB that delivers statistics about data packets and network traffic to an SNMP console. Abbreviated as RMON MIB.

Repeater

(1.) A device that boosts a signal from one network lobe or trunk and continues transmission to another similar network lobe or trunk. Protocols must match on both segments. The repeater provides service at level 1 of the OSI reference model. (2.) A device that regenerates and propagates electrical signals between two network segments.

Repertory Dialing

Sometimes known as *memory dialing* or *speed-calling*. A feature that allows you to recall phone numbers from a phone's memory with the touch of just one, two, or three buttons.

Replication

A process by which a set of distinct physical backups of a database, set, or table is kept in synchrony by a distributed database. Replication is typically used for distributing routing tables, domain maps and tables, yellow page-type network information, mail addresses, user access rights, and distributed databases.

Replication Server

A process or device which duplicates and synchronizes databases, sets, or tables to other servers through a network environment.

Request for Comments

Documents used as the primary means for communicating information about the Internet. Some RFCs are designated by the IAB or IETF as Internet standards. Most RFCs document protocol specifications such as telnet and FTP but some are humorous and/or historical. RFCs are available from Internet Network Information Centers. Abbreviated as RCF.

Request to Send

An RS-232 control signal that requests a data transmission on a communications line. Abbreviated as RTS.

Resistance
The measurement of the electrical properties of the coaxial cable and network hardware that describes their ability to hinder the passage of electrons.

RCF
Acronym for Request for Comments (q.v.).

RFI
Acronym for Radio Frequency Interference (q.v.).

Ring
(1.) The connection and call signaling method on POTS. (2.) A network topology that has stations in a circular configuration.

Ring Topology
Topology in which the network consists of a series of repeaters connected to one another by unidirectional transmission links to form a single closed loop. Each station on the network connects to the network at a repeater.

RJ-11
Standard four-wire connectors for phone lines.

RJ-22
Standard four-wire connectors for phone lines with secondary phone functions (such as call forward, voice mail, or dual lines).

RJ-45
Standard eight-wire connectors for networks. Also used as phone lines in some cases. *See also* ISO/IEC 8877.

RJ-48
See RJ-45.

RJ-11 jack
A six pin modular jack.

RJ-45 jack
An eight pin modular jack.

RMON
Acronym for Remote Monitoring (q.v.).

RMON MIB
Acronym for Remote Monitoring MIB (q.v.).

Route
A path through an internetwork.

Routed Protocol
A protocol that can be routed by a router. To route a routed protocol, a router must understand the logical internetwork as perceived by that routed protocol. Examples of routed protocols include DECnet, AppleTalk, and IP.

Router
(1.) A device that interconnects networks that are either local or wide area. (2.) A device providing intercommunication with multiple protocols. 3 A device providing service at level 3 of the OSI reference model. *See also* Bridge and Gateway. (4.) A router examines the network address of each packet. Those packets that contain a network address different from the originating PC's address are forwarded onto an adjoining network. Routers also have network-management and filtering capabilities, and many newer routers incorporate bridging capabilities as well.

Routing
The process of finding a path to the destination host. Routing is very complex in large networks because of the many potential intermediate destinations a packet might traverse before reaching its destination host.

RS-232
EIA protocol for interface between data terminal equipment and data communications equipment employing serial binary data interchange.

RS-232-C
Popular physical layer interface for serial connections. Virtually identical to the V.24 specification.

RS-422
A balanced electrical implementation of RS-449 for high-speed data transmission.

RS-449
EIA standard for general-purpose 37- and 9-pin interface for data terminal equipment and data circuit-terminating equipment employing serial binary data interchange. Popular physical layer interface. Essentially a faster (up to 2 Mbps) version of RS-232-C capable of longer cable runs.

RTI
Acronym for Receiver Terminal Identification (q.v.).

RTS
Acronym for Request to Send (q.v.).

s
Abbreviation for a second.

S/T-interface
A 4-wire ISDN circuit. The S/T interface is the part of an ISDN line that connects to the terminal equipment.

SAR
Acronym for Segmentation And Reassembly (q.v.).

Satellite Communication
Use of geostationary orbiting satellites to relay data between multiple earth-based stations. Satellite communications offer high bandwidth at a cost which is not related to distance between earth stations, relatively long propagation delays, and broadcast capability.

SC-PAM
Acronym for Simple Coded-Pulse Amplitude Modulation (q.v.).

Scanner
A testing tool which verifies the integrity and performance of network wiring and cable. It tests for electrical breaks, shorts, impedance, capacitance, signal crosstalk, and signal attenuation. These tools are sometimes called *pair* or *ring scanners*.

SMDS
Acronym for Switched Multimegabit Digital Service (q.v.).

Segmentation And Reassembly
The process, ISO layers, and software that converts data into frames or packets and converts these frames or packets back into the base data. Abbreviated as SAR.

Serial Line Internet Protocol
A simple TCP/IP networking extension for low-speed serial (modem and telephone circuits) interface that does not handle noise on the line, lacks checksums, cannot negotiate any parameters for the connection itself, and cannot dynamically alter IP addresses. Abbreviated as SLIP.

Service
The facility provided to the user of an entity.

Service Profile Identification
The ISDN profile assignment for a particular channel. Abbreviated as SPID.

Session
A logical connection with a host system. The session begins when you establish the communications link and ends when you terminate emulation and return to the operating system.

SFTP
Acronym for Shielded Foil Twisted Pair (q.v.).

Shadowing
A technique which writes the same data to multiple disks simultaneously to minimize the chance that data is lost due to a malfunction.

Shield
A barrier, usually metallic, within a wiring bundle that is intended to contain the high-powered broadcast signal within the cable. The shield reduces Electromagnetic interference and Radio-frequency interference, and signal loss.

Shielded Foil Twisted Pair
A pair of insulated wires which are twisted together in a spiral manner. In addition, the pair is wrapped with metallic foil or braid, designed to insulate (shield) the pair from electromagnetic interference. Possible medium for Category 6 wiring specification under consideration.

Shielded Twisted Pair
Pairs of 22 to 26 AWG gauge wire clad with a metallic signal shield. Abbreviated as STP.

Short
A physical discontinuity (usually electrical, rarely optical) such that one or more signal conductors in the network media leaks the signal into other conductors. Usually, a short refers to an electric short circuit between a signal conductor and the shield or ground. It can also refer to a short between receive or transmit pairs.

SID
Acronym for System Identification (q.v.).

Sideband
The use of other frequencies or other channels (than the transmission channel) for measurement, control, and management of a transmission channel. Also called *out-of-band*.

Signal
A transmission broadcast. The electrical or optical pulse that conveys information.

Signaling
The process of sending a transmission signal over a physical medium for purposes of communication.

Signaling System Number 7
An LEC connection method that provides fast call setup and remote database interactions. Abbreviated as SS7.

Simple Coded-Pulse Amplitude Modulation
A signal transmission technology used to extend the ISDN BRI signal beyond 4587 m (15,000 ft) up to 10091 m (33,000 ft).

Simple Mail Transfer Protocol
An Internet standard for transferring E-mail from host to host. It can handle only ASCII text mail messages. Abbreviated as SMTP.

Simple Network Management Protocol
An IETF-defined protocol that runs natively on TCP/IP networks. Considered the de facto standard for network management, it is used to monitor the status of devices but, because of its lack of security, is rarely used to control network devices. Abbreviated as SNMP.

Simple Network Management Protocol, Version 2
The next generation of SNMP. It adds security (encryption and authentication), supports a hierarchical management scheme, and runs on network transports other than TCP/IP, including AppleTalk, IPX, and OSI. Abbreviated as SNMP-II or SNMP v2.

Single Line Service
Telephone service where only one pair of wires is brought to the home or small business.

Slave Controller
A specialized ISDN controller which utilizes the normal CPU access process and bus access controller to move segments of memory between RAM and the device which it controls.

SLIC
Acronym for Subscriber Line Interface Card (q.v.).

Sliding Window Flow Control

Method of flow control in which a receiver gives transmitter permission to transmit data until a window is full. When the window is full, the transmitter must stop transmitting until the receiver advertises a larger window. TCP, other transport protocols, and several link-layer protocols use this method of flow control.

SLIP

Acronym for Serial Line Internet Protocol (q.v.).

Slot Time

(1.) The time during which the protocol allows a node or station to transmit. (2.) A multipurpose parameter to describe the contention behavior of the data link layer.

SMTP

Acronym for Simple Mail Transfer Protocol (q.v.).

Sneakernet

A slang term for computer data exchange by floppy disk or other portable media physically carried from point-to-point by a person.

SNMP

Acronym for Simple Network Management Protocol (q.v.).

SNMP-II and SNMPv2

Acronym for Simple Network Management Protocol, Version 2 (q.v.).

Socket

(1.) An IP socket is created when an IP port on a host machine makes a connection to an IP port on a client machine. IP sockets are ephemeral by nature. When a standard request port has formed a socket connection to a standard services port, additional requests for access to that port might be denied. (2.) The 68-pin PCMCIA-defined, connector module in which PC cards are inserted. (3.) Software structure operating as a communications end point within a network device. A functional socket is important for enabling laptop ISDN connectivity with a PCMCIA (or PC card) ISDN adapter.

SONET

Acronym for Synchronous Optical Network (q.v.).

Source Address

The transmitting station's logical address.

Spectrum

The electromagnetic spectrum. Includes radio and microwaves, infrared. visible light, ultraviolet. x-, and gamma rays. The radio portions of the spectrum are licensed and regulated by the FCC in the United States.

SPID

Acronym for Service Profile Identification (q.v.).

Spike

An in-phase impulse causing spontaneous increases in voltage. Spikes are impulses less than 100 microseconds of high-voltage electricity ranging from 400 to 5600 volts superimposed on the normal AC electrical sine wave.

SPNI

Acronym for Service Provider Network Identifier (q.v.).

Spoofing

The process of sending a false or dummied acknowledgment signal in response to a request for status or receipt. Spoofing is typically applied for host transmission over LAN-type transmission networks or routers so that processes are not falsely terminated for lack of message response activity.

SPX

Acronym for System Packet Exchange (q.v.).

SS7

Acronym for Signaling System Number 7 (q.v.).

Stable

Describes a network protocol, infrastructure, or application subject to error condition and traffic loading levels.

Standard

A commonly used or officially specified set of rules or procedures.

Star Wiring Configuration

A wiring configuration in which each jack or end point is directly connected to a central distribution point.

Station

(1.) A logical, nonphysical interconnection to the network that supports computer workstations or other types of physical devices on a network. Alternatively, a station might connect to a wiring concentrator providing network access for many devices. A device might be a terminal server or a shared peripheral such as a file server, printer, or plotter. (2.) A single addressable device on FDDI, generally implemented as a stand-alone computer or a peripheral device such as a printer or plotter. *See also* Node and Workstation. (3.) A station might be a terminal server or a shared peripheral such as a file server, printer, or plotter.

Statistical Multiplexer
Multiplexing equipment that dynamically allocates trunk capacity only to active input channels, allowing more devices to be connected than with a traditional multiplexer. Also referred to as a statistical time division multiplexer or a stat mux.

STN
Acronym for Switched Telephone Network (q.v.).

STP
Acronym for Shielded Twisted Pair (q.v.).

Structured Wiring
Adherence to the concept of installing a general-purpose and multipurpose copper (and fiber) plan for telecommunications and data communications. *See also* EIA/TIA 568/569.

Subchannel
In broadband terminology, a frequency-based subdivision creating a separate communications channel.

Subnet
For routing purposes, IP networks can be divided into logical subnets by using a subnet mask. Values below those of the mask are valid addresses on the subnet.

Subnetwork
Sometimes used to refer to a network segment. In IP networks, a network sharing a particular subnet address. A single administrative domain using a single network access protocol.

Surge
An increase in line voltage that lasts longer than one cycle of the line frequency of 60 Hz, the North American frequency (50 Hz in Europe, Asia, and Africa). Surges are the increase flow of current through an electrical device brought about by an instantaneous change in its resistance or impedance.

Surge Suppresser
A device that filters excess power (surges and spikes) through power or phone lines and shunts them away from sensitive equipment to a ground.

Sustainable Rate
The maximum throughput bursty traffic can achieve within a given virtual circuit without risking data loss.

SVC
Acronym for Switched Virtual Circuit (q.v.).

Switch
The hardware at a central office that directs and connects incoming and outgoing ISDN transmissions.

Switched 56
Digital service at 56 kbps provided by local telephone companies and long-distance carriers. Similar to ISDN, Switched 56 traffic can travel over the same physical infrastructure that supports ISDN. Switched 56, however, is an older technology with decreasing significance.

Switched Multimegabit Digital Service
Any of a variety of switched digital services ranging from 1.544 to 44.736 Mbps (T-1 to T-3 speeds). Abbreviated as SMDS.

Switched Service
A connection service where users pay only for the time during which voice or data transmission service occurs.

Switched Telephone Network
A standard wired telephone system at the LEC central office. Abbreviated as STN.

Switched Virtual Circuit
A dial-up digital connection or switched delivery. This compares to the permanent virtual circuit or leased line. *See also* Committed Information Rate and Frame Allocated Bandwidth. Abbreviated as SVC.

Switching Uninterruptible Power Supply
Power Supply that is active when power levels drop below 10 percent of normal.

Symbol
(1.) Six consecutive bits of a Reed-Solomon block forming an element of a Galois field.
(2.) The instantaneous frequency of a signal waveform that defines a single bit.

Symbol Error Rate
The average number of symbols in error in a Reed Solomon block in a defined period of time. The symbol error rate can only be measured using Reed-Solomon blocks that con-

tain a correctable number of symbol errors. Up to six bits might be in error in a single symbol error.

Synchronization
(1.) The process of achieving a common interpretation of a transmitted bit stream between more than one entity at the same point in the bit stream. Synchronization might be required at each layer of a protocol stack. (2.) The event occurring when transmitting and receiving stations operate in unison for very efficient (or inefficient) utilization of the communications channel. (3.) Establishing common timing between sender and receiver.

Synchronization Error
A cell, frame, or packet that is framed improperly by the receiving station or overruns the controller.

Synchronous
Refers to synchronous transmission methods where the data transmission is circumscribed to occur within a specified time interval known to both sending and receiving nodes. Transmission is moderated by a clock or negotiation.

Synchronous Optical Network
A common carrier fiber-optic transmission link providing basic bandwidth in blocking units of 45 Mbps. Multiple streams can support bandwidths up to 18 billion bits per second. Used with ATM protocols. Abbreviated as SONET.

System Identification
Issued by the FCC for each cellular market. Abbreviated as SID. However, each market can apply to change or consolidate SIDs from adjoining markets. The SID is electronically generated by the cellular system computer and received by the cellular mobile unit. Its main function is to operate the HOME and ROAM light in the cellular mobile unit. *See also* BID. Abbreviated as SID.

System Packet Exchange
Novell routing specification for OSI level-4 data exchange applying RIP. Abbreviated as SPX.

T-1
(1.) Bell technology referring to a 1.544-Mbps communications circuit provided by long-distance carriers for voice or data transmission through the telephone hierarchy. Since the required framing bits do not carry data, actual T-1 throughput is 1.536 Mbps. T-1 lines might be divided into 24 separate 64-kbps channels. This circuit is common in North America. Elsewhere, the T-1 is superseded by the ITU/TTS designation DS-12. A 2.054-Mbps communications circuit provided by long-distance carriers in Europe for voice or data transmission.

T-3
An AT&T standard for dial up or leased line circuits with a signaling speed of 44.736 Mbps. Superseded in Europe by the ITU (ITU/TTS) DS-3 designation.

T-Carrier
Time-division multiplexed transmission method usually referring to a line or cable carrying a DS-1 signal.

TA
Acronym for Terminal Adapter (q.v.).

TAPI
Acronym for Telephone Application Programming Interface (q.v.).

Tariff
Requirement that the rates, terms and conditions at which customers are charged for services be published and that the state review those rates, terms and conditions. The document might contain these rules, terms, and conditions.

T-Carrier
Time-division multiplexed transmission method usually referring to a line or cable carrying a DS-1 signal.

TCEP
Acronym for Transport Connection End Point (q.v.).

TCP
Acronym for Transmission Control Protocol (q.v.).

TCP/IP
Acronym for Transaction Control Protocol/Internet Protocol. *See also* TCP and IP.

TDM
Acronym for Time-Division Multiplexing (q.v.).

TDR
Acronym for Time Domain Reflectometer (q.v.).

Tear Down
Phrase referring to disconnection of a logical or physical WAN service, such as an ISDN channel connection.

Teflon®
Trade name for fluorinated ethylene propylene. A nonflammable material used for cable foam and jacketing.

TEI
Acronym for Temporary Equipment Identifier (q.v.) and Terminal End Point Identifier (q.v.).

Telco
A reference to modular telephone wiring.

Telecommunications
Term referring to communications (usually involving computer systems) over the telephone network.

Telephone Application Programming Interface
A Microsoft-designed application development interface for Windows 95 and Windows NT for uniform control functionality of telephone equipment. Abbreviated as TAPI.

Telephony Services Application Programming Interface
A Novell application development interface for Novell fileservers to provide uniform control functionality of telephone equipment. Abbreviated as TSAPI.

Teleservices
Services offered across communications links. Includes e-mail and facsimile features.

Telex
Teletypewriter service allowing subscribers to send messages over the PSTN.

Telnet
The virtual terminal protocol used by the Internet. It is part of the TCP/IP protocol suite that allows managers to control network devices, such as routers, remotely.

Temporary Equipment Identifier
A temporary identifier assigned to a data link to distinguish it from other data links in a domain (for example, a domain of ISDN daisy chained devices and directory numbers). Abbreviated as TEI.

Term ID
Also called XID, an SNA cluster controller identification. TermID is only meaningful for switched lines.

Terminal Adapter

An ISDN phone or a PC card that emulates the phone. It connects a non-ISDN terminal to an ISDN network. Abbreviated as TA.

Terminal Equipment

A telecommunications or input/output device which is at the end of the communications circuit. Includes such equipment as telephones, personal computers and answering machines.

TSAPI

Acronym for Telephony Services Application Programming Interface (q.v.).

Throughput

(1.) A measurement of work accomplished. (2.) The volume of traffic that passes through a pathway or intersection. Typically, this refers to data communications packets or cells, and is measured in packets, cells, or bits per second. (3.) Rate of information arriving at. and possibly passing through, a particular point in a network system.

TIA

See Electronics Industry Association/Telecommunication Industry Association.

Time Division Multiplexer

A method using specific time slots to access a communication link. This is accomplished by combining data from several devices into one transmission. Abbreviated as TDM.

Time Domain Reflectometer

Test equipment that verifies proper functioning of the physical components of the network with a sequence of time-delayed electrical pulses. Primarily, this tool checks for contiguity and isolation of STP and UTP wiring. Abbreviated as TDR.

Time-Division Multiplexer

A method using specific time slots to access a communication link. This is accomplished by combining data from several devices into one transmission. Abbreviated as TDM.

Tip and Ring

Two conductors carry speech from the central telephone office to a home or business. One is the tip and the other is the ring.

Token-Ring

(1.) An IBM network protocol and trademark. (2.) A popular example of a local area network from which the IEEE 802.5 standard was derived from original IBM working papers. Token-Ring applies the IEEE 802.2 MAC protocols and uses the nonpersistent token protocol on a logical ring, although in a physical star topology. Transmission rate is a 4 Mbps. with upgrades to 16 Mbps and options to release a token upon completion of frame transmission, early token release, and burst mode option.

Topology
Layout of a network. This describes how the nodes are physically joined to each other.

TOS
Acronym for Type of Service (q.v.).

TP
Acronym for Twisted Pair (q.v.).

TPS
Acronym for Transactions per Second (q.v.).

TR41.8.1
EIA/TIA subcommittee delegated with the task of standardizing cable testing specifications.

Trace Trigger
An RMON MIB filter set to create an alarm based upon an upper-bound threshold exceeded or a lower-bound threshold reached.

Traffic
(1.) The communications carried by a system. (2.) A measure of network load which refers to the frame transmission rate (frames per second or frames per hour).

Transaction Control Protocol/Internet Protocol
(1.) A complete implementation of this networking protocol includes Transaction Control Protocol (TCP), Internet Protocol (IP), Internetwork Control Message Protocol (ICMP). User Datagram Protocol (UDP), and Address Resolution Protocol (ARP). Standard applications are File Transfer Protocol (FTP), Simple Mail Transfer Protocol (SMTP), and telnet which provides a virtual terminal on any remote network system. (2.) Common communication protocol servicing the network and transport layers that provides transmission routing control and data transfer. This represents logical connectivity at levels 2 and 3 of the OSI reference model, although the protocol does not conform in fact to this model. Abbreviated as TCP/IP.

Transaction Log
(1.) A list of activities or events on the enterprise network. (2.) A list of transactions performed against a Database Management System.

Transactions per Second

The number of discrete data entry, data update, and data requests processed by a system each second; a measurement of performance capacity or measure of work accomplished generally applied to on-line transaction processing environments. Abbreviated as TPS.

Transmission

Any electronic or optical signal used for telecommunications or data communications to send a message.

Trunk

Transmission channel connecting two switching devices.

Twisted Pair

Telephone wire twisted over its length to preserve signal strength and minimize electromagnetic interference. Abbreviated as TP.

Type Error

A packet that is improperly labeled with protocol information.

Type Of Service Routing

Routing scheme where the choice of a path through the internetwork depends on the characteristics of the subnetworks and packet involved, as well as the shortest path to the destination.

U-Interface

Two-wire ISDN circuit—essentially today's standard one-pair telephone company local loop made of twisted wire. The U interface is the most common ISDN interface and extends from the central office.

U25 bis dialing

ITU/TSS standard for in-band dialing on bit-synchronous (HDLC) serial lines. Supports Addressed Call Mode.

UART

See Universal Asynchronous Receiver/Transmitter (q.v.).

UDP

Acronym for User Datagram Protocol (q.v.).

UDP/IP

Acronym for User Datagram Protocol/Internet Protocol (q.v.).

UL
Acronym for Underwriters Laboratories, Inc. (q.v.).

UL 1449 Rating
A rating given to surge suppressers with a clamping level of 330 V. *See* Surge Suppresser.

ULP
Acronym for Upper-Layer Protocol (q.v.).

Undercarpet Ribbon
A specialized cable designed for installation in walkways and undercarpet applications, as its name implies. This cable is best used for lengths not too exceed 2 m (5 ft).

Underwriters Laboratories, Inc.
An organization funded by manufacturers that tests the safety of products.

Uniform Resource Locator
The Internet address to a World Wide Web page or specific document by file transfer protocol. Abbreviated as URL.

Unimodem
A specification from Microsoft Corporation describing Windows 95 (and beyond) services and communication drivers supporting the transmission of digitized voice and data.

Uninterruptible Power Supply
A backup power supply in case the main electrical source fails. Abbreviated as UPS.

Universal Asynchronous Receiver/Transmitter
The chip in a PC that allows the computer to communicate through the serial port with mice, modems, and light pens, and directly with other computers. Abbreviated as UART.

Unshielded Twisted Pair
Pairs of 22- to 26-gauge wire usually in bundles of 2, 4, or 25 pairs installed for telephone service and occasionally for data networks. Referred to as voice-grade twisted pair or voice-grade wiring. Abbreviated as UTP.

Upper-Layer Protocol
A protocol higher in the OSI reference model than the current reference point. ULP is often used to refer to the next-highest protocol in a particular protocol stack. Abbreviated as ULP.

UPS
Acronym for Uninterruptible Power Supply (q.v.).

URL
Acronym for Uniform Resource Locator (q.v.).

User Datagram Protocol
A connectionless transport-layer protocol belonging to the Internet protocol family of the Transmission Control Protocol for application-level data. An in-band, connectionless transmission for congestion management and information collection typically employed by bridges, routers, gateways, and SNMP. Abbreviated as UDP.

User Datagram Protocol/Internet Protocol
A connectionless transport-layer protocol belonging to the Internet protocol family of the Transmission Control Protocol for application-level data. An in-band, connectionless transmission for congestion management and information collection typically employed by bridges, routers, gateways, and SNMP. Abbreviated as UDP/IP.

UTP
Acronym for Unshielded Twisted Pair (q.v.).

V Series Protocol
A set of standards published by the ITU/TTS for data communication over telephone networks.

V.10
ITU/TSS electrical characteristics for unbalanced double-current interchange circuits for general use with integrated circuit equipment in the field of data communications. V.10 is also known as ITU/TSS X.26; EIA RS-423-A; and FED-STD 1030A.

V.11
ITU/TSS electrical characteristics for balanced double-current interchange circuits for general use with integrated circuit equipment in the field of data communications V.11 is also known as ITU/TSS X.27; EIA RS-422-A; and FED-STD-1020A.

V.120
ITU protocol for data transmission rate adaptation.

V.21
ITU/TSS standard for 300 bps duplex modem standardized for use on the general switched network.

V.22
ITU/TSS standard for 1200 bps duplex modem standardized for use on the general switched telephone network and on leased circuits [FED-STD 1008].

V.22 bis
ITU/TSS standard for 2400 bps duplex modem using the frequency division technique standardized for use on the general switched telephone network and on point-to-point, two-wire leased, telephone-type circuits [FED-STD 1008].

V.23
ITU/TSS standard for 600/1200 bps modem standardized for use on the general switched network.

V.24
ITU/TSS list of definitions for interchange circuits between data terminal equipment and data circuit-terminating equipment. Also known as EIA RS-232-C, RS-449, RS-449.1. and RS-266-A.

V.25
ITU/TSS standard for automatic calling and/or answering equipment on the general switched telephone network, including disabling of echo suppressers on manually established calls. Also known as EIA RS-366-A.

V.26
ITU/TSS standard for 2400-bps modem standardized for use on four-wire leased circuits.

V.26 bis
ITU/TSS standard for 2400/1200-bps modem standardized for use in the general switched telephone network. Also known as FED-STD 1005.

V.27
ITU/TSS standard for 4800-bps-modem with manual equalizer standardized for use on leased telephone circuits.

V.27 bis
ITU/TSS standard for 4800/2400-bps modem with automatic equalizer standardized for use on leased telephone-type circuits. Also known as FED-STD 1006.

V.27 ter
ITU/TSS standard for 4800/2400-bps modem standardized for use in the generalized switched telephone network. Also known as FED-STD 1006.

V.28

ITU/TSS standard for electrical characteristics for unbalanced, double-current interchange circuits. Also known as EIA RS-232-C.

V.29

ITU/TSS standard for 9600-bps modem standardized for use on point-to-point, leased, telephone-type circuits. Also known as FED-STD 1007.

V.3

ISO International Alphabet No. 5. Also known as ISO 646; ANSI X(3.)4; and FIPS 1-1.

V.31

ITU/TSS standard for electrical characteristics for single-current-interchange circuits controlled by contact closure.

V.32

ITU/TSS standard for a family of two-wire duplex modems operating at data signaling rates of up to 9600 bps for use on the general switched telephone network and on leased telephone-type circuits.

V.32 bis

ITU/TTS V Series protocol with transmission at 14,400 bps with fall back to 12,000, 9600, 7200, and 4800 bps.

V.34 diagnostic

The most exhaustive phone-line quality diagnostic test sequence available. This is a 23-tone V.8 sequence specified by the ITU for testing phone quality.

V.35

ITU/TSS standard for data transmission at 48,000 bps using 60- to 180-kHz group band circuits.

V.36

ITU/TSS standard for modems for synchronous data transmission using 60- to 180-kHz group band circuits.

V.4

ITU/TSS general structure of signals of International Alphabet No. 5 code for data transmission over public telephone networks. Also known as ITU/TSS X.4; ISO 1155, 1177; ANSI X3.15, X3.16; FED-STD 1010, 1011; and FIPS 16-1, 17-1.

V.42
ITU/TSS standard for data compression technique for modems using LAPM error-control protocol.

V.42 bis
ITU/TTS V Series protocol with transmission applying data compression using a Lempel-Ziv-Welch or Shannon-Fano technique, which detects frequently occurring character strings and replaces them with tokens. Typical compression for text is 50 percent or better, with a nearly 20 percent gain from synchronous conversion that reduces transmission time by almost 60 percent.

V.5
ITU/TSS standardization of data signaling rates for synchronous data transmission in the general switched telephone network. Also known as ANSI X3.1; EIA-RS-269-B; FED-STD 1013, and FIPS 22-1.

V.54
ITU/TSS standard for loopback devices for modems. Also known as EIA RS-449.

V.6
ITU/TSS standardization of data signaling rates for synchronous data transmission of leased telephone-type circuits. Also known as ANSI X3.1; EIA RS-269-B; FED-STD 1013, and FIPS 22-1.

Variable Bit Rate
A user network interface congestion control method designed for ATM to limit traffic to a guaranteed level of service (that is, establishing a dedicated bandwidth) based upon a selection of a committed information rate and a burst (overflow) rate when unused virtual channel bandwidth exists. Nodes which exceed negotiated bandwidth are monitored by a leaky bucket algorithm for each transmission speed. Abbreviated as VBR.

VBR
Acronym for Variable Bit Rate (q.v.).

VC
Acronym for Virtual Circuit (q.v.).

VCFC
Acronym for Virtual Circuit Flow Control (q.v.).

VCI
Acronym for Virtual Channel Identifier (q.v.).

Very-Small-Aperture Terminal
The earth-based antenna used for satellite data communications. Abbreviated as VSAT.

Video
Digital video is a precalculated sequence of images that are displayed to show an animation; audio is usually present. Analog video is common TV signals; audio is almost always present.

Video Capture
The process of digitizing an analog video image.

Video Compression
A method for reducing the amount of information required to store and recall a frame of video. Compression is critical to delivering digital full motion video to the user in the most effective manner (in terms of performance and storage costs).

Virtual Channel
A defined route between two end points in an ATM network that might physically trace several different virtual paths.

Virtual Channel Identifier
The unique numerical tag used to identify every virtual channel across an ATM network provided by a 1-bit field in the ATM cell header. Abbreviated as VCI.

Virtual Circuit
A portion of a virtual path or a virtual channel that is used to establish a single virtual connection between two end points. Abbreviated as VC.

Virtual Circuit Flow Control
A credit-based congestion control method designed for ATM. Every link connecting every node in the network is monitored. Credits are exchanged between nodes to indicate available cell buffer space, and this flow control method requires that nodes count incoming and outgoing cells and send frequent status messages. Abbreviated as VCFC.

Virtual Path
A group of virtual channels that can support multiple virtual circuits.

VN3/VN4
The French ISDN standard.

Voice Channel
The channel on which a voice conversation occurs and on which brief messages might be sent from a base station to a mobile station or from a mobile station to a base station.

Voice Grade
EIA/TIA category 1, 2, or 3 twisted pair wiring best utilized for analog lines rather than high-speed data communication lines. *See* Twisted Pair.

Voice Line
A communications link designed for communication with a frequency of 4 to 12 Khz, typically the human voice.

Voltage Rating
The highest voltage that might be continuously applied to a wire in conformance with standards or specifications.

VSAT
Acronym for Very-Small-Aperture Terminal (q.v.).

WAN
Acronym for Wide Area Network (q.v.).

W4WG
Acronym for Microsoft's Windows for Work Groups.

WFWG
Acronym for Microsoft's Windows for Work Groups.

Wide Area Network
A network that spans cities, states, countries, or oceans. Command carrier services usually supply the network links. Abbreviated as WAN.

WinSock
Microsoft Windows Sockets application programming interface (API) that defines a means by which IP sockets are mapped to the Windows environment.

Wire Line
The local cellular carrier in each market that was also the land line earner when licenses were given out in the 1980s. The tag does not indicate any difference in quality of service, although these carriers tend to cooperate in roaming agreements.

Wire Speed
A term used to describe full utilization of the available bandwidth for a particular medium. This is also the rated transmission speed for a medium. For example, Ethernet wire speed is 16 Mbps on coaxial cable or UTP, whereas ISDN is a multiple of 56 or 64 kbps per channel.

Wiring Closet
Specially designed room used for wiring data and voice networks. Wiring closets serve as a central junction point for wiring and wiring equipment that is used for interconnecting devices.

Wiring Hub
A central wiring concentrator for a series of ATM, Ethernet, FDDI, and Token-Ring nodes that may provide remote ISDN ports too.

Workstation
(1.) Any computer device. (2.) Any device on a network. (3.) A workstation can also be a single addressable site on FDDI that is generally implemented as a stand-alone computer or a peripheral device, connected to the ring with a controller. *See also* Node.

X.1
ITU/TSS standard for intentional user classes of service in public data networks. Also known as ANSI X3.1, X3.36; EIA RS-269-B; FED-STD 1001, 1013; FIPS 22-1, 37.

X.10
ITU/TSS categories of access for data terminal equipment to public data transmission services provided by PDNs and/or ISDN through on-terminal adapters.

X.121
ITU/TSS standard for international numbering plan for public data networks.

X.2
ITU/TSS standard for international user services and facilities in public data networks. Also known as FED-STD 001041.

X.200
ITU/TSS standard for Reference Model.

X.21
ITU/TSS recommendation that defines a synchronous protocol for communication between a circuit-switched network and user devices.

X.21 bis
ITU/TSS standard for interface on public data networks of data terminal equipment which is designed for interchange to synchronous V series modems.

X.22
ITU/TSS procedures for the exchange of control information and user data between a PAD and a packet mode DTE or another PAD.

X.24

ITU/TSS definitions for interchange circuits between a DTE and a DCE on public data networks.

X.25

ITU/TTS standard that describes how data is handled in a packet-switched network and how commercial links are established by packet switches across such a network. The designation also defines the packet format for data transfers in a public data network. Typical X.25 speeds range from 9.6 to 56 or 64 kbps with ISDN or frame relay.

X.28

ITU/TSS standard for the DTE/DCE interface for start-stop mode data terminal equipment accessing the PAD in a public data network situated in the same country.

X.29

ITU/TSS recommendation that defines the packet assembly/disassembly-(PAD-) computer interface.

X.3

ITU/TSS standard for packet assembly/disassembly (PAD) facility in a public data network.

X.400

ITU/TSS recommendation specifying a standard for electronic mail transfer.

X.430

ITU/TSS standard for Message Handling Systems (MHS) access protocol for teletype terminals.

X3.44

ANSI protocol for determination of the performance of data communications systems.

Zero Code Suppression

Coding scheme to substitute a 1 in the seventh bit (eighth position) of a string of eight consecutive zeros.

Index